CLINICAL EVALUATION OF MEDICAL DEVICES

D1292645

CLINICAL EVALUATION
OF MEDICAL DEVICES

Principles and Case Studies

Second Edition

Edited by

KAREN M. BECKER, PhD

Becker & Associates Consulting Inc., Washington, DC

JOHN J. WHYTE, MD, MPH

Discovery Health Channel, Silver Spring, MD

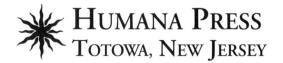

HUMANA PRESS
TOTOWA, NEW JERSEY

© 2006 Humana Press Inc.
999 Riverview Drive, Suite 208
Totowa, New Jersey 07512

humanapress.com

For additional copies, pricing for bulk purchases, and/or information about other Humana titles, contact Humana at the above address or at any of the following numbers: Tel.: 973-256-1699; Fax: 973-256-8341; E-mail: orders@humanapr.com; or visit our Website: www.humanapress.com

Due diligence has been taken by the publishers, editors, and authors of this book to assure the accuracy of the information published and to describe generally accepted practices. The contributors herein have carefully checked to ensure that the drug selections and dosages set forth in this text are accurate and in accord with the standards accepted at the time of publication. Notwithstanding, as new research, changes in government regulations, and knowledge from clinical experience relating to drug therapy and drug reactions constantly occurs, the reader is advised to check the product information provided by the manufacturer of each drug for any change in dosages or for additional warnings and contraindications. This is of utmost importance when the recommended drug herein is a new or infrequently used drug. It is the responsibility of the treating physician to determine dosages and treatment strategies for individual patients. Further it is the responsibility of the health care provider to ascertain the Food and Drug Administration status of each drug or device used in their clinical practice. The publisher, editors, and authors are not responsible for errors or omissions or for any consequences from the application of the information presented in this book and make no warranty, express or implied, with respect to the contents in this publication.

Production Editor: Melissa Caravella
Cover design by Patricia F. Cleary
Cover Illustration: From Fig. 1B in Chapter 18, "Role of Device Retrieval and Analysis in the Evaluation of Substitute Heart Valves," by Frederick J. Schoen and Figs. 4 and 5 in Chapter 17, "Polyurethane Pacemaker Leads: *The Contribution of Clinical Expertise to the Elucidation of Failure Modes and Biodegradation Mechanisms,*" by Ken Stokes.

This publication is printed on acid-free paper. ∞
ANSI Z39.48-1984 (American National Standards Institute) Permanence of Paper for Printed Library Materials.

Printed in the United States of America. 10 9 8 7 6 5 4 3 2 1
eISBN 1-59745-004-9
Library of Congress Cataloging-in-Publication Data
Clinical evaluation of medical devices / edited by Karen M. Becker,
John J. Whyte. -- 2nd ed.
p. ; cm.
Includes bibliographical references and index.
ISBN 1-58829-422-6 (alk. paper)
1. Medical instruments and apparatus--Evaluation. 2. Clinical trials.
I. Becker, Karen M. II. Whyte, John J. (John Joseph), 1966- .
[DNLM: 1. Equipment and Supplies--standards. 2. Clinical Trials
--standards. 3. Device Approval--standards. 4. Evaluation Studies.
W 26 C641 2006]
R856.C545 2006
610'.28--dc22
                                2005011142

PREFACE

The original edition of this text, *Clinical Evaluation of Medical Devices: Principles and Case Studies*, provided the first overview of key principles and approaches to medical device clinical trials, illustrated with a series of detailed, real-world case studies. The book is designed as a resource for clinical professionals and regulatory specialists working in the field of new medical device development and marketing. Since the first edition of this text was published in 1997, the rapid pace of innovation in health care technologies continues to yield exciting and important new products. The regulatory landscape has also evolved, reflecting some of the changes and needs within the medical device industry.

The purpose of *Clinical Evaluation of Medical Devices: Principles and Case Studies, Second Edition* is to provide an updated and expanded presentation of the scientific methods and regulatory requirements applied to the study of new significant risk medical devices. The text now includes (1) new information on the requirements and process for gaining reimbursement of new products from Medicare and private insurers, with case studies of research specifically designed for this purpose as well as health care technology assessment methods; (2) information on new statistical methodologies applied to medical device trials; and (3) all new case studies, including examples of combination products, three-phase development models (i.e., feasibility, FDA approval, Medicare reimbursement), and novel study designs. This second edition builds on the strength and foundation of the first, and would not have been possible without those colleagues who graciously contributed their expertise in the form of chapters and ideas to the final product.

Karen M. Becker, PhD
John J. Whyte, MD, MPH

CONTENTS

PART II   CASE STUDIES

CONTRIBUTORS

KAREN M. BECKER, PhD, *President and CEO, Becker & Associates Consulting Inc., Washington, DC*

MICHAEL D. BEDNAREK, JD, *Senior Partner, Intellectual Property Group, Pilsbury Winthrop Shaw Pittman LLP, McLean, VA*

ROBIN BOSTIC, BS, *Vice President, Reimbursement Affairs, Thoratec Corporation, Pleasanton, CA*

SHARON D. BROOKS, JD, MPH, *Associate, Alston & Bird LLP, Washington, DC*

MITCHELL I. BURKEN, MD, MPP, *Medical Director, TrailBlazer Health Enterprises, LLC, Timonium, MD*

GUY CHAMBERLAND, MSc,PhD, *Program Director, Drug Development Programs, MDS Pharma Services, Montreal, Quebec, Canada*

JULIE C. CHOE, MPH, MS, *Clinical Research Coordinator and Project Leader, International Center for Health Outcomes and Innovation Research (InCHOIR), Columbia University, New York, NY*

SAMANTHA R. COOK, PhD, *Post-Doctoral Fellow, Department of Statistics, Harvard University, Cambridge, MA*

DEBORAH V. DAVIS ASCHEIM, MD, *Assistant Professor of Medicine, International Center for Health Outcomes and Innovation Research (InCHOIR), Columbia University, New York, NY*

MICHAEL P. DIAMOND, MD, *Associate Chair and Kamran S. Moghissi Professor of Obstetrics and Gynecology; Director, Division of Reproductive Endocrinology and Infertility, Wayne State University, Detroit, MI*

LISA M. DWYER, JD, *Associate, Patton Boggs LLP, Washington, DC*

JANET M. FAULS, BS, *Vice President, Regulatory, Quality, and Clinical Affairs, CardioGenesis Corporation, Foothill Ranch, CA*

ADAM K. GALEON, MBA, *Vice President, Medical Technology and Hospital Supplies, Credit Suisse First Boston, New York, NY*

ANNETINE C. GELIJNS, PhD, *Co-director and Professor of Surgical Sciences and Public Health, International Center for Health Outcomes and Innovation Research (InCHOIR), Columbia University, New York, NY*

TELBA Z. IRONY, PhD, *Chief, General and Surgical Devices Branch, Division of Biostatistics, Center for Devices and Radiological Health, Food and Drug Administration (FDA), Rockville, MD*

PETER M. KAZON, ESQ., *Senior Counsel, Alston & Bird LLP, Washington, DC*

Daniel A. KRACOV, JD, *Partner and Deputy Director of Public Policy and Regulatory Department, Patton Boggs LLP, Washington, DC*

STUART S. KURLANDER, JD, MHA, *Partner, Local Chair, and Global Co-chair, Health Care and Life Sciences Practice Group, Latham & Watkins, LLP, Washington, DC*

FARIBORZ MOAZZAM, PhD, JD, MBA, *Founding Partner, Moazzam & Latimer LLP, Reston, VA*

ALAN J. MOSKOWITZ, MD, *Co-director and Associate Professor of Medicine, International Center for Health Outcomes and Innovation Research (InCHOIR), Columbia University, New York, NY*

JAN ERIK NORDREHAUG, MD, PhD, *Professor and Director, Department of Heart Disease, Haukeland University Hospital, Bergen, Norway*

SUZAN ONEL, JD, *Partner, Medical Device, Food, Drug, and Cosmetic Law, Kirkpatrick & Lockhart, LLP, Washington, DC*

MICHAEL PARIDES, PhD, *Associate Professor of Biostatistics, Department of Biostatistics and International Center for Health Outcomes and Innovation Research (InCHOIR), Columbia University, New York, NY*

DONALD B. RUBIN, PhD, *John L. Loeb Professor of Statistics, Harvard University, Cambridge, MA*

ESTHER R. SCHERB, DMD, JD, *Of Counsel, Health Care & Life Sciences Practice Group, Latham & Watkins, LLP, Washington, DC*

URSULA MARIA SCHMIDT-OTT, MD, *Clinical Research Coordinator and Project Leader, International Center for Health Outcomes and Innovation Research (InCHOIR), Columbia University, New York, NY*

FREDERICK J. SCHOEN, MD, PhD, *Professor of Pathology and Health Sciences and Technology, Harvard Medical School, Boston, MA; Executive Vice Chairman, Department of Pathology, Brigham and Women's Hospital, Boston, MA*

RICHARD SIMON, DSc, *Chief, Biometric Research Branch, National Cancer Institute, Rockville, MD*

KEN STOKES, PhD, *Consultant, Medtronic Inc., Minneapolis, MN*

JULIE K. TAITSMAN, MD, JD, *Senior Counsel, Industry Guidance Branch, Office of Counsel to the Inspector General, Washington, DC*

JOHN J. WHYTE, PhD, *Vice President, Continuing Medical Education, Discovery Health Channel, Silver Spring, MD*

I

Fundamentals of Clinical Study Design and Evaluation

1

Clinical Trials in Development and Marketing of Medical Devices

Karen M. Becker

1. Introduction

Medical devices are health care products distinguished from drugs for regulatory purposes in most countries based on mechanism of action. Unlike drugs, medical devices operate via physical or mechanical means and are not dependent on metabolism to accomplish their primary intended effect. As defined in the federal Food, Drug, and Cosmetic (FD&C) Act , the term *medical device*:

> ...means an instrument, apparatus, implement, machine, contrivance, implant, in vitro reagent, or other similar or related article...intended for use in the diagnosis of disease or other conditions, or in the cure, mitigation, treatment, or prevention of disease...or intended to affect the structure or any function of the body...and which does not achieve its primary intended purposes through chemical action within or on the body....[1]

This broad definition of medical devices encompasses literally tens of thousands of different types of health care products, including in vitro diagnostics.

Developing new medical devices and extending the scope of what is known about the performance of already marketed products often requires clinical investigations. Designing well-controlled prospective clinical trials of medical devices presents unique challenges that differ from those faced in studies of pharmaceuticals. For example, clinical outcomes observed in medical device studies are influenced not only by the product under evaluation and the patient, but also by the skill and discretion of the user, who is typically a health care professional but may be the patient. The impact of this third parameter—the medical device user—is a variable unique to medical device studies and can be responsible for the greatest degree of variability in the clinical outcomes. Being aware of and controlling for the user's influence on device performance is a critical variable that requires attention in designing a clinical study. Other

From: *Clinical Evaluation of Medical Devices: Principles and Case Studies, Second Edition*
Edited by: K. M. Becker and J. J. Whyte © Humana Press Inc., Totowa, NJ

critical features typically considered in the design of well-controlled studies, such as the choice of a control group, the need to reduce bias, and the need to control for confounders, are common to both drug and device trials; however, the nature of the difficulties presented and the approaches used to successfully address these challenges often differ.

When designing a clinical trial for a medical device, it is useful to consider both regulatory requirements and the manufacturer's established business goals. The Food and Drug Administration (FDA) is concerned with the safety and efficacy of a product, whereas health care providers and payors are interested in comparative performance, superiority or product differentiation claims, and/or economic data. Therefore, an optimal clinical research program provides not only the data needed for marketing authorization in the United States or abroad but also information to obtain coverage and reimbursement in the targeted markets and support competitive claims.

This chapter provides a detailed discussion of the features of medical devices that can pose challenges in the design of well-controlled clinical studies as well as methods for addressing these design challenges. It also presents an overview of the role of clinical research in the lifecycle of medical device product development and marketing along with the essential elements of a clinical investigational plan for a prospective medical device clinical trial.

2. Drugs vs Devices: Is There a Difference?

The Medical Devices Amendments to the FD&C signed into law in 1976 provided the FDA with broad jurisdiction and authority over the commercialization of medical devices. Before the law's development, the Secretary of Health and Human Services assembled a task force to consider the nature of the medical device industry in the United States, the extent to which the products of this industry should be subject to regulation, and the best mechanisms for protecting the public health without applying an undue burden to industry or preventing innovation. The task force, known as the Cooper Committee, (after Dr. Theodore Cooper, who had been the director of the National Heart, Lung, and Blood Institute at the time), submitted a report in 1970, the conclusions of which formed the framework of the 1976 legislation.[2]

The Cooper Committee concluded that medical devices differ significantly from pharmaceuticals, and as such, direct application of the "drug model" of regulation to these products was neither desirable nor feasible. Instead, the committee recommended a novel regulatory approach that was based on the extent of risk posed to the patient from the use of the device. Among the "inherent differences between drugs and devices"[2] noted by the Cooper Committee were that medical devices are an extremely diverse group of products varying widely in their intended uses and principles of operation, they are often designed by physicians and subject to frequent innovation in both design and use, they are used primarily by health care professionals rather than patients, they are most often developed by small companies, and their annual sales are only a fraction of that for a typical pharmaceutical product on a per product basis.

Relying on the Cooper Committee's recommendations and subsequent testimony, Congress developed legislation incorporating a regulatory pathway for medical devices based on the consideration of the relative risk posed by each product and an apparent acceptance of the principle that the user's skill and clinical judgment ultimately has a major role in the performance of any particular medical device, regardless of federal regulatory requirements. According to the premarket approval regulations for medical devices, sponsors must provide valid scientific evidence of safety and efficacy. Unlike requirements for drug approvals, this evidence can come from sources other than well-controlled clinical investigations, such as "partially controlled studies, studies and objective trials without matched controls, well-documented case histories conducted by qualified experts, and reports of significant human experience with a marketed device."[3] However, the regulation specifically precludes reports of clinical experience that are not adequately supported by data such as anecdotal reports or opinion.

The standard for approval of medical devices is also more flexible than for drugs in that the regulations require "reasonable assurance" of safety and effectiveness, rather than the more onerous burden of "substantial evidence" specified for drugs.[4,5] Although the FDA's Center for Devices and Radiological Health has taken responsible and aggressive steps to ensure the rigor of clinical research required to support registration of new medical devices, it has recently reaffirmed the differential standard of evidence for approval of devices vs drugs.[6,7] As noted by the FDA, the primary practical consequence of the regultaions is that the approval requirements for drugs require replication of clinical findings (i.e., more than one clinical trial), but for devices a single pivotal clinical trial is sufficient for approval because, "for medical devices, where the mechanism of action is a result of product design and substantially verified by in vitro performance testing, the agency has routinely relied on single studies evaluated for internal and across-center consistency."[6]

The differences between drugs and medical devices identified by the Cooper Committee 35 years ago remain valid today, and some are key to understanding features of medical devices that influence product development plans and clinical study design (*see* Table 1).

2.1. Devices Are Primarily Used by Health Care Professionals

In studies of pharmaceuticals, the two principal interacting variables are the drug and the patient. Given that investigators can control for other variables, such as concomitant exposures, pre-existing conditions, and progression of dis-

Table 1
Characteristics of Medical Devices That Affect Clinical Trial Design

Characteristic	Clinical study design issue
Devices are primarily used by health care professionals.	Product performance is influenced by user.
Devices are subject to frequent incremental innovation.	The user often cannot be blinded to the study intervention. Bench testing and animal models alone may validate new designs.
Some devices are implanted.	Ethical considerations may preclude comparative trials. Results from long-term clinical studies may no longer be relevant to current products and medical procedures. Exposure to the product is not readily terminated or without irreversible consequences to patient. Placebo control groups (sham-surgery) are not possible. Medically appropriate alternative treatment regimens may not be available to provide randomized, concurrent controls. Long-term performance evaluations primarily rely on design controls and failure analysis.
Devices are often developed by small companies; sales on a per-product basis are less than that for average pharmaceuticals.	Practical considerations (regulations, financial constraints) limit new product development and testing.

ease, the outcomes measured are ultimately a function of the interaction of the drug and the patient. In contrast, the interaction of a medical device and the patient is usually mediated by a third party, the product's user, who is typically a health care professional. Thus, the clinical outcomes measured in the study of a medical device's safety and effectiveness are a function of the user's skill as well as the interaction between the device and the patient. The user, as an intermediary, poses two major difficulties in the design of a clinical trial for devices that are not commonly encountered in studies of pharmaceuticals: Users can rarely be blinded to the treatment intervention and can impact on the product's performance. Indeed, the user is an integral variable in the performance of the product. A device performs better in the hands of an experienced user than in the hands of a naïve user, a phenomenon typically called the *learning curve*. Training in the use of a device is a key part of the investigation of its clinical performance and eventual marketing. If the variability in the user's proficiency is not accurately assessed or minimized in importance, the device's performance may be inaccurately estimated.

The inability to blind the user, and often the patient, to the intervention under study can introduce bias into the assessment of clinical performance if the clinical investigator is jointly responsible for treatment and assessment of performance. Thus, whenver possible, blinded evaluators are preferred to clinical investigators in assessing efficacy.

2.2. Devices Are Subject to Frequent, Incremental Innovations

Frequent innovations in the design and use of medical devices are standard practice in the industry. These are often minor modifications that enhance safety, reliability, patient comfort, or ease of use and do not require regulatory approval or premarket notification in most cases. Bench testing and/or evaluations in animal models are often sufficient to validate the suitability of a design change without the need for controlled clinical trials. It is not uncommon for clinicians and Institutional Review Boards (IRBs) to find the results of an in vitro performance evaluation of a new design sufficiently compelling that they are reluctant to proceed with a comparative clinical trial, because continued use of the older design may be deemed unethical. However, clinical studies are usually necessary for design innovations intended to significantly improve performance parameters (efficacy) or to expand indications for use.

2.3. Some Devices Are Implanted

It is estimated that 20 to 25 million people in the United States alone have some type of implanted medical device, such as pacemakers, intraocular lenses, and artificial joints.[8] Some consequences of implanting a medical device are irreversible for the patient, regardless of how long the device is used. Unlike a clinical drug experiment, an implanted medical device trial is not readily terminated. Given that clinical studies of implanted devices are surgical trials, the use of placebo- or sham-operated control groups is usually precluded. Because an ethically appropriate alternative treatment group may be difficult to identify, the use of historical controls in the trial or patients as their own controls (pre- and postsurgery) may be required to evaluate outcomes. Controlled, prospective, long-term performance evaluations of implanted devices (>2 year) are uncommon because they pose logistical constraints (e.g., large sample sizes are required to mitigate against loss to follow-up, inability to identify a sufficient number of suitable patients, expense of a large trial in relation to market size). Instead, information to track rare complications, identify failure modes, and contribute to enhanced designs is most often collected from the analysis of case series, registries, failure analysis of retrieved devices, bench testing, and formal design reviews.

2.4. Device Manufacturers May Be Small Companies

Medical devices are often products developed by small companies that generate annual sales revenue that is only a fraction of that generated by the average pharmaceutical. A responsible manufacturer conducts whatever testing is required to develop a safe and reliable product, but practical constraints experienced by this particular segment of the health care industry are particularly influential in driving product development and testing decisions.

3. The Role of Clinical Studies in Product Development

Most commonly, the impetus for conducting a clinical study is to demonstrate the safety and effectiveness of an investigational device before marketing, a requirement for registration of implants and other significant risk (SR) devices in the United States and most international markets. An SR device is a product that presents a potential for serious risk to the health, safety, or welfare of a subject and is most commonly an implant or life-sustaining product.[9] The goal of the clinical study is to confirm, validate, or supplement data from bench and/or animal testing. Clinical studies are commonly performed to support a novel design, new technology, and/or new indications for use. However, carefully conceived clinical research also has a role in enhancing product development and marketing for nonsignificant risk products, despite most devices and diagnostics reaching the market after only safety and performance testing in animal models and in vitro. Postmarketing studies can yield information on safety (e.g., long-term safety and performance, uncommon complications), enhance product design, extend labeling claims, and provide data on comparative effectiveness and support for cost–benefit claims. Table 2 summarizes a typical classification scheme for clinical investigations of medical devices.

3.1. Pilot Studies

Pilot studies, or feasibility studies, are usually single-center studies of a limited number of patients designed to accomplish any number of objectives within a clinical-testing program. Pilot studies are not usually designed as hypothesis-testing studies; rather they are intended to generate data in support of the design of rigorous analytical (i.e., hypothesis-testing) trials. The first study of a novel investigational device in humans is usually a small pilot study undertaken to evaluate safety under carefully controlled conditions and to provide data supporting broader performance testing in a larger population. Pilot studies are the

Table 2
Typical Classification of Medical Device Clinical Studies

Pilot studies of safety, performance, and/or design before marketing
Pivotal trials of safety and effectiveness before marketing
Postmarketing studies
    In support of expanded labeling claims
    In support of comparative performance claims
    Pharmacoeconomic studies
    Observational or analytical studies of specific safety or performance issues
    Explant retrieval and failure analysis investigations

---

first opportunity to evaluate the role of the user in device performance under actual clinical conditions and gather information on design features that may be modified to optimize proper use of the device. Before designing a pivotal clinical trial to evaluate device safety and effectiveness, pilot studies allow the sponsor to collect data on a series of patient outcomes that may be related to device performance, thereby contributing to the identification and selection of clinically significant measures for use as effectiveness endpoints in a subsequent pivotal trial. Frequently, the selection of measures of safety and effectiveness requires the development of validated assessment methods. More extensive pilot studies can incorporate validation of assessment tools and be used to generate enough data on the interpatient variability of endpoints to support sample-size calculations for use in a hypothesis-testing study.

3.2. Pivotal Trials of Safety and Effectiveness

A single, well-controlled clinical trial of device performance remains sufficient for FDA approval of a SR device. These prospective, analytical studies provide objective evidence of effectiveness based on single or multiple clinical outcomes of significance. In combination with data from bench testing and animal studies, results from a single trial are adequate to establish "reasonable evidence" of safety and effectiveness. When direct comparisons are made to alternative treatment options, effectiveness of the new device is expected to be *not worse than* that of other available devices or treatment. With rare exceptions, pivotal trials in support of successful FDA premarket approval applications (PMAs) are multicentered.

Clinical research conducted on an investigational device before marketing creates the foundation for claims that will appear on its label once marketing authorization is accomplished. This point is especially critical in the United States, where the expectations for reliable data in support of all aspects of the label are the most rigorous. For this reason, the clinical portion of the product development plan should never be considered in isolation from the ultimate marketing goal. In some cases, bench testing and animal studies can provide additional performance data to augment the clinical research and support expanded label claims.

3.3. Postmarketing Studies

Two categories of postmarketing studies can be distinguished: mandated postapproval studies and postmarket surveillance studies. It has become increasingly common for the FDA to require sponsors of Class III devices to conduct a postapproval clinical study as a mandatory condition of PMA approval. These studies are usually narrow in scope and focus on generating additional data to expand on results of the pivotal trial(s) in support of product approval. The objectives of postapproval studies, whether mandated by a regulatory agency or the state of the science, typically include longer-term follow-up, additional data on the incidence and time-course to appearance of adverse events, and additional data in support of broader label claims (e.g., indications for use, duration of effectiveness, product benefits). Postapproval studies may be undertaken as an extension of a pivotal trial via protocol amendments to extend follow-up or they may be conceived as an entirely separate study. The trend toward mandatory postapproval studies reflects the FDA's commitment to work with sponsors whose investigational plans were finalized before 1993, when the agency shifted to a more rigorous standard of clinical trial requirements.[6,10,11] Careful consideration of pivotal trial design and good communication with the FDA in the design of investigational plans before initiation of pivotal clinical trials will likely minimize the need for mandatory postapproval studies.

Distinct from postapproval studies are the various types of postmarketing studies undertaken by a manufacturer to accomplish a variety of goals. These studies may be sponsor-initiated or required by a regulatory agency in response to a perceived safety concern. The federal Safe Medical Devices Act of 1990 and the FDA Modernization Act of 1997 (FDAMA) empowered the FDA to require mandatory postmarket surveillance trials for certain types of devices as well as discretionary postmarket surveillance trials. Under FDAMA, the FDA can require companies to conduct postmarket surveillance studies for Class II or Class III devices under the following conditions:

1. The failure of the device would be reasonably likely to result in serious health concerns.
2. The device is intended for implantation in the human body for a period greater than 1 year.
3. The device is a life-sustaining or life-supporting product that is used outside a device user facility.

Mandatory postmarket surveillance trials to date have been directed at developing systematic data on either long-term failure modes and/or the potential for serious adverse events occurring in a small number of patients receiving devices (e.g., heart valves, injectable collagen, polyurethane breast implants, pacemaker leads). Compliance with Quality System Regulations for medical devices requires manufacturers to engage in postmarket surveillance monitoring of marketed products. This includes requirements to evaluate and act on complaints, product failures, and adverse events associated with product use. This is not a passive process; the responsible manufacturer maintains routine procedures for systematic evaluation of postmarket experience directed toward investigating product failures and successes and may include research (bench testing or clinical) to improve product performance and safety.

In addition to meeting regulatory requirements, other goals of postmarket research, including comparative studies with alternative or competitive treatments and/or devices, may be to provide support for pharmacoeconomic claims, comparative effectiveness claims, or expanded label claims. As previously noted, implanted device studies in which patients are followed prospectively for more than 2 years are not generally practical because of the loss of follow-up, enormous expense, and rapidly progressing changes in medical practice. Carefully considered programs to exploit explant analysis and failure investigations, coupled with design controls before marketing, are the most common, effective, and efficient means of gathering data on long-term performance. Sponsors are also beginning to use observational studies effectively for retrospective studies of clinical experience, especially when the goal is to gather data on patient- or device-specific factors that may contribute to long-term performance failures. As with pharmaceuticals, vigilance in the evaluation of adverse events reports, returned goods, and complaints are important sources of information on clinical experience. Although anecdotal, these data are the foundation for research into design deficiencies and strengths and may lead to products that perform better.

4. Elements of a Clinical Investigational Plan

Clinical research studies can be categorized as either observational or analytical. An observational study is designed to collect and analyze data on subject exposure or treatment interventions and does not include a control group. The investigators collect, record, and analyze data to generate or test a hypothesis. Examples of observational studies that are designed to collect data include daily recording of rainfall and temperature, surveys of dietary intake, and data on national cancer incidence. Observational studies are valuable when an experimental approach is not practical or is otherwise unfeasible.

Table 3
Elements of the Investigational Plan

Device description
Study objective
Study design
Study population
Treatment regimen
Control group
Endpoints evaluated
    Effectiveness
    Safety
Definition of trial success (if hypothesis-testing study)
Study procedures and duration
    Screening and assignment to treatment
    Assessments and follow-up
    Training procedures (if appropriate)
Sample-size calculations
Data analysis plan
Risk analysis
Case report forms
Informed consent forms
Investigational site(s) and IRB information
Data safety monitoring board (optional)
Monitoring plan

IRB, Institutional Review Board.

Analytical studies are hypothesis-testing trials comprised of a cohort exposed to a specific intervention, the impact of which is subsequently assessed. The "gold standard" of analytical clinical trial design is the prospective, randomized study with concurrent control group(s). The reader is referred to several excellent sources for detailed discussions of the principles of good clinical study design, conduct, and analysis. Although generally written from the point of view of pharmaceuticals, the principles of good clinical study design apply equally to both drugs and devices.[12–16]

Clinical protocols for medical device trials typically incorporate a device description and a patient-risk analysis along with information describing the design, conduct, and analysis of the planned trial. The essential elements of a clinical investigational plan are listed in Table 3. Each of these is considered in the following discussion, with particular emphasis on features that may be problematic for medical devices.

4.1. Device Description

A description of the product under investigation is provided with sufficient information for the clinical investigators to understand the design of the device, the rationale in support of the product design (which may include references to preclinical testing), device performance specifications, a statement of intended use, and the instructions for use.

4.2. Study Objective

Before designing a clinical study, it is necessary to clearly formulate the question(s) to be answered by the research effort. It should be possible to prepare a summary statement of the objective for any protocol by noting four features for the study:

1. The product tested.
2. The indication for use (treatment or condition to be affected).
3. The primary outcome measure.
4. The subject characteristics (e.g., disease stage).

4.3. Study Design

The clinical study design specifies whether the study is prospective or retrospective, open-label (nonblinded) or controlled, and single-center or multicenter. Although the simplest controlled design includes two groups of subjects, controlled trials may have many variations (e.g., crossover studies, sequential studies). More than two groups of subjects result in a multiple-arm trial, a design that may be selected to incorporate a sham- as well as an active-control group. It is generally best to choose the least complicated design required to successfully address the trial objective. For more information on clinical trial design, refer to the Spilker textbook.[12]

4.4. Study Population

In addition to articulating the clinical condition of the subjects, investigators often include demographic criteria specifying patient age, sex, and race. Economic status or educational level may be relevant; for instance, this data may useful in trials evaluating the labeling and instructions for use of over-the-counter in vitro diagnostics. Defining specific subject inclusion and exclusion criteria is an important means of narrowing the range of subjects studied, thereby minimizing the impact of uncontrolled variables and variability in the effectiveness and safety endpoints observed. For a pivotal clinical trial intended to support FDA approval, the definition of the patient population is significant because the approved PMA will generally have a label, which only supports use in the study population evaluated.

4.5. Treatment Regimen

The nature of the intervention under study encompasses a description of the device, instructions for use, and any other ancillary or related procedure or treatments. For surgical trials, it is especially critical to work with clinical investigators at all sites to develop a consensus to the greatest extent possible on surgical procedures to be used. Uniformity of procedures across sites, coupled with training of clinical investigators and their staff, are the two most critical techniques used to reduce variability and site-specific bias. For devices that represent a truly novel innovation and for which investigators are not expected to have significant first-hand experience, trials can include a prespecified run-in period to stratify sequential procedures as a function of investigator experience and thereby evaluate the impact of the learning curve on device performance.

4.6. Control Group

The control group serves as a benchmark for gaging safety and effectiveness of the device and allows investigators to estimate the clinical significance of device's effect in a defined patient population. By definition, the control group comprises a set of patients or subjects who are not exposed to the intervention or device.

There are two broad categories of control groups: concurrent controls and nonconcurrent controls. Concurrent controls are subjects assigned to a control exposure and observed contemporaneously with the experimental group. The control exposure may be no treatment at all, treatment with a placebo, or treatment with an alternative therapy or device. Concurrent controls are always the preferred choice. Using a concurrent control group allows trial subjects to be randomized between control and treatment groups, thereby eliminating selection bias and controlling for confounding variables.

Nonconcurrent controls are subjects who are not observed contemporaneously with the experimental group. The most common nonconcurrent control is the historical control group, comprising a cohort previously observed, treated, studied, or reported on. Historical control groups must be used with caution, because the quality of the historical data set is often variable and may be unreliable. The subjects in the historical cohort may not be comparable to those in the treatment group in terms of demographics or disease status, and confounding variables cannot be controlled through randomization. Furthermore, because the nature and success of medical treatments tend to improve over time, historical cohorts are more likely to bias the results of a trial toward a positive outcome.

Although the choice of the proper control group is driven by the objectives of the trial and should be supported by a sound rationale, ethical and practical constraints impact the decision. It is not uncommon to discover that the ideal control group from a scientific point of view is not always feasible for logistical or ethical reasons (e.g., a sham-operated control in a surgical trial). Well-controlled studies of some implants can be conducted with the patients as their own controls if an alternative device or procedure is not available. Historical controls should be used only if the quality of the data set is deemed to be reliable and valid, and if the pathogenesis of the disease under study is well understood. They may also be used when the objectives of the trial are limited (e.g., feasibility study).

4.7. Effectiveness and Safety Endpoints

The protocol should specify the clinical endpoint to be assessed in the evaluation of device effectiveness. The selection of the primary effectiveness endpoint is critical to the success of the trial. An appropriate endpoint is a clinically significant outcome that can be measured using a validated method. For a hypothesis-testing study, the extent of change compared to control should be projected to be large enough to have a clinically significant impact on the patient in terms of quality of life, disease progression, diagnosis, or mitigation. Although it is most common to use a single, primary endpoint, some trials may incorporate multiple primary endpoints, with the goal being to demonstrate a clinically significant response through several measures. Secondary endpoints can provide additional data to support product claims, confirm effectiveness, and aid in establishing the reimbursement strategy.

Data on safety and adverse events associated with the studied intervention should be collected as broadly as possible. Previous clinical experience with the device and/or data from nonclinical studies highlight specific safety issues that should be considered in a clinical trial; however, these targeted evaluations do not preclude the need to capture all data on observed adverse events, regardless of whether they are deemed to be device-related at the time.

4.8. Definition of Trial Success

If the clinical trial is a hypothesis-testing study, it is powered to detect a clinically meaningful difference in treatment and control. This is usually expressed as the magnitude of some clinical outcome deemed to be sufficient to conclude that the device is effective in comparison to a baseline value and/or the control group response. In the United States, demonstrations of device effectiveness do not require that the new device be superior to existing products; instead, it is acceptable for product performance to be *not worse than* the standard alternative treatment(s).

4.9. Study Procedures and Duration

This section of the protocol describes the conduct and duration of the study, including screening procedures, scope of the initial patient work-ups, schedules and procedures for follow-up visits, and procedures for discontinuations. The procedures for assigning subjects to the intervention (i.e., randomization scheme) must also be included.

Standardized training procedures for using the device are important features of well-controlled studies of medical devices. Training programs enhance the rigor of a trial by minimizing bias and the impact of confounding variables on device performance that often result from variability in individual user skill and discretion. The extent of training provided to users in a clinical trial varies, and extensive training may appropriately raise questions regarding the extent of training that a manufacturer should provide once marketing commences. In certain trials (e.g., for over-the-counter in vitro diagnostics), it is appropriate not to provide training in the use of the device in order to simulate the actual conditions of use.

4.10. Sample-Size Calculation and Data Analysis Plan

The number of subjects selected for inclusion in a hypothesis-testing study should be justified by sample-size calculations that specify the statistical power of the study. To calculate sample sizes, it is necessary to define the number of subjects specified by the underlying statistical model selected. Usually these parameters include the magnitude of the anticipated outcome resulting from the intervention under study, the variability in the measure of that outcome, the desired power of the study (usually 80%), and the statistical significance (most often 95%). The statistical approach to be used in the analysis of the data collected is articulated in the data-analysis plan. Consultation and collaboration with an expert statistician is necessary to address these aspects of the protocol.

4.11. Risk Analysis

The risk analysis describes the potential risks and benefits to the study subjects in sufficient detail to provide adequate informed consent and support the conclusion that the risks to the patients are not unreasonable. The risk analysis includes a description of alternative procedures available, a consideration of potential failure modes, the steps taken in the design of the device and the trial to minimize risks to the trial subjects, and the rationale for the anticipated benefit to the patient.

4.12. Informed Consent Forms

Informed consent materials given to study subjects should provide complete information on the procedure and its potential risks in a format that is easy to read and understand. The informed consent form should be explicit regarding the voluntary nature of the subject's participation, the known and potential risks and benefits of participation, and the subject's willingness to participate in all required aspects of the study. Regulatory authorities throughout the world require adequate informed consent from clinical trial subjects.[17–19]

4.13. Case Report Forms

Each trial must include case report forms to record the data collected on each subject in accordance with the study protocol. These forms contain information on subject screening, operative information, postoperative and follow-up visit data, adverse event reports, and subject withdrawal forms.

4.14. Investigational Site(s) and Institutional Review Board Approvals

The investigational plan specifies the investigational site(s) involved in the study and includes information on the qualifications of the clinical investigators, the IRB procedures, and records of IRB approvals.

4.15. Investigator's Brochure

The goal of the investigator's brochure is to provide the investigator with a thorough understanding of the risks and adverse reactions associated with the device, as well as a detailed description of the device and how it functions. The topics addressed in the investigator's brochure include:

1. A summary of the literature.
2. A summary of previous clinical research.
3. A description of the device.
4. Device hazards and risk analysis.
5. Device performance and preclinical testing.
6. Anticipated risks in human subjects.
7. Reported adverse events.

4.16. Data Safety Monitoring Board

In certain circumstances, a sponsor will elect to establish a Data Safety Monitoring Board comprised of experts who continually evaluate the safety data accrued in a study. This is most common for early clinical experience with an investigational device or studies in which significant morbidity or mortality is expected. In either case, the monitoring of safety data while a study is ongoing is blinded and intended to minimize risks to the patients by providing a system to alert investigators to unexpected adverse outcomes that may require premature termination of a trial. Alternatively, if a treatment is unexpectedly robust in terms of efficacy, a trial may be terminated early on the grounds that continuing the study would be unethical given the benefit provided to the treatment group.

4.17. Monitoring Plan

A clinical trial monitoring plan defines the procedures that the sponsor or his or her representative will undertake to provide quality assurance in the study's conduct. The monitoring plan ensures that the study is conducted in accordance with the procedures specified in the protocol. The importance of a carefully designed and implemented monitoring plan cannot be overstated. This aspect of study conduct ensures that the data generated from the trial are valid, serves to identify early in the study any significant problems that may arise, and helps ensure that the eventual audit of the clinical study by regulatory authorities is satisfactory.

5. Conclusions

Well-controlled clinical trials of medical devices have become the standard in the industry for the premarket evaluation of new products and systematic evaluations of performance for products already on the market. The basic principles of good clinical study design developed for trials of pharmaceuticals provide the best foundation for the design of trials of medical devices. Although it is common to discover in the design of a prospective, analytical study of a medical device that there are unexpected sources of bias and confounders that require adjustments to the clinical trial strategy, rigorous studies can be developed using innovative or alternative methods. A successful clinical investigation requires careful planning to clearly delineate a testable hypothesis and to select:

1. A study design that can support the required analysis.
2. A suitable control group.
3. Primary outcome measures that are objective, validated, and clinically relevant.

It is important to remember that some questions regarding safety and effectiveness of medical devices are not readily answered in prospective, controlled clinical studies—especially for implanted devices. Most notably, it is necessary to rely on a combination of data from bench testing, animal studies, device-retrieval analysis, and observational research to ascertain with some degree of confidence the anticipated durations of in vivo performance, failure modes, and long-term fate of implanted materials.

References

1. U.S. Congress. *Federal Food, Drug, and Cosmetic Act of 1938*. Public Law Number 75-717, 52 Stat. 1040 (1938) 21 USC §201.
2. Study Group on Medical Devices. 1970. *Medical Devices: A Legislative Plan*. Department of Health, Education, and Welfare, Washington, DC.

3. Food and Drug Administration. 2004. *Determination of Safety and Effectiveness.* 21 CFR §860.7(c)(2).
4. U.S. Congress. *Federal Food, Drug, and Cosmetic Act of 1938.* Public Law Number 75-717, 52 Stat. 1040 (1938) 21 USC §505.
5. Hutt, P.B., Merrill, R.A. and Kirschenbaum, A.M. 1992. Devices are not drugs: the standard of evidence required for premarkert approval under the 1976 medical device amendments. Unpublished.
6. Food and Drug Administration. 1993. *Final Report of the Committee for Clinical Review. "The Temple Report."* FDA Report. March 1993. pp. 1–45.
7. Food and Drug Administration. 1995. Notice. Statement regarding the demonstration of effectiveness of human drug products. *Fed. Regist.* 60:39,180–39,181.
8. Marwick, C. 2000. Implant Recommendations. *JAMA.* 283:869.
9. Food and Drug Administration. 2004. *General Provisions. Definitions.* 21 CFR §812.3(m).
10. Food and Drug Administration, Center for Devices and Radiological Health. 1995. *Clinical Trial Guidance for Non-Diagnostic Medical Devices.* Rockville, MD, pp. 1–25.
11. Food and Drug Administration, Center for Devices and Radiological Health. 1996. *Statistical Guidance for Clinical Trials of Non-Diagnostic Medical Devices.* Rockville, MD, pp. 1–32.
12. Spilker, B. and Schoenfelder, J. 1990. *Presentation of Clinical Data.* Raven, New York.
13. Food and Drug Administration, Center for Drug Evaluation and Research. 1998. *Guideline for the Format and Content of the Clinical and Statistical Sections of an Application.* Rockville, MD, pp. 1–125.
14. Minert, C.L. 1986. *Clinical Trials. Design, Conduct, and Analysis.* Oxford University Press, New York.
15. Food and Drug Administration. 1995. International conference on harmonization. Draft guidelines on good clinical practice. *Fed Regist.* 60:42,948–42,957.
16. Committee for Proprietary Medicinal Products. 1994. *Biostatistical Methodology in Clinical Trials in Applications for Marketing Authorization for Medicinal Products.* Document No. III/3630/92-EN. European Commission, Directorate-General III, Brussels.
17. Food and Drug Administration. 2004. *Protection of Human Subjects.* 21 CFR §50.
18. Declaration of Helsinki. World Medical Assembly, Helsinki, Finland, 1964; 1975; 1983.
19. CEN. European Committee for Standardization. 1993. *Clinical Investigation of Medical Devices for Human Subjects.* Document No. EN 540; 1993 E. European Standard, Central Secretariat, Brussels.

2

Regulatory Requirements for Clinical Studies of Medical Devices and Diagnostics

Daniel A. Kracov and Lisa M. Dwyer

1. Introduction

The investigational device exemption (IDE) provides an release for medical devices from various sections of the federal Food, Drug, and Cosmetic (FD&C) Act.[1] Without the exemption, medical devices would have to comply with performance standard, premarket approval, or notification requirements to be lawfully shipped and used for investigational purposes.[2] Furthermore, itwould be exceedingly difficult—if not impossible—to conduct clinical trials for devices to support premarket approval applications (PMAs) or 510(k) premarket notifications without violating the act, if the IDE did not exist. Indeed, according to Congress and the Food and Drug Administration (FDA), the twin objectives of the exemption are "to encourage discovery and development of useful devices for human use"[3] and "to protect the public health by requiring safeguards for human subjects of investigations, sound ethical standards, and procedures to assure development of reliable scientific data."[4]

1.1. Brief History of Clinical Trial Regulation of Devices

The IDE was enacted as part of the Medical Device Amendments of 1976 ("the Amendments").[5] As the FDA had regulated the investigational use of new drugs and biologics for years,[6] it was widely recognized that the new framework for medical device regulation required a corresponding—if not equivalent—framework for controlling investigational devices. The statutory provision granting the exemption, which can be found in Section 520(g) of the FD&C Act,[7] expressly authorized the Secretary of what was then called the Department of Health, Education and Welfare to regulate the investigational use of medical devices.

From: *Clinical Evaluation of Medical Devices: Principles and Case Studies, Second Edition*
Edited by: K. M. Becker and J. J. Whyte © Humana Press Inc., Totowa, NJ

Notably, Section 520(g) required the agency to issue regulations within 120 days of the effective date of the 1976 Amendments.[8] These regulations are now found in 21 C.F.R. Part 812. Although the IDE regulations vary from the investigational new drug regulations, they are substantively similar in many respects.[9] One fundamental difference, however, is that Section 520(g) of the act expressly permits FDA to vary the procedures and conditions of the regulation of investigational devices based on the nature of the device, the scope and duration of the trial, the number of human subjects involved, the need for changes to be made in the device during the investigation, and whether the purpose of the data is to obtain approval to commercially distribute the device.[10]

1.2. Investigational Devices Subject to Regulation

Clinical investigations for most new devices or for new uses of devices are subject to the IDE regulations in 21 C.F.R. Part 812. However, FDA exempted clinical investigations for the following types of devices:[11]

1. Pre-1976 Amendment devices that currently are being investigated for the same indications that existed before the 1976 Amendments.
2. Post-1976 Amendment devices that are being investigated for purposes that have been cleared through the 510(k) premarket notification process.
3. Diagnostic devices that meet certain requirements.
4. Devices undergoing consumer preference testing and other similar types of testing if the testing is not used to determine safety and efficacy and does not put subjects at risk.
5. Devices intended solely for veterinary use.
6. Devices being tested solely on or with laboratory animals.
7. Custom devices.[12]

1.3. Structure of FDA Regulation of the Investigational Device Exemption

As mentioned, the IDE regulations permit sponsors to conduct clinical studies on new devices or new device indications to collect the requisite safety and efficacy data to support PMAs and, in some instances, 510(k) premarket notifications. The IDE regulations establish minimum requirements that must be met before an investigational device study begins. For example, the sponsor of any clinical device investigation must obtain approval from an Institutional Review Board (IRB) and informed consent from study subjects before the study begins.

If the investigational device poses a serious risk to the health, safety, or welfare of a subject (i.e., a significant risk [SR] device),[13] the sponsor also must obtain FDA's approval of the IDE application. The IDE application must contain information concerning the study's investigational plan, prior investigations, device manufacture, IRB actions, investigator agreements, the subjects' informed consent forms, device labeling, the cost of the device, and other matters related to the study. FDA has 30 calendar days from the date it receives the application to approve or disapprove the application.[14]

2. Regulations Relating to People and Institutions Engaged in Clinical Trials

The FDA has promulgated a number of regulations delineating the responsibilities of the key players involved in clinical trials for devices (i.e., sponsors, investigators, IRBs).

2.1. Sponsors

A *sponsor* is a person, an institution, or a company that initiates, but does not actually conduct, an investigation.[15] Importantly, a *sponsor* is distinct from a *sponsor-investigator*, who both initiates and actually conducts or oversees an investigation.[16] Sponsor-investigators must comply with FDA's regulations for both sponsors and investigators.

FDA's regulations provide that sponsors generally are responsible for the following: (1) selecting investigators (i.e., individuals who actually conduct an investigation),[17] (2) providing investigators with necessary information to conduct the investigation, (3) ensuring proper monitoring of the investigation, (4) ensuring that the IRB review and approval are obtained, (5) submitting an IDE application to FDA, and (6) ensuring that any reviewing IRB and FDA are promptly informed of any significant new information about an investigation.[18] Sponsors must also comply with the labeling, reporting, and record-keeping requirements established in 21 C.F.R. Part 812 and refrain from engaging in promotional activities and the other prohibited activities enumerated in 21 C.F.R. §812.7 (e.g., commercializing an investigational device).[19]

FDA regularly issues warning letters to sponsors who fail to comply with these general responsibilities. For example, on October 3, 2003, FDA sent a warning letter to an orthopedic device company that sponsored a device investigation regarding its failure to comply with its responsibilities in Part 812 of the regulations.[20] Among violations mentioned in the warning letter, the company failed to submit an IDE application, failed to ensure that investigators received IRB approval before use of the investigational device, and failed to ensure proper monitoring of the study. As a result, 31 devices (the name of which has been redacted from the warning letter) were implanted in research subjects without FDA and IRB approval. Of those 31 devices, 12 devices were implanted using a procedure or instrument that was not part of the investigational plan, and nine devices were implanted in patients not enrolled in the study. Moreover, the warning letter cited a number of additional deviations from the IDE regulations. This warning letter demonstrates how failure to comply with the IDE regulations can expose patients—even those not enrolled in the study—to uncontrolled situations that may present unnecessary risks.

Table 1
Information Required for Transfer of Sponsorship

Minimum information required for transfer[a]

- Identification of the new sponsor (e.g., name, address, contact person, telephone number)
- Effective date of transfer
- Certification that all relevant records will be transferred to the new sponsor by the effective date of the transfer
- An agreement from the new sponsor, stating that the new sponsor will assume all sponsor responsibilities for the study
- An agreement from the new sponsor, stating that the new sponsor will comply with any terms or outstanding conditions of approval of the investigation

Additional information requested by FDA after transfer[a]

- A statement: that there are no changes in the investigation caused by the transfer or requesting specific approval for changes in the investigation that could affect the scientific soundness of the investigation or the rights, safety, and welfare of the subjects
- Acknowledgment that all investigators and associated IRBs will be informed of the sponsorship change by the effective date
- Certification that the new sponsor will not permit investigators to participate in the investigation until they have signed the investigator agreement.

[a]See id.

2.1.1. Transfer of Sponsorship

To transfer sponsorship to another person, a sponsor must submit a minimum amount of information in the form of an IDE supplement to FDA.[21] Once FDA has acknowledged the transfer, the agency must request additional information. To streamline the process, original and new sponsors may want to consider submitting the minimum information required for acknowledgement of the transfer and the information in response to the required follow-up questions in the initial IDE supplement notifying the agency of the transfer. Table 1 summarizes the information required for the transfer of sponsorship in the initial IDE supplement and in response to FDA's follow-up questions.

2.2. Investigators

An *investigator* is an individual who actually conducts a clinical investigation (i.e., the person who directly oversees the administration of a device to a test subject). In situations in which there is a team of researchers, the investigator is the team leader.[22]

An investigator's general responsibilities include:[23] (1) ensuring that an investigation is conducted in accordance with a signed agreement, the investigational plan, and FDA regulations, (2) protecting the rights, safety, and welfare of subjects under the investigator's care, (3) controlling devices under the investigation (e.g., supervising device use and disposal),[24] and (4) ensuring that informed consent is obtained from subjects.[25] Investigators, like sponsors, are also subject to certain record-keeping and reporting requirements.[26]

The FDA does not hesitate to issue warning letters to investigators who fail to comply with these general responsibilities. For example, on July 26, 2004, FDA sent a warning letter to an investigator, citing him for conducting an investigation that made numerous deviations from the study protocol, including failing to exclude subjects who did not meet study criteria; failing to measure body mass index, as required by the study protocol; and failing to perform follow-up visits in a timely manner, also as specified by the protocol.[27]

2.2.1. Financial Disclosure Requirements

When reviewing studies included in marketing applications, the FDA may consider clinical studies inadequate if it determines that the studies are biased. Bias in clinical studies can arise if investigators have a financial interest in the outcome of the study, a proprietary interest in the product, or an equity interest in the sponsor of the study.[28]

Thus, the FDA requires all investigators to disclose accurate financial information to the sponsor, so that the sponsor is able to submit to the FDA either: (1) certification that no suspect financial arrangements (e.g., compensation affected by the outcome of the study, significant equity interest in the sponsor of the study, or a proprietary interest in the tested product) exist, or (2) a disclosure statement disclosing any suspect arrangements and the steps that have been taken to minimize bias.[29]

In addition, FDA requires all investigators to promptly update their financial information if any relevant changes occur during the course of the investigation.[30] FDA uses this financial information, along with information about the design and purpose of the study, to assess whether the data in the study is reliable.[31]

2.2.2. Disqualification of Investigator

The FDA may initiate the process to disqualify investigators and render them ineligible to receive investigational devices if it has information that: (1) the investigator has deliberately failed to comply with regulations governing IDEs,[32] protection of human subjects,[33] or IRBs;[34] or (2) the investigator has repeatedly or deliberately submitted false information either to the sponsor of the investigation or in any required report.[35]

Once the FDA has received such information, the Center for Devices and Radiological Health (CDRH) will give the investigator written notice of the matter and offer the investigator an opportunity to explain it in an informal conference. If the CDRH accepts the explanation, it will stop the disqualification process; however, if CDRH does not accept the explanation, it will give the investigator an opportunity for a formal hearing on whether the investigator should be entitled to receive investigational devices.

If the process progresses to a formal hearing, the FDA Commissioner will consider all available evidence on the matter. If the commissioner then determines that the investigator should be disqualified, he or she notifies the investigator, the sponsor of any study in which the investigator is a participant, and the IRB that the investigator is not entitled to receive investigational devices.

Notably, the FDA rarely disqualifies investigators. In fact, from 1966 through June 2004, the FDA's list of disqualified investigators shows that only about 95 clinical investigators have been disqualified based on their investigations of drugs, biologics, and medical devices. Of the 95 disqualified investigators on FDA's list, only one was investigating medical devices.[36]

2.3. Institutional Review Boards

An *IRB* is any board, committee, or other group formally designated by an institution to review biomedical research involving human subjects,[37] and it serves an important role in safeguarding the health and welfare of the subjects.

Under the FDA regulations, IRBs must review investigations, and continually monitor them throughout the length of the study.[38] IRBs also have the authority to approve, disapprove, or require modification of investigations.[39] IRBs typically use a group process to review investigation protocols and related materials, such as informed consent documents.

Notably, most established research institutions have an IRB and/or established researchers that are affiliated with a certain IRB; however, in the absence of an IRB affiliated with an institution or a researcher, independent IRBs can be contracted to act as an IRB for a given investigation or IRBs can be established in accordance with 21 C.F.R. Part 56. The FDA regulations set forth specific organization and membership requirements as well as requirements relating to review procedures, approval criteria, and record-keeping requirements.[40]

3. Pre-Investigation Device Exemption Process

FDA's Office of Device Evaluation (ODE) reviews IDE applications. FDA generally encourages applicants to begin communicating with the ODE before they submit an original IDE application via a pre-IDE meeting and/or a pre-IDE submission.[41] Pre-IDE meetings are beneficial for the ODE reviewing division and the applicant because they increase the applicant's familiarity with the review process, increase the ODE's familiarity with the technology at issue, expedite the regulatory process, and minimize delays.

3.1. Pre-Investigational Device Exemption Meetings

The FDA encourages applicants to contact the FDA for a pre-IDE meeting before they submit their IDE application. Pre-IDE meetings are particularly beneficial for applicants who do not have previous experience with the agency and for those who are planning to study new technologies or new uses for existing technologies.

3.1.1. Informal Guidance Meetings

Informal guidance meetings provide applicants with the opportunity to meet with the ODE reviewing division before they submit an IDE application.[42] Typically, the meetings serve as forums for the reviewers to provide advice regarding the development of pre-clinical data that would support an IDE application or the development of an investigational plan for the IDE application.

These meetings may be conducted via telephone, videoconference, or in person. During the meeting, the ODE reviewing division is responsible for recording the meeting. The minutes from the meeting generally include: (1) the date of the meeting, (2) the attendees, (3) whether material was submitted before the meeting for discussion or review, (4) a summary of the discussion, and (5) all of the ODE's recommendations for the sponsor. Applicants may schedule informal guidance meetings simply by contacting the ODE reviewing division or the IDE staff.

3.1.2. Formal Guidance Meetings

The Food and Drug Modernization Act of 1997 expressly provided for two types of formal pre-IDE meetings to provide applicants with clear direction on the testing and development of devices requiring clinical investigations: *determination meetings* and *agreement meetings*.[43]

The main objective of a determination meeting is to provide anyone submitting a PMA or a product development protocol[44] with a determination regarding the type of valid scientific evidence that will be necessary to demonstrate that the device is effective for its intended use. Typically, determination meetings focus on deciding whether clinical studies are necessary to establish effectiveness. The meetings also focus on developing a broad outline of a clinical trial design that represents the least burdensome method of evaluating device efficacy and has a reasonable likelihood of success.[45] By statute, the FDA is required to provide a determination to the applicant in writing within 30 days following the meeting[46] and the determinations are binding.[47] Applicants should submit requests for determinations in the form of a pre-IDE submission and identify the request as a "determination meeting request."[48]

The main objective of the agreement meeting is to establish the parameters of the investigational plan for the device, including the clinical protocol. Investigational plans include a written protocol describing the trial methodology, a risk analysis, and monitoring procedures.[49] Applicants should submit requests for agreement meetings in the form of a pre-IDE submission and identify them as an "agreement meeting request."[50] By statute, FDA must meet with the applicant no later than 30 days after receipt of the request. The written request should include: (1) a detailed description of the device, (2) a detailed description of the proposed conditions of use of the device, (3) a proposed plan (including a clinical protocol) for determining whether there is a reasonable assurance of effectiveness, and, if available, (4) information regarding the expected performance of the device.[51] Agreements reached during these meetings are binding, and the agreements, which must be put in writing, become part of the administrative record.[52]

3.2. Pre-Investigation Device Exemption Submissions

In addition to informal and formal guidance meetings, sponsors may submit pre-IDE submissions to the ODE reviewing division if they would like informal guidance on certain parts of the IDE application (e.g., clinical protocol design, the preclinical testing proposal, preclinical test results, protocols for foreign studies used to support the FDA marketing applications). Pre-IDE submissions must be clearly identified as such and submitted in duplicate.

Once the FDA receives pre-IDE submissions, it logs them in its pre-IDE tracking system, where they are attached to a tracking sheet. The submission is then forwarded to the appropriate review division, which will send an acknowledgment letter to the sponsor. The review division must then issue a response via a letter, a meeting, or a telephone call to the sponsor within 60 days of receiving the submission. If the response is issued during a meeting or a telephone call, the division must prepare a memorandum memorializing the conversation.

4. Investigational Device Exemption Application

A study sponsor must have an approved IDE application before it can conduct clinical studies on an investigational device, which may be subject to certain exemptions (e.g., certain diagnostic devices, devices intended solely for veterinary use).[53] Furthermore, clinical investigations are subject to different levels of regulation, depending on the level of risk.

Table 2
Examples of Significant Risk and Nonsignficant Risk Devices

Significant risk devices[a]	Nonsignificant risk devices[b]
Cardiac pacemakers	Daily wear contact lenses and solution
Orthopedic implants	Ultrasonic dental scalers
Sutures	Foley catheters
Hydrocephalus shunts	Denture repair kits
Respiratory ventilators	External insulin monitors

[a]*See* Information Sheets. Guidance for Institutional Review Boards and Clinical Investigators, 1998 Update. Available from: www.fda.gov/oc/ohrt/irbs/devices.html#risk;[b] *see also* Device Advice. Clinical Trial and IDE. Approval Process. Available from: www.fda.gov/cdrh/devadvice/ide/approval.shtml.

4.1. Significant Risk Devices vs Nonsignificant Risk Devices

There are two paths to obtaining an IDE approval, depending on whether the product is an SR device or nonsignificant risk (NSR) device (*see* Table 2).[54] An SR device poses serious risk to the health, safety, or welfare of a subject.[55] An investigational device should be classified as an SR device if its studies pose life-threatening harm, could cause permanent physical damage or impairment, or would require medical intervention to prevent such damage.[56] An NSR device is any device that does not meet the SR definition.[57]

The determination of whether a device is SR or NSR has important implications for both research sponsors and regulators because SR device studies are subject to more stringent regulatory requirements.[58] SR and NSR device procedures differ primarily in the approval process as well as in record-keeping and reporting requirements.[59] If a sponsor determines that an investigational product is an SR device, the clinical investigation cannot begin until FDA approves the IDE application and the IRB approves the study.[60] Furthermore, the study sponsor and investigators for an SR device must comply with the full list of IDE requirements in 21 C.F.R. Part 812.[61]

In contrast, the FDA established the NSR device category to "avoid delay and expense where the anticipated risk to human subjects [does] not justify the [agency's] involvement."[62] As a result, an NSR device sponsor follows abbreviated IDE regulation requirements.[63] For instance, the FDA does not require an IDE application for an NSR device study, and the sponsor need only obtain approval from an IRB, not from the FDA. The IRB serves as FDA's proxy for the NSR review and approval process.[64]

FDA considers an NSR device study to have an approved IDE application if the sponsor complies with the abbreviated requirements,[65] and the sponsor may begin clinical trials once the IRB approves its study.[66] If the sponsor believes that the device it is studying is an NSR, it must explain this finding to the reviewing IRB and include other information that will help the IRB assess any risk posed by the investigation.[67] If the IRB agrees with the sponsor's NSR determination and approves the study, the sponsor may begin clinical trials immediately without submitting an IDE application to the FDA;[68] however, if the IRB concludes that the device poses a significant risk, the sponsor cannot begin investigating the device until it has received the agency's approval.[69]

4.2. Nonsignificant Risk Devices: Abbreviated Requirements

When an investigational device is designated as an NSR device, the sponsor participates in a streamlined approval and application process. In addition to bypassing FDA review of the device study, the sponsor of an NSR device is also subject to less stringent IDE regulations.[70] Section 812.2(b) of the FDA's IDE regulations outlines the abbreviated IDE requirements for an NSR device investigation.[71]

The abbreviated IDE regulations merely require the sponsor to: (1) label the NSR device in accordance with the regulations,[72] (2) obtain the reviewing IRB's approval for the NSR device study (and maintain that approval for the study's duration), (3) ensure that each investigator obtains and documents participating subjects' informed consent, (4) monitor the investigation to protect subjects and ensure compliance, (5) maintain certain records and make certain reports,[73] (6) ensure that participating investigators maintain certain records and make certain reports,[74] and (7) comply with the prohibitions against commercialization, promotion, and test marketing, among other things, in 21 C.F.R. §812.7.

4.3. Significant Risk Devices: Application Requirements

Once a device has been designated an SR device, its sponsor must comply with all the IDE regulations in 21 C.F.R. Part 812, including the application requirements. Section 812.20 of the IDE regulations provides a detailed list of the information a sponsor must include in the IDE application.[75]

As specified by the IDE regulations, the sponsor of an SR device study must provide the following information for the IDE application, in the following order:[76]

1. The sponsor's name and address.
2. A comprehensive report of previous investigations for the device and a complete investigational plan.[77]

3. A detailed description of the methods, facilities, and controls used for the device's manufacture, processing, packing, storage, and installation.
4. An example of the agreements that the investigators will sign to comply with their regulatory obligations, and a complete list of the names and addresses of the investigators who signed the agreement.[78]
5. A certification that all participating investigators have signed the agreement and that no investigator will be added without signing the agreement.
6. A list of the names, addresses, and chairpersons of each IRB that has been—or will be—asked to review the investigation, as well as a certification concerning each IRB's action in the investigation.
7. The name and address of any institution where part of the investigation may occur that has not yet been identified.
8. If the device will be sold, the price of the device and an explanation of why the sale does not qualify as commercialization.
9. An environmental assessment or a claim for categorical exemption from the requirement.
10. Copies of all labeling for the device.
11. Copies of all informed consent forms and informational materials that will be given to study subjects.
12. Any other relevant information that FDA requests to review the IDE application.[79]

Notably, the FDA provides an administrative checklist that sponsors can use to ensure that their applications are complete. It is available from www.fda.gov/cdrh/devadvice/ide/application.shtml#modifications. Once the sponsor of an SR device study is sure that the application is complete, it must submit three copies of the signed IDE application and a cover letter to the FDA.[80]

4.4. Modifications to Investigation Device Exemption Applications

Although sponsors of SR device studies can make some modifications to IDE applications without prior approval, any changes to the investigational plan affecting the study design's scientific soundness or data validity require regulatory review.[81] Before making such a change to the investigational plan, the sponsor must submit a supplemental IDE application to FDA and obtain IRB approval.[82] Investigational plan modifications requiring a supplemental IDE include:

1. A change in indication.[83]
2. A change in type or nature of study control.[84]
3. A change in primary endpoint.[85]
4. A change in the method of statistical evaluation.[86]
5. Early termination of the study.[87]
6. Increasing the number of study subjects.[88]
7. Increasing the number of investigational sites.[89]

Under certain circumstances, a sponsor may modify a device or the investigational study without submitting a supplemental IDE application, but the FDA must be notified of the change within 5 working days.[90] For instance, if the sponsor departs from investigational protocol for emergency use (i.e., to protect the subject's life or welfare), it does not need FDA approval before implementing the change.[91] Similarly, notice of 5 working days is sufficient for developmental changes that do not significantly alter the device's design or operation and for certain changes to clinical protocol, as long as the protocol changes do not affect the data validity, information on the approved protocol, or the risk/benefit ratio for subjects; the investigational plan's scientific soundness, or the study subjects' rights, safety, or welfare.[92] Notably, the changes to the development and protocol have to meet the exceptions for submitting a supplemental IDE application on the basis of "credible information," which is specifically defined in the regulations.[93]

Notably, for extremely minor changes, sponsors need not submit an IDE supplement or provide the agency with 5 day notice. Sponsors need only report changes in the annual progress report if the changes are in the following areas:

1. The purpose of the study.
2. Risk analysis.
3. Monitoring procedures.
4. Labeling.
5. Informed consent materials.
6. IRB information, if the changes do not affect the data validity, information on the approved protocol, or the risk–benefit ratio for subjects; the investigational plan's scientific soundness; or the study subjects' rights, safety, or welfare.[94]

4.5. Disapproval or Withdrawal of Approval of Investigation Device Exemption Applications

The FDA can approve an IDE application as submitted or with modification, or it can disapprove or even withdraw approval of an IDE application.[95] The IDE regulations list the various grounds for IDE disapproval or withdrawal.[96] The FDA may disapprove or withdraw an IDE application if:

1. The sponsor has not complied with the applicable IDE regulatory requirements, statutes or regulations, or any condition imposed by the IRB or the FDA.
2. The application or report contains false statements or misleading information.
3. The sponsor does not respond to the FDA's request for additional information in a timely fashion.
4. The benefits of the study or its importance do not outweigh the risks, informed consent is inadequate, the investigation is scientifically unsound, or the device is used ineffectively.

5. It is unreasonable to begin or continue the investigation because of how the device is used.
6. There is an inadequacy in the investigational plan; the methods, facilities, and controls used to manufacture, process, package, store, and install the device; or the monitoring of the investigation.[97]

Both a disapproval order and a proposed withdrawal of approval of an IDE application must state the reasons for disapproval or withdrawal, and they must give the sponsor notice that it has the opportunity to request a hearing,[98] although the FDA should communicate with the sponsor in other manners before issuing a proposed withdrawal.[99] For instance, the ODE should send correspondence to the sponsor stating its concerns and referencing the basis for its concerns.[100] If the ODE believes that a meeting may alleviate its concerns, it may request a meeting with the sponsor.[101] Under certain circumstances, the Office of Compliance (OC) should send warning letters specifying any issues that need to be addressed.[102]

4.6. Closing an Investigation Device Exemption

The complexity of closing procedures depends on when the sponsor decides to close the IDE.[103] If the decision comes early in the IDE process, the sponsor has fewer requirements to satisfy. For example, if the FDA has not yet approved the IDE, the sponsor can simply request to withdraw the IDE application.[104] Moreover, even if the FDA has already approved the IDE application, the sponsor may still request withdrawal if it has not yet enrolled any subjects.[105]

Once a sponsor has enrolled subjects in a study, the sponsor must conform to stricter requirements. Although the sponsor can stop enrolling subjects in the study at this stage, the sponsor must follow-up with all enrolled subjects in accordance with the investigational plan before closing the IDE.[106] The IDE is not officially closed until the sponsor has followed up with all study subjects and has submitted a final report to the FDA and the reviewing IRBs within 6 months thereafter.[107]

5. Investigation Device Exemption Study Requirements

After the necessary approval for the IDE application is obtained, clinical studies authorized under the application must be conducted in accordance with the requirements and regulations collectively referred to as Good Clinical Practices (GCPs). The GCPs require, among other things, oversight by an IRB (*see* Section 2.3), informed consent, and monitoring of the clinical study as well as submission of reports and records maintenance. Additionally, during the clinical studies, certain behavior, such as promotion or commercialization of the investigational device, is prohibited.

5.1. Informed Consent

Generally, investigators must obtain a written informed consent form from the patient or the patient's representative before the patient participates in a clinical study under an IDE application.[108] Because the FDA established the informed consent regulations in part to protect patients, the regulations require informed consents to contain adequate information in understandable language to allow patients to determine whether to participate in the study. They also require that the patients have adequate time to consider participation in the study, to minimize the possibility of coercion or undue influence. In addition, they prohibit exculpatory language that in any way appears to require the patient to waive any legal rights or release the investigator, sponsor, or institution from liability for negligence.

The following basic elements must be included in the informed consent:[109]

1. A study summary (i.e., a statement that the study involves research, an explanation of the purposes of the research, a statement regarding the expected duration of the subject's participation, and a description of the procedures to be followed and identification of any procedures that are experimental).
2. A description of any reasonably foreseeable risks or discomfort to the patient.
3. A description of any benefits to the subject or to others that may reasonably be expected from the research.
4. A disclosure of appropriate alternative procedures or courses of treatment, if any, that might be advantageous to the subject.
5. A statement describing the extent, if any, to which confidentiality of records identifying the subject will be maintained and that notes the possibility that the FDA may inspect the records.
6. For research involving more than minimal risk, an explanation regarding any compensation and whether any medical treatments are available if injury occurs and, if so, what they consist of or where further information may be obtained.
7. An explanation of whom to contact for answers to pertinent questions about the research and research subjects' rights and whom to contact in the event of a research-related injury to the subject.
8. A statement that participation is voluntary, that refusal to participate will involve no penalty or loss of benefits to which the subject is otherwise entitled, and that the subject may discontinue participation at any time without penalty or loss of benefits.

The informed consent form must also include any of the following provisions, as appropriate:[110]

1. A statement that the particular treatment or procedure may involve currently unforeseeable risks to the subject (or to the embryo or fetus, if the subject is or may become pregnant).

2. Anticipated circumstances under which the subject's participation may be terminated by the investigator without regard to the subject's consent.
3. Any additional costs to the subject that may result from participation in the research.
4. The consequences of a subject's decision to withdraw from the research and procedures for orderly termination of participation by the subject.
5. A statement that significant new findings developed during the course of the research that may be related to the subject's willingness to continue participation will be provided to the subject.
6. The approximate number of subjects involved in the study.

As part of the IDE process, the FDA must review and approve sample informed consent documents to ensure that they contain the necessary information. The IRB must also review the documents and may modify the language and format to be consistent with the institution's policies and requirements. The informed consent form must be signed by the patient (or legal representative) and maintained by the clinical investigator, and a copy must be given to the patient.

Importantly, the FDA will issue warning letters to investigators who fail to comply with the informed consent regulations. For example, on July 30, 2003, the FDA sent a warning letter to an investigator who failed to include several basic elements in his informed consent forms, such as the expected duration of the study and an identification of the procedures that were experimental.[111] The letter also cited the investigator for failing to obtain signed and dated copies of the informed consent forms and for not providing the person signing the consent form with a copy of the form. According to the letter, all of the informed consent forms were signed by the investigator and/or a witness who had read the forms to the subjects over the telephone—which was insufficient to meet the requirements of the informed consent regulations.

5.2. Exception to Informed Consent: Emergency Research

The FDA provides exceptions to the informed consent requirements,[112] including an exception for emergency research.[113] The exception for emergency research addresses circumstances in which the human subject is in a life-threatening situation[114] and it is not feasible to obtain informed consent.[115] The emergency research exception is appropriate only if:[116]

1. Participation in the research holds out the prospects of direct benefit to the subjects (for the reasons enumerated in the regulations).[117]
2. The clinical investigation could not practicably be carried out without the waiver.[118]

3. The proposed investigational plan defines the length of the potential therapeutic window based on scientific evidence.[119]
4. The IRB has reviewed and approved informed consent procedures and an informed consent document consistent with the elements of informed consent.[120]
5. Additional protections of the rights and welfare of the subjects will be provided, including the minimum protections enumerated in Section 50.24(a)(7) of the regulations.[121]

5.3. Monitoring Clinical Investigations

Under the IDE regulations, sponsors are responsible for ensuring that clinical investigations are appropriately monitored.[122] Accordingly, the regulations require sponsors to identify the name and address of a monitor who has appropriate training and experience.[123] Notably, the FDA's published guidelines on acceptable monitoring approaches[124] require that sponsors provide written monitoring procedures for all studies involving more than one investigator.

Sponsors are also responsible for securing compliance with the signed agreement, the investigational plan, the IDE regulations, and other conditions imposed by the FDA or an IRB. Sponsors who discover that an investigator is not complying must discontinue shipments of the device to the investigator promptly and terminate the investigator's participation in the study. Sponsors must also require the investigator to return or otherwise dispose of the device, unless it would jeopardize a research subject.[125]

Sponsors who fail to secure compliance are subject to warning letters. For instance, the FDA sent Silimed, Inc., a warning letter on August 21, 2003. The agency cited Silimed for failing to document that the occurrence of a study initiation visit, which was needed ensure that investigators were complying with the regulations and study protocol, in accordance with the IDE monitoring plan.[126]

The regulations also require sponsors to immediately conduct evaluations of any unanticipated adverse effect of a device. If the sponsor determines that the device presents an unreasonable risk to research subjects, the sponsor must terminate all investigations (or the parts of the investigations that present the risk) immediately (i.e., not later than 5 working days after the sponsor makes the determination and not later than 15 working days after the sponsor receives notice of the effect).[127] If the study is terminated because the sponsor has determined that the device presents an unreasonable risk, the sponsor cannot resume the investigation without express approval of the FDA and IRB(s), regardless of whether the study involves an SR or NSR device.[128]

5.4. Records and Reports

During the IDE study, sponsors and investigators have significant responsibilities to maintain accurate, complete, and current records and to make timely required reports. Record-keeping responsibilities for sponsors and investiga-

Table 3
Record-Keeping Requirements for Sponsors[a]

SR devices

- All correspondence with another sponsor, an investigator, a monitor, an IRB, and the FDA, including required reports
- Shipment records, including name and address of consignee, type and quantity of the device, date of shipment, and batch number or code
- Disposition records, such as batch number or code of any devices returned to the sponsor and devices repaired or disposed of in other ways and reason for the method of disposal
- Signed investigator agreements including financial disclosure information
- Records concerning complaints and adverse effects, regardless of whether they are anticipated
- Any other records that the FDA requires to be maintained by regulation or specific requirement for a category of investigation or a particular investigation

NSR devices

- Records concerning the following: (1) name and intended use of the device; (2) objectives of the investigation; (3) a brief explanation of why the device is not an SR device; (4) name and address of each investigator; (5) name and address of each IRB; and (6) statement of the extent to which the good manufacturing practices will be followed in manufacturing the device
- Records concerning complaints and adverse device effects, regardless of whether they are anticipated

[a]*See id.* §812.140.
SR, significant risk; NSR, nonsignificant risk; IRB, Institutional Review Board; FDA, Food and Drug Administration.

tors vary depending on whether the device being investigated presents a significant risk or a nonsignificant risk (*see* Tables 3 and 4).

The FDA regularly issues sponsors warning letters record-keeping violations. For example, on August 30, 2000, the FDA sent a warning letter to Paradigm Medical Industries, Inc., citing the company for failure to maintain financial disclosure information for investigators and records of correspondence with investigators, monitors, and IRBs.[129] Regarding records of correspondence, the FDA found that the company had no documentation showing that the sites were sent revised protocols and given permission to deviate from the protocol requirement of using independent laboratories to perform certain functions. Moreover, the FDA found that there were no copies of progress reports from the investigational sites and no records to show when the data was received from the sites and entered into the database.

Table 4
Record-Keeping Requirements for Investigators[a]

SR devices

- All correspondence, including required reports, with another investigator, an IRB, the sponsor, a monitor, or the FDA
- Records of receipt, use, or disposition of the investigational device, including type and quantity of device; date of receipt; batch number or code; name of person who received, used, or disposed of each device; and why and how many units of the device have been returned to the sponsor, repaired, or otherwise disposed of
- Records of each subject's case history and exposure to the device, including signed and dated consent forms, condition of each subject upon entering the study, relevant previous medical history, record of the exposure to the investigational device (date and time of each use), observations of adverse device effects, medical records, results of diagnostic tests, case report forms, and any other supporting information
- Protocol and documentation (date and reason) for each deviation from the protocol
- Any other records that the FDA requires to be maintained by regulation or specific requirement for a category of investigation or a particular investigation

NSR devices

- Records of each subject's case history and records of exposure to the device. Such records must include documents evidencing informed consent and, for any use of a device by the investigator without informed consent, any written concurrence of a licensed physician and a brief description of the circumstances justifying the failure to obtain informed consent. The case history for each individual must document that informed consent was obtained prior to participation in the study

[a]21 C.F.R. §812.140.
SR, significant risk; NSR, nonsignificant risk; IRB, Institutional Review Board; FDA, Food and Drug Administration.

The FDA takes all record-keeping requirements for investigators seriously and regularly issues warning letters to those who fail to comply. For example, on Nov. 7, 2003, the FDA sent a warning letter to an investigator, noting that the investigator had failed to maintain records of receipt and disposition of the investigational devices as well as each subject's case history and exposure to the device, including pre- and postoperative X-ray reports, X-ray reports for 6- and 12-month study visits, and a signed and dated informed consent form for one subject.[130]

5.4.1. Maintenance of Records

Sponsors and investigators must maintain the required records for 2 years after the date the investigation is completed or terminated or until the records are no longer required to support a PMA or product development protocol, whichever date is later.[131] However, an investigator or sponsor may transfer custody of the records to another person to maintain them during that period.[132] Investigators or sponsors who transfer custody of the records to another person must notify the FDA within 10 working days after the transfer occurs.

Sponsors, IRBs, and investigators are required to permit authorized FDA employees reasonable access at reasonable times to inspect and copy all investigation records.[133] After giving notice, FDA may inspect and copy records that identify subjects. FDA has authority to inspect facilities at which investigational devices are being held, including any establishments where devices are manufactured, packed, installed, used, or implanted.

5.4.2. Sponsor and Investigator Reports to FDA

To keep the FDA and/or IRBs up to date with the progress of the clinical study, the sponsor and investigators are responsible for submitting reports and notifications on various aspects of the study.[134] Table 5 lists the required reports and notifications.

5.5. Prohibitions of Promotion and Other Practices

The IDE regulations prohibit the promotion and commercialization of an SR or NSR device that has not been cleared or approved for marketing by FDA.[135] This prohibition applies to sponsors and investigators (or any person acting on behalf of a sponsor or investigator). These individuals may *not*: (1) promote or test market an investigational device; (2) charge subjects or investigators for the device a price larger than is necessary to recover the manufacturing, research, development, and handling costs; (3) prolong an investigation beyond the point needed to collect data required to determine whether the device is safe and effective; or (4) represent that the device is safe or effective for the purposes for which it is being investigated.

Importantly, the prohibition against promotion does not proscribe efforts to publicize an investigational device for the purpose of obtaining clinical investigators or subjects to participate in a clinical study. The FDA has published a guidance document providing the parameters for solicitations that is intended to ensure that such recruitments efforts are bona fide.[136]

The FDA monitors solicitation efforts and issues warning letters to individuals who do not comply with its regulations. On January 25, 2001, the FDA issued a warning letter to a device manufacturer for making particularly blatant

Table 5
Sponsor/Investigator Reports and Notifications[a,b]

Sponsor

- *Unanticipated adverse device effects*: The sponsor must report the results of an evaluation of an unanticipated adverse device effect to the FDA, all reviewing IRBs, and investigators within 10 working days after the sponsor first receives notice of the adverse effect[c]
- *Withdrawal of IRB approval*: The sponsor must notify the FDA, all reviewing IRBs, and participating investigators of the withdrawal of IRB approval of an investigation (or any part of an investigation) within 5 working days of receipt of the withdrawal of approval[c]
- *Withdrawal of FDA approval*: The sponsor must notify all reviewing IRBs and participating investigators of any withdrawal of FDA approval within 5 working days after receipt of the notice[c]
- *Current list of investigators*: Every 6 months the sponsor must submit to the FDA a current list of the names and addresses of all investigators participating in an SR device investigation
- *Progress reports (or annual reports)*: At regular intervals and at least yearly, the sponsor must provide progress reports to all reviewing IRBs. For an SR device, the sponsor must also submit the progress report to the FDA[c]
- *Recalls and device disposition*: The sponsor must notify the FDA and all reviewing IRBs of any request that an investigator return, repair, or dispose of any unit of an investigational device. The notice must be made within 30 working days after the request is made and must state why the request was made[c]
- *Final report*: For an SR device, the sponsor must notify FDA and all reviewing IRBs within 30 working days of the completion or termination of the investigation. The sponsor must also submit a final report to the FDA, all reviewing IRBs, and participating investigators within 6 months after the completion or termination of the investigation. For an NSR device, the sponsor must submit a final report to all reviewing IRBs within 6 months after completion or termination[c]
- *Informed consent*: Sponsors must submit a copy of any investigator's report of the device's use without first obtaining informed consent. The report must be made to FDA within 5 working days after receipt of the notice of such use[c]
- *SR device determination*: If an IRB determines that the device is an SR device and not an NSR device as the sponsor had proposed to the IRB, a report must be submitted to FDA within 5 working days after the sponsor learns of the IRB's determination[c]
- *Other reports*: The sponsor must provide accurate, complete, and current information about any aspect of the investigation on request from the reviewing IRB or FDA[c]

(continued)

Table 5 (Continued)

Investigator

- *Unanticipated adverse device effects*: The investigator must submit to the sponsor and the reviewing IRB a report of any unanticipated adverse device effect as soon as possible, but no later than 10 working days, after the investigator first learns of the effect[c]
- *Withdrawal of IRB approval*: The investigator must report to the sponsor a withdrawal of approval of the reviewing IRB within 5 working days[c]
- *Progress reports*: The investigator must submit progress reports to the sponsor, the monitor, and the reviewing IRB at regular intervals but no less than on a yearly basis
- *Deviations from the investigational plan*: The investigator must notify the sponsor and the reviewing IRB of any deviation from the investigational plan to protect the life or physical well-being of a subject in an emergency. The notice must be provided as soon as possible but no later than 5 working days after the emergency occurred. If it is not an emergency, prior approval from the sponsor is required for changes in or deviations from the investigational plan. If the change or deviation may affect the scientific soundness of the investigational plan or the rights, safety, or welfare of the subject, the sponsor is required to obtain prior IRB approval and obtain FDA approval for an SR device investigation by submitting a supplemental application
- *Informed consent*: If an investigator uses a device without obtaining informed consent, he or she must report the use to the sponsor and the reviewing IRB within 5 working days after the use occurs[c]
- *Final report*: The investigator must submit a final report to the sponsor and to the reviewing IRB within 3 months after termination or completion of the investigation
- *Other reports*: The investigator must provide accurate, complete, and current information about any aspect of the investigation on request from the reviewing IRB or FDA

[a]*See id.* §812.150(b).
[b]*See id.* §812.150(a).
[c]These reports and notifications are required for NSR devices as well as SR devices.
FDA, Food and Drug Administration; IRB, Institutional Review Board; SR, significant risk; NSR, nonsignificant risk.

claims that a particular device was safe and effective for the purpose for which it was being investigated.[137] In that case, the company claimed in a brochure that the device was able to "alleviate a wide variety of illnesses: neurodegenerative diseases like Alzheimer's, multiple sclerosis, and Parkinson's; cerebral palsy, autism, and epilepsy; and migraine headaches" during studies at multiple clinical facilities in the United States and Europe. The company also claimed that the

FDA had determined that the devices generally presented "no significant risk," despite the fact that the FDA's determination of NSR applied only when the device was used to treat a certain disease. In addition, the company's advertisements to recruit subjects expressly stated that the treatment was safe.

6. Patient Access to Unapproved Medical Devices

Typically, a patient cannot access a medical device that has not been approved or cleared with a PMA or a 510(k), unless the device has an approved IDE application and the patient meets the criteria to participate in the clinical trials for the device.[138] In certain instances, however, physicians may wish to use unapproved devices to save a patient's life or to help a patient who is suffering from a serious disease or condition for which there is no alternative approved therapy. FDA regulations and guidance documents provide the following exceptions to the typical IDE and device approval requirements: (1) emergency use, (2) compassionate use, (3) treatment use, and (4) continued access during PMA preparation and review.

These exceptions can provide patients crucial access during different, although sometimes overlapping, periods of device development. Table 6 summarizes the criteria for each exception, the period during which each exception may be used, and whether the exception requires prior FDA approval. In addition, more detailed discussions of each exception follow.

6.1. Emergency Use

The need for emergency use of an unapproved device may arise any time before or after the initiation of a clinical trial, when an IDE does not exist, when the physician wants to use the device in a way that is outside the scope of the IDE, or when the physician is not a part of the clinical study.[139] The FDA will permit such use, as long as three criteria are met: (1) the patient has a life-threatening condition that needs immediate treatment, (2) there is no alternative treatment, and (3) there is no time to obtain FDA approval.[140]

Of note, physicians who have patients who are potential candidates for emergency use can contact the ODE to discuss a patient's condition, and the ODE may act in an advisory capacity. However, the final decision of whether a situation warrants emergency use lies with the physician—no prior approval by the FDA is required.[141]

The FDA expects that any physician providing a patient with access to an unapproved device under the emergency use exemption would follow as many patient protection procedures as possible (e.g., obtaining an independent assessment from another physician, obtaining an informed consent).[142] In addition, any IDE sponsor shipping an unapproved device to a physician for emergency use should notify the FDA within 5 working days after the shipment is made.[143] An unapproved device may not be shipped in anticipation of an emergency.

Table 6
Early/Expanded Patient Access

	Criteria	Time period	Prior FDA approval required
Emergency use	(1) Life-threatening condition that needs immediate treatment (2) No alternative (3) No time for FDA approval	Before the IDE, market approval	No
Compassionate use	(1) Serious disease or condition (2) No alternative	After clinical trials have begun, market approval	Yes
Treatment use	(1) Serious or immediately life-threatening disease (2) No alternative (3) Device is under investigation in a controlled clinical trial for the same use under an approved IDE (or the clinical trials have been completed) (4) Sponsor of the study is actively pursuing marketing approval or clearance with due diligence	Serious diseases: after the completion of the clinical trial, market approval Life-threatening diseases: immediately after the clinical trial has begun and looks promising, market approval	Yes
Continued access (extended investigation)	(1) Public health need (2) Preliminary evidence shows that the device will be effective and there are no significant safety concerns	After IDE investigation completion, market approval	Yes

FDA, Food and Drug Administration; IDE, investigational device exemption.

43

As discussed, an IRB may also approve emergency research without an informed consent when obtaining the informed consent is not feasible (e.g., the subject cannot give informed consent as a result of his or her medical condition).[144] Sponsors must monitor the progress of all investigations involving emergency research and file certain information with the FDA.[145]

6.2. Compassionate Use

After IDE clinical trials have begun, the FDA recognizes a compassionate use exception for patients who do not have life-threatening conditions and do not meet study criteria.[146] Candidates for the compassionate use exemption must have a serious disease or condition and no available treatment alternative.

Unlike the emergency use exception, the compassionate use exception may be used only with FDA approval, which requires an IDE supplement requesting protocol deviation. The request must include (1) a description of the patient's condition and the circumstances necessitating treatment, (2) a discussion of why alternative therapies are unsatisfactory and why the probable risk of using the investigational device is no greater than the probable risk from the disease or condition, (3) the identification of any deviations in the approved clinical protocol that are needed to treat the patient, and (4) the patient protection measures that will be followed.[147]

If the FDA approves the request, the attending physician should devise an appropriate schedule for monitoring the patient to detect any problems associated with the device. A follow-up report must be submitted to FDA as an IDE supplement.

The compassionate use exception may also be used for a small number of patients. To use the exemption for more than one patient, the physician should speak to the sponsor to request access to the investigational device, and the sponsor should submit to the FDA an IDE supplement with the same information that would be included in the request for a single patient. If the FDA approves the request, sponsors should implement the same procedures that they would follow for a single patient.

6.3. Treatment Use

The FDA also provides a *treatment use* exception that provides expanded access to an investigational device after a clinical trial has already begun and before any final decision is made on the PMA.[148] If an ongoing clinical trial yields promising results and indicates that a device is effective, it may be expanded to include other people with serious or immediately life-threatening diseases. The purpose of this exception is to expedite the availability of promising new devices to desperately ill patients as early in the device development process as possible.[149] This exception covers both diagnostic and therapeutic devices.[150]

The treatment use exception distinguishes between *serious diseases* and *immediately life-threatening diseases* (i.e., a stage of disease in which there is a reasonable likelihood that death will occur within a matter of months or in which premature death is likely without early treatment). For serious diseases, the device generally would not be made available to additional patients until all clinical trials have been completed (but before a final decision is made on the PMA), whereas the device would be made available to patients with immediately life-threatening diseases while trials are underway.[151]

The FDA would consider the use of an investigational device under the treatment use exception if:

1. The device is intended to treat or diagnose a serious or immediately life-threatening disease.
2. There is no alternative therapy.
3. The device is under investigation in a controlled clinical trial for the same use under an approved IDE (or the clinical trials have been completed).
4. The sponsor of the study is actively pursuing marketing approval or clearance with due diligence.[152] Treatment use of an investigational device is also contingent on sponsors and investigators following the safeguards of the IDE process, informed consent regulations, and the IRB procedures.[153] Those applying for treatment use exceptions must complete applications that contain the detailed information required by FDA regulations.[154]

Sponsors may begin using the device 30 days after the FDA receives the treatment use application, unless the FDA notifies the sponsor earlier that treatment may or may not begin.[155] The FDA may disapprove or withdraw treatment use IDEs, after following certain procedures,[156] for the following reasons:

1. The criteria for the treatment use exception have not been met.
2. The device is intended for a serious disease or condition, and there is insufficient evidence of safety and efficacy.
3. The device is intended for an immediately life-threatening disease or condition, and the available scientific evidence, taken as a whole, fails to provide a reasonable basis for concluding that the device is safe and effective.[157]
4. There is reasonable evidence that the treatment use has impeded enrollment or otherwise interfered with the controlled investigation of the same or another investigational device.
5. The device has received marketing approval or clearance or another comparable device or therapy has become available.
6. The sponsor of the clinical trial has not pursuing marketing approval or clearance with due diligence.
7. Approval of the IDE for the controlled clinical trial has been withdrawn.
8. The clinical investigators named in the treatment IDE are not qualified based on their scientific training and/or experience.

9. There has been a failure to comply with any applicable legal requirement.
10. The application contains an untrue statement of material fact or omits material information required by the FDA's regulations.
11. The sponsor has not responded to a request for additional information.
12. There is reason to believe that the risks associated with the device's benefits do not outweigh its risks or that the device is ineffective.
13. It is otherwise unreasonable to begin or continue the use because of the way the device is used or the inadequacy of certain components of the investigation.[158]

Until a manufacturer has filed a marketing application, sponsors of treatment IDEs must submit progress reports to the FDA and all reviewing IRBs on a semiannual basis.[159]

6.4. Continued Access Policy During Premarket Approval Preparation and Review

The FDA may permit subjects to continue using investigational devices after the completion of a clinical trial under certain circumstances, providing there is a public health need or preliminary evidence that the device will be effective and there are no significant safety concerns.[160] Periods when a subject might use the device beyond the investigation include extended investigations, while the sponsor is preparing the marketing application, and while FDA is reviewing the application.

To request an extended investigation, a sponsor should submit an IDE supplement that includes the following:

1. A justification for the extension.
2. A summary of the preliminary safety and efficacy data generated under the IDE;
3. A brief discussion of the risks posed by the device.
4. The proposed rate of continued enrollment (i.e., number of sites and subjects).
5. The clinical protocol (if it differs from the initial clinical trial) and the proposed objectives of the extended study.
6. A brief discussion of the sponsor's progress in obtaining market approval or clearance.[161]

The treatment use exception and the continued access policy overlap in that they are both intended to provide individuals with continued access to a promising device during the market clearance or approval process. Although the treatment use exception can be helpful during the course of a clinical trial (whereas the continued access policy is useful only after the trial has been completed), the treatment use exception is actually somewhat narrower than the continued access policy, since it provides access only to individuals with a serious or an immediately life-threatening disease.[162]

7. In Vitro Diagnostic Device Studies

Investigational in vitro diagnostic devices (IVDs) are subject to less stringent regulation than other investigational devices are. IVDs are defined as:

> reagents, instruments, and systems intended for use in the diagnosis of disease or other conditions, including a determination of the state of health, in order to cure, mitigate, treat, or prevent disease or its sequelae. Such products are intended for use in the collection, preparation, and examination of specimens taken from the human body.[163]

Although in certain circumstances, IVDs must comply with IRB and informed consent requirements,[164] IVDs are exempt from IDE requirements if the testing is noninvasive, does not require invasive sampling presenting significant risk, does not introduce energy into a subject, and is not used diagnostically without confirming the diagnosis with another medically established device or procedure.[165]

Although requirements for IVDs are less stringest, these devices must bear one of the following labeling statements, whichever is applicable, during shipment or delivery: "For Research Use Only. Not for use in diagnostic procedures," or "For Investigational Use Only. The performance characteristics of this product have not been established."[166] In addition, the provisions concerning investigator disqualification apply to IDE-exempt IVDs.[167]

Typically, if IVDs are truly exempt from the IDE requirements and comply with the labeling requirement, the FDA's Division of Bioresearch Monitoring will refrain from issuing warning letters for IDE requirement violations, except in unusual circumstances. Even in unusual circumstances, it would not issue a warning letter without first consulting the Office of Chief Counsel, the OC, and the ODE management.

If the IVD testing fails to meet the criteria for exemption from IDE requirements, the IDE, IRB, and informed consent requirements apply.

8. International Issues

8.1. European Standards for Device Studies

Medical devices were first regulated in a harmonized manner in Europe in 1990, with the Active Implantable Medical Device Directive,[168] and in 1992, with the Medical Device Directive.[169] The following year, the European standard, EN 540: Clinical Investigation of Medical Devices for Human Subjects,[170] was issued to provide medical device manufacturers with guidelines for conducting clinical trials in Europe. The countries subject to the standard included all European Union members, as well as Austria, Finland, Iceland, Norway, Sweden, and Switzerland.[171] The primary purpose of EN 540 was to protect human subjects participating in clinical trials.

Two new harmonized standards were published in 2003 by the International Organization for Standardization (ISO): ISO 14155-1: Clinical Investigation of Medical Devices for Human Subjects—Part 1: General Requirements[172] and ISO 14155-2: Clinical Investigation of Medical Devices for Human Subjects—Part 2: Clinical Investigation Plans.[173] These standards include most of the guidelines that were in EN 540, although they are more comprehensive. The ISO standards essentially replace EN 540.[174]

ISO 14155 Part 1 establishes procedures for conducting investigational device studies and specifies general requirements to protect human subjects.[175] Although the first part does not apply to IVDs, ISO 14155 Part 2 applies to all medical devices being tested for safety and efficacy in human subjects. Part 2 specifically requires that devices in clinical studies be used in the manner that they would be used during normal clinical use, and attempts to ensure that adverse events that would be experienced under normal conditions of use will be observed during the clinical trial. It also attempts to make sure that the investigation will permit sponsors and investigators to assess the risks associated with the investigational device and contains requirements regarding clinical trial organization, conduct, monitoring, data collection, and documentation.[176]

Unlike the FD&C Act and IDE regulations in the United States, the ISO standards are strictly voluntary and serve only as guidelines.[177] Before conducting an investigational device study in a European country, sponsors should consult the specific laws of that country, as well as the ISO 14155 guidelines.

8.2. Clinical Studies Conducted Outside the United States

Generally, the FDA will accept data from a clinical study conducted outside of the United States if the study was conducted under an IDE and complies with all applicable regulations.[178] The FDA will also recognize data that constitute valid scientific evidence[179] and was collected from studies conducted pursuant to the Declaration of Helsinki or the laws and regulations of the country in which the research is conducted, whichever provides greater protection to human subjects.[180] The FDA defines *valid scientific evidence* as:

> [E]vidence from well-controlled investigations, partially controlled studies, studies and objective trials without matched controls, well-documented case histories conducted by qualified experts, and reports of significant human experience with a marketed device, from which it can fairly and responsibly be concluded by qualified experts that there is reasonable assurance of the safety and effectiveness of a device under its conditions of use. The evidence required may vary according to the characteristics of the device, its conditions of use, the existence and adequacy of warnings and other restrictions, and the extent of experience with its use.[181]

The FDA's regulations further state that "[i]solated case reports, random experience, reports lacking sufficient detail to permit scientific evaluation, and unsubstantiated opinions" do not constitute valid scientific evidence, although they may be considered in identifying a device with questionable safety and efficacy.[182]

Regarding the second element of the test, an applicant who chooses to follow the laws of an individual country must state the differences between the Declaration of Helsinki and the country's standards in detail and explain why the country's standards offer more protection.[183]

In addition, the FDA will approve a PMA that otherwise meets the criteria for approval, based *solely* on foreign clinical data if the foreign data are applicable to the US population and medical practice,"[t]he studies have been performed by clinical investigators of recognized competence," and the data may be considered valid without the need for an on-site inspection by the FDA. The FDA can perform an on-site investigation if it deems it necessary or can validate data through other appropriate means.[184] The FDA encourages applicants for PMAs based solely on foreign clinical data to meet with the FDA in a presubmission meeting.[185]

8.3. Exportation of Unapproved Devices for Investigational Use

Although the FDA does not have jurisdiction over how investigational studies are conducted outside the United States, it does have jurisdiction over unapproved devices that are being exported for use in foreign studies. The FDA has a two-tier system for the exportation of unapproved investigational devices, depending on the recipient country. For countries in the first tier, unapproved investigational devices may be exported under Section 802(c) of the FD&C Act[186] without FDA approval; however, unapproved investigational devices may only be exported to countries that are not in Tier I with FDA approval under Section 801(e)(2)[187] of the FD&C Act.[188]

Tier I countries include Australia, Canada, Israel, Japan, New Zealand, Switzerland, South Africa, and member countries of the European Union or the European Economic Area.[189] Section 802(c) of the FD&C Act permits unapproved investigational devices to be exported to these countries without FDA approval, if the unapproved device is exported in accordance with the laws of that country.[190] The device must also:

1. Be in compliance with the specifications of the foreign purchaser.
2. Not be in conflict with the laws of the country to which it is intended for export.
3. Be in a shipping packaged labeled on the outside that it is intended for export.
4. Not be sold or offered for sale in the US commerce.
5. Be in substantial conformity with the current good manufacturing practices or be certified as meeting international standards by a recognized organization.
6. Not be adulterated (other than by lack of marketing approval).
7. Not present an imminent hazard to public health.

8. Be labeled and promoted in accordance with the laws of the country to which the device is being exported.[191]

Devices that are exported under Section 802(c) need not comply with the other IDE requirements in 21 C.F.R. Part 812. When a foreign purchaser requests proof of compliance with US law before export, the FDA will provide a certificate of exportability.[192]

Similar to the rules for Tier I countries, exporting unapproved investigational devices to non-Tier I countries requires that the devices comply with the specifications of the foreign purchaser, not be in conflict with the laws of the country to which it is intended for export, be shipped in a package labeled that it is intended for export, and not be sold or offered for sale in the US commerce. The FDA must also approve these devices for export under Section 801(e)(2). FDA will approve the device for export only if it determines that exporting the device is "not contrary to public health and safety" and "has the approval of the country to which it is intended for export."[193]

Exporters shipping unapproved devices to Tier I countries must also provide the FDA with simple notification when the exporter first begins to export the device to another country; similarly, exporters shipping unapproved devices to non-Tier I countries must notify the FDA of the device and the country involved. All exporters must maintain records regarding exported devices and the countries to which they were exported.[194]

References

1. *See* 21 U.S.C. §320j(g) (Supp. 2003); *see also* 21 C.F.R. § 812.1(a) (2003) (exempting investigational devices from various sections of the Act, such as those governing misbranding, registration, listing, premarket notification, performance standards, and premarket approval).
2. *See* 21 C.F.R. § 812.1(a) (2003).
3. Proposed Investigational Device Exemptions; Cross-Reference Amendments, 41 Fed. Reg. 35282 (Aug. 20, 1976); *see also* 21 U.S.C. §320j(g) (Supp. 2003); 21 C.F.R. §812.1(a) (2003).
4. Proposed Investigational Device Exemptions; Cross-Reference Amendments, 41 Fed. Reg. 35282 (Aug. 20, 1976); *see also* 21 U.S.C. §320j(g) (Supp. 2003); 21 C.F.R. §812.1(a) (2003).
5. Medical Device Amendments of 1976, P.L. 94-205 (1976).
6. *See* 41 Fed. Reg. at 35283.
7. 21 U.S.C. §320j(g) (Supp. 2003).
8. *See* 21 U.S.C. §320j(g)(2)(A) (Supp. 2003).
9. *See* 41 Fed. Reg. at 35283
10. *See* 21 U.S.C. §320j(g)(2)(C) (Supp. 2003).
11. *See* 21 C.F.R. §812.2(c) (2003).

12. A "custom device" is a device that: (1) necessarily deviates from an applicable performance standard or approval requirement in order to comply with the order of an individual physician or dentist, (2) is not generally available to, or generally used by, other physicians or dentists, (3) is not generally available in finished form for purchase or for dispensing upon prescription, (4) is not offered for commercial distribution through labeling or advertising, and (5) is intended for use by an individual patient named in the order of the physician or dentist, and is to be made in a specific form for that patient, or is intended to meet the special needs of the physician or dentist in the course of professional practice. *See id.* §812.3(b) (2003).
13. *See* 21 C.F.R. §812.3(m) (2003); Device Advice, Clinical Trial and IDE, Approval Process, http://www.fda.gov/ cdrh/devadvice/ide/approval.shtml.
14. *See generally* 21 C.F.R. pt. 812 (2003); *see also* Device Advice, Clinical Trial and IDE, Introduction, http://www.fda.gov /cdrh/devadvice/ide/index.shtml#IDE_Over.
15. *See* 21 C.F.R. §812.3(n) (2003).
16. *See id.* §812.3(n).
17. *See id.* §812.3(i).
18. *See* 21 C.F.R. pt. 812, subpt. C (2003).
19. *See infra*, discussion at Section V.
20. *See* FDA Warning Letter to Plus Orthopedics, dated Oct. 3, 2003.
21. *See* Guidance on IDE Policies and Procedures, IDE Staff, Office of Device Evaluation, Center for Devices and Radiological Health, FDA (Jan. 20, 1998) ("Guidance on IDE Policies and Procedures"), at 6-7.
22. *See* 21 C.F.R. §812.3(i) (2003).
23. *See id.* at pt. 812, subpt. E.
24. *See id.* §812.110.
25. *See also* 21 C.F.R. at pt. 50 (2003).
26. *See id.* §§812.140, 812.150; *see infra*, discussion at Section V(D).
27. *See* FDA Warning Letter to Hans C. Kioschos, M.D., dated July 26, 2004.
28. *See* 21 C.F.R. §54.1 (2003).
29. 21 C.F.R. §§54.4, 812.110(d); *see also* Guidance: Financial Disclosure by Clinical Investigators, FDA (Mar. 20, 2001); Device Advice, Responsibilities, http://www.fda.gov/cdrh/devadvice/ide/ responsibilities.shtml#ResponInvestigators.
30. *See* 21 C.F.R. §812.110(d) (2003).
31. *See id.* §54.1.
32. *See id.* at pt. 812.
33. *See id.* at pt. 50.
34. *See id.* at pt. 56.
35. *See id.* §812.119.
36. *See* Disqualified/Totally Restricted List for Clinical Investigators, FDA, http://www.fda.gov/ora/compliance_ref /bimo/disqlist.htm.
37. *See* 21 C.F.R. §812.3(f) (2003).
38. *See id.* §§812.62, 812.64.

39. *See id.* §812.62(a).
40. *See id.* at pt. 56.
41. Guidance on IDE Policies and Procedures, at 1.
42. *See id.*
43. 21 U.S.C. §§360c(a)(3)(D), 360j(g)(7) (Supp. 2003).
44. A product development protocol is an alternative to the PMA process for Class III devices, which allows sponsors to come to early agreement with the FDA as to what needs to be done to demonstrate the safety and effectiveness of a new device. *See* Draft Guidance for Industry – Contents of a Product Development Protocol, ODE (Jan. 27, 1998).
45. *See* Early Collaboration Meetings Under the FDA Modernization Act (FDAMA); Final Guidance for Industry and for CDRH Staff, at 1.
46. *See* 21 U.S.C. §§360c(a)(3)(D) (Supp. 2003).
47. *See* Early Collaboration Meetings Under the FDA Modernization Act (FDAMA); Final Guidance for Industry and for CDRH Staff, at 3.
48. *See* Device Advice, Clinical Trial and Investigational Device Exemption ("IDE"), http://www.fda.gov/cdrh/devadvice/ide/approval.shtml#pre_ide
49. *See* 21 C.F.R. §812.25 (2003).
50. *See* 21 U.S.C. §§360j(g)(7) (Supp. 2003); Device Advice, Clinical Trial and Investigational Device Exemption ("IDE"), Approval Process, http://www.fda.gov/cdrh/devadvice/ide/approval.shtml#pre_ide
51. *See* 21 U.S.C. §§360j(g)(7) (Supp. 2003); Device Advice, Clinical Trial and IDE, Approval Process, http://www.fda.gov/cdrh/devadvice /ide/ approval.shtml#pre_ide; *see supra*, discussion at Section I(C).
52. *See* 21 U.S.C. §§360j(g)(7) (supp. 2003); Early Collaboration Meetings Under the FDA Modernization Act (FDAMA), Final Guidance for Industry and for CDRH Staff, Feb. 28, 2001, at 3, http://www.fda.gov/cdrh/ode/guidance/310.html.
53. *See* 21 U.S.C. §360j(g) (Supp. 2003); 21 C.F.R. §812.1 (2003); FDA, Device Advice, Clinical and IDE Approval Process, Introduction, http://www.fda.gov/cdrh/devadvice/ide/index.shtml.
54. *See* Information Sheets, Guidance for Institutional Review Boards and Clinical Investigators, 1998 Update, http:// www.fda.gov/oc/ohrt/irbs/devices.html#risk.
55. *See* 21 C.F.R. §812.3 (2004); Device Advice, Clinical Trial and IDE, Approval Process, http://www.fda.gov/ cdrh/devadvice/ide/approval.shtml. The full regulatory definition states that a significant risk device is "an investigational device that: (1) is intended as an implant and presents a potential for serious risk to the health, safety, or welfare of a subject; (2) is for use in supporting or sustaining human life and represents a potential for serious risk to the health, safety, or welfare of a subject; (3) is for a use of substantial importance in diagnosing, curing, mitigating, or treating disease or otherwise preventing impairment of human health and presents a potential for serious risk to the health, safety, or welfare of a subject; or (4) otherwise presents a potential for serious risk to a subject." 21 C.F.R. §812.3(m) (2003).

56. *See* 21 C.F.R. §812.3(m) (2003); Device Advice, Clinical Trial and IDE, Approval Process, http://www.fda.gov/cdrh/devadvice/ide/approval.shtml.

57. *See* Device Advice, Clinical Trial and IDE, Approval Process, http://www.fda.gov/cdrh/devadvice/ide/approval.shtml.

58. Information Sheets, Guidance for Institutional Review Boards and Clinical Investigators, 1998 Update, http:// www.fda.gov/oc/ohrt/irbs/devices.html#risk; *see also* Device Advice, Clinical Trial and IDE, Approval Process, http:// www.fda.gov/ cdrh/devadvice/ide/approval.shtml; David G. Adams, *et al.*, 2 Fundamentals of Law and Regulations: An In-Depth Look at Therapeutic Products 280 (1997).

59. *See* Information Sheets, Guidance for Institutional Review Boards and Clinical Investigators, 1998 Update, http:// www.fda.gov/oc/ohrt/irbs/devices.html#risk.

60. *See id.*

61. *See* 21 C.F.R. pt. 812 (2003); *see also* Information Sheets, Guidance for Institutional Review Boards and Clinical Investigators, 1998 Update, http:// www.fda.gov/oc/ohrt/irbs/devices.html#risk

62. Information Sheets, Guidance for Institutional Review Boards and Clinical Investigators, 1998 Update, http:// www.fda.gov/oc/ohrt/irbs/devices.html#risk

63. *See id.*

64. *See id.*

65. FDA will otherwise notify the sponsor if the IDE application has been disapproved. *See* Device Advice, Clinical Trial and IDE, Approval Process, http:// www.fda.gov/ cdrh/devadvice/ide/approval.shtml.

66. *See id.* In contrast, a sponsor cannot begin trials on a SR device until FDA has approved the application. FDA considers the IDE application approved 30 days after its receipt, unless the agency notifies the sponsor otherwise.

67. *See id.*

68. Once the IRB agrees the device poses a non-significant risk and approves the study, FDA considers the device approved. *See id.*

69. If the IRB thinks the device poses a significant risk, the sponsor must report this finding to the FDA within five working days. *See* 21 C.F.R. §812.150(b)(9) (2003); *see also* Device Advice, Clinical Trial and IDE, Approval Process, http://www.fda.gov/ cdrh/devadvice/ide/approval.shtml.

70. *See* 21 C.F.R. §812.2(b) (2003).

71. *See id.*

72. *See id.* §812.2(b)(1)(i). The label must include the name and business address of the manufacturer, packer, or distributor. If applicable, the label must list content quantity. The label also must contain the statement "CAUTION—Investigational Device. Limited by Federal (or United States) law to investigational use." *Id.* §812.5. The label must describe the relevant contraindications, hazards, adverse effects, interfering substances or devices, warnings, and precautions. The label cannot contain any false or misleading statements and cannot imply that the device is safe or effective for the uses being investigated. Finally, the sponsor should

provide detailed information on labeling in the investigational protocol. *See id.*
See also Device Advice, Clinical Trial and IDE, Application, http://www.fda.gov/
cdrh/devadvice/ide/application.shtml#modifications.

73. *See* 21 C.F.R. §§812.140(b)(4),(5), 812.150(b)(1)-(3), (5)-(10) (2003).
74. *See id.* §§812.140(a)(3)(i), 812.150(a)(1),(2),(5), and (7).
75. *See* 21 C.F.R. §812.20 (2003).
76. *See id.* §812.20(b).
77. *See id.* §812.20(b)(2). For a list of the information that must be included in the report of previous investigations, *see id.* §812.27(2003); for the investigational plan requirements, *see id.* §812.25.
78. *See id. See also* 21 C.F.R. §812.43 (listing the information that the agreement must include).
79. *See id.* §812.20(b) (listing all of the application requirements); *see also* Device Advice, Clinical Trial and IDE, Application, http://www.fda.gov/cdrh/devadvice/ide/application.shtml#modifications (providing a detailed guide for preparing the IDE application).
80. *See* 21 C.F.R. §812.20(a)(3) (2003). *See generally* Device Advice, Clinical Trial and IDE, Approval Process, http://www.fda.gov/ cdrh/devadvice/ide/approval.shtml (with suggested content for the cover letter).
81. *See* 21 C.F.R. §812.35 (2003).
82. *See id.* §812.35; *see also* Device Advice, Clinical Trial and IDE, Application, http://www.fda.gov/cdrh /devadvice/ide/application.shtml#modifications.
83. *See* Device Advice, Clinical Trial and IDE, Application, http://www.fda.gov/cdrh/ devadvice/ide /application.shtml#modifications.
84. *See id.*
85. *See id.*
86. *See id.*
87. *See id.*
88. *See id.*
89. *See id*
90. *See id.*
91. *See id.* (noting that the sponsor still must report the change to FDA within five working days).
92. *See* 21 C.F.R. §812.35(a)(3) (2003); *see also* Device Advice, Clinical Trial and IDE, Application, http://www.fda.gov/cdrh /devadvice/ide/application.shtml# modifications.
93. *See* 21 C.F.R. §812.35(a)(3)(iii) (2003).
94. *See* 21 C.F.R. §812.150(b)(5) (2003); *see also* Device Advice, Clinical Trial and IDE, Application, http://www.fda.gov/cdrh /devadvice/ide/application.shtml# modifications.
95. *See* 21 C.F.R. §812.30 (2003).
96. *See id.* §812.30.
97. *See id.* §812.30(b). *See also* Device Advice, Clinical Trial and IDE, Application, http://www.fda.gov/cdrh/devadvice/ ide/application.shtml#modifications.

98. *See* 21 C.F.R. §812.30(c) (2003); Device Advice, Clinical Trial and IDE, Approval Process, http://www.fda.gov/ cdrh/devadvice/ide/approval. shtml.

99. *See* Guidance on IDE Policies and Procedures, at 8.

100. *See id.*

101. *See id.*

102. *See id.*

103. *See id.* at 7.

104. *See id.*

105. *See id.*

106. *See id.*

107. *See id. See also* 21 C.F.R. §812.150 (2003); Device Advice, Clinical Trial and IDE, Reports, http://www.fda.gov/cdrh/devadvice/ide /reports.shtml

108. *See* 21 C.F.R. §50.20 (2003).

109. *See id.* §50.25(a).

110. *See id.* §50.25(b).

111. *See* FDA Warning Letter to James Yanney, DDS, MD, dated July 30, 2003.

112. *See* 21 C.F.R. §50.23 (2003).

113. *See id.* §50.24.

114. *See id.* §50.24(a)(1) (requiring that the "human subjects are in a life-threatening situation, available treatments are unproven or unsatisfactory, and the collection of valid scientific evidence, which may include evidence obtained through randomized placebo-controlled investigations, is necessary to determine the safety and effectiveness of particular interventions").

115. *See id.* §50.24(a)(2) (providing that obtaining informed consent must be infeasible because: (1) subjects will not be able to give their informed consent as a result of their medical condition, (2) the intervention under investigation must be administered before consent from the legally authorized representatives is feasible, and (3) there is no reasonable way to identify prospectively the individuals likely to become eligible for participation in the clinical investigation).

116. *See id.* §50.24(a).

117. *See id.* §50.24(a)(3) (providing that there must be a direct benefit because: (1) subjects are facing a life threatening situation that necessitates intervention; (2) appropriate animal and other preclinical studies have been conducted, and the information derived from those studies and related evidence support the potential for intervention to provide a direct benefit to the subjects, and (3) risk associated with the investigation are reasonable in relation to what is known about the medical condition of the potential class of subjects, the risks and benefits, if any, and what is known about the risks and benefits of the proposed intervention or activity).

118. *See id.* §50.24(a)(4).

119. *See id.* §50.24(a)(5) (requiring the investigator to contact the legally authorized representative during that window, among other things).

120. *See id.* §50.24(a)(6).

121. *See id.* §50.24(a)(7).

122. *See id.* §812.40.
123. *See id.* §§812.43(d), 812.46.
124. *See* FDA, Guidelines for Monitoring of Clinical Investigations (Jan. 1988).
125. *See* 21 C.F.R. §812.46(a) (2003).
126. *See* FDA Warning Letter to Silimed, Incorporated, dated August 21, 2003.
127. *See* 21 C.F.R. §812.46(b) (2003).
128. *See id.* §812.46(c).
129. FDA Warning Letter to Paradigm Medical Industries, Inc., dated Aug. 30, 2000.
130. *See* FDA Warning Letter to Wesley Kinzie, M.D., dated Nov. 7, 2003.
131. *See* 21 C.F.R. §812.140(d).
132. *See id.* §812.140(e).
133. *See id.* §812.145.
134. *See id.* §812.150.
135. *See id.* §812.7.
136. *See* Guidance for Industry and FDA Staff, Preparing Notices of Availability of Investigational Medical Devices and for Recruiting Study Subjects, FDA (March 19, 1999).
137. *See* FDA Warning Letter to Jacobson Resonance Enterprises, Inc., dated Jan. 25, 2001.
138. *See* 21 U.S.C. §360j(g) (Supp. 2003); 21 C.F.R. §812.1 (2003); FDA, Device Advice, Clinical Trials & Investigational Device Exemption (IDE), Early/Expanded Access, http://www.fda.gov/cdrh/devadvice/ide/print/ early.html#emergencyuse.
139. *See* 21 C.F.R. §§50.25, 812.35, 812.47 (2003); Guidance for Emergency Use of Unapproved Medical Devices (Oct. 22, 1985), http://www.fda.gov/cdrh/manual/unappr.html.
140. *See* Guidance for Emergency Use of Unapproved Medical Devices (Oct. 22, 1985), http://www.fda.gov/cdrh/manual/unappr.html; *see also* FDA, Device Advice, Clinical Trials & Investigational Device Exemption (IDE), Early/Expanded Access, http://www.fda.gov/cdrh/devadvice/ide/print/ early.html#emergency use.
141. *See* Guidance on IDE Policies and Procedures, at 17-19.
142. *See* Guidance for Emergency Use of Unapproved Medical Devices (Oct. 22, 1985), http://www.fda.gov/cdrh/manual/unappr.html.
143. *See* 21 C.F.R. §812.35 (2003); Guidance on IDE Policies and Procedures, at 17-19.
144. *See* 21 C.F.R. §50.24 (2003); *see supra*, discussion at Section V(B).
145. *See* 21 C.F.R. §812.47 (2003); *see also* FDA, Device Advice, Clinical Trials & Investigational Device Exemption (IDE), Early/Expanded Access, http://www.fda.gov/cdrh/devadvice/ide/print/ early.html#emergencyuse.
146. *See* Guidance on IDE Policies and Procedures, at 19-20; FDA, Device Advice, Clinical Trials & Investigational Device Exemption (IDE), Early/Expanded Access, http://www.fda.gov/cdrh/devadvice/ide/print/ early.html#emergencyuse.

147. *See* 21 C.F.R. §812.35 (2003); *see also* Guidance on IDE Policies and Procedures, at 19-20; FDA, Device Advice, Clinical Trials & Investigational Device Exemption (IDE), Early/Expanded Access, http://www.fda.gov/cdrh/devadvice/ ide /print/early.html#emergencyuse.
148. *See* 21 C.F.R. §812.36 (2003); Guidance on IDE Policies and Procedures, at 20-22; FDA, Device Advice, Clinical Trials & Investigational Device Exemption (IDE), Early/Expanded Access, http://www.fda.gov/cdrh/devadvice/ide /print/ early.html#emergencyuse.
149. *See* 21 C.F.R. §821.36(a) (2003).
150. *See id.*
151. *See id.*
152. *See id.* §812.36(b)(1)-(4).
153. *See id.* §812.36(e).
154. *See id.* §812.36(c); *see also* FDA, Device Advice, Clinical Trials & Investigational Device Exemption (IDE), Early/Expanded Access, http://www.fda.gov/ cdrh/devadvice/ide /print/early.html#emergencyuse.
155. *See* 21 C.F.R. §812.36(d) (2003).
156. *See id.* §§812.30(c); 812.36(d)(3).
157. *See id.* §812.36(d)(2)(iv) (establishing a specific test).
158. *See id.* §§812.30(b); 812.36(d)(2); *see also* FDA, Device Advice, Clinical Trials & Investigational Device Exemption (IDE), Early/Expanded Access, http:// www.fda.gov/cdrh/devadvice/ide /print/early.html#emergencyuse.
159. *See* 21 C.F.R. §812.36(f) (2003); *see also* Suggested Format for IDE Progress Report, http://www.fda.gov/cdrh/dsma /311.html.
160. *See* FDA, Device Advice, Clinical Trials & Investigational Device Exemption (IDE), Early/Expanded Access, http://www.fda.gov/cdrh/devadvice/ide/print/ early.html#emergencyuse.
161. *See id.*
162. *See id.*
163. 21 C.F.R. §809.3 (2003).
164. *See id.* at pts. 50 and 56; Guidance for FDA Staff Regulating *In Vitro* Diagnostic Device ("IVD") Studies, Office of Compliance, Division of Bioresearch Monitoring, CDRH, FDA (Dec. 17, 1999).
165. *See* 21 C.F.R. §812.2(c)(3) (2003).
166. *Id.* §§809.10(c)(2); 812.2(c)(3) (2003).
167. *See* Guidance for FDA Staff Regulating *In Vitro* Diagnostic Device ("IVD") Studies, Office of Compliance, Division of Bioresearch Monitoring, CDRH, FDA (Dec. 17, 1999).
168. Council Directive 90/385/EEC on Active Implantable Medical Devices (1990).
169. Council Directive 93/43/EEC on Medical Devices (1992).
170. Clinical Investigation of Medical Devices for Human Subjects, Doc. No. 540, European Commission of Standardization (1993).
171. *See id.*

172. *See* ISO 14155-1:2003 Clinical Investigation of Medical Devices for Human Subjects – Part 1: General Requirements.
173. *See* ISO 14155-2: Clinical Investigation of Medical Devices for Human Subjects – Part 2: Clinical Investigation Plans.
174. *See* Danielle Giroud, *A Revised Guideline for Medical Device Clinical Investigations: ISO 14155 part 1 and 2: 2003*, Qual. Assur. J. (2004).
175. *See* ISO 14155-1:2003 Clinical Investigation of Medical Devices for Human Subjects – Part 1: General Requirements.
176. *See* ISO 14155-2: Clinical Investigation of Medical Devices for Human Subjects – Part 2: Clinical Investigation Plans.
177. *See* ISO 14155-1:2003 Clinical Investigation of Medical Devices for Human Subjects – Part 1: General Requirements; ISO 14155-2: Clinical Investigation of Medical Devices for Human Subjects – Part 2: Clinical Investigation Plans; *see also* ISO Website, About Us, http://www.iso.org/iso/en/aboutiso/introduction/index.html.
178. *See* 21 C.F.R. pt. 812 (2003).
179. *See id.* §860.7.
180. *See id.* §814.15(b); *see also* Guidance for Industry Acceptance of Foreign Clinical Studies, FDA (Mar. 2001).
181. 21 C.F.R. §860.7(c)(2) (2003).
182. *See id.*
183. *See id.* §814.15(b).
184. *Id.* §814.15(d).
185. *See id.* §814.15(e).
186. 21 U.S.C. §382(c) (Supp. 2003).
187. *Id.* §381(e)(2) (Supp. 2003).
188. *See* 21 C.F.R. §812.18(b) (2003).
189. *See* 21 U.S.C. §382(b)(1)(A)(i)-(ii) (Supp. 2003).
190. *See id.* §382(c).
191. *See id.* §382(f).
192. *See* FDA Guidance to Industry on: Exports and Imports Under the FDA Export Reform and Enhancement Act of 1996 (Draft) Feb. 1998, http://www.fda.gov/opacom/fedregister/frexport.html#general; *see also* Device Advice, Import and Export of Investigational Devices, http://www.fda.gov/cdrh/devadvice/ide/import_export.shtml; Device Advice, Exporting Medical Devices, http://www.fda.gov/cdrh/devadvice/39.html.
193. 21 U.S.C. §381(e)(2) (Supp. 2003). For information regarding the precise materials that should be filed with a request for approval under Section 801(e)(2), *see* Device Advice, Import and Export of Investigational Devices, http://www.fda.gov/cdrh/devadvice/ide /import_export.shtml.
194. *See* 21 U.S.C. §383(g) (Supp. 2003); 64 Fed. Reg. 15944 (Apr. 2, 1999); *see also* Device Advice, Import and Export of Investigational Devices, http://www.fda.gov/cdrh/devadvice/ide/import_export.shtml.

3

Requirements for Medicare Coverage and Reimbursement for Medical Devices

Esther R. Scherb and Stuart S. Kurlander

1. Introduction

More device manufacturers are recognizing the importance of considering Medicare coverage and reimbursement principles during the development of their new technologies. By adopting long- and short-term strategies that incorporate these principles, manufacturers can avoid certain foreseeable delays in bringing devices to market. Although the roadmap for demonstrating to the Food and Drug Administration (FDA) the safety and efficacy of a device has been fairly well established, the steps for obtaining third-party payors' approvals have been less apparent. This chapter highlights Medicare coverage and reimbursement principles and considerations that might be incorporated into a company's business plan during and after the product development stage.

From the time it was first created, the Medicare program, a federal health insurance program and the nation's single largest payor, has become progressively more expansive and more complex. Established in 1965 as Title XVIII of the Social Security Act,[1] the program now provides benefits to more than 40 million beneficiaries ages 65 years and older, certain disabled individuals, and individuals with end-stage renal disease. Medicare's claims administrators annually process more than 900 million claims for health care services and Medicare benefit expenditures for fiscal year 2003 reached approximately $275 billion.[2] Because it has been lauded as having one of the most stringent review standards for health care coverage, other insurers—government and commercial alike—look to policies and procedures developed by Medicare regulators as a guide in establishing their own. If an item or service is not covered by Medicare, there is a strong likelihood that other payors also will decide not to cover the item or service. Likewise, Medicare's reimbursement methodologies

From: *Clinical Evaluation of Medical Devices: Principles and Case Studies, Second Edition*
Edited by: K. M. Becker and J. J. Whyte © Humana Press Inc., Totowa, NJ

are often followed by other payors. A sufficient payment amount, of course, is as important as a coverage decision, because inadequate payments can render even a seemingly favorable coverage decision meaningless.

Medicare's coverage principles may be understood by examining how the program's "reasonable and necessary" standard has been interpreted. This standard differs from the FDA's "safe and effective" standard, which is at the core of the FDA's regulatory approvals. Companies that design appropriate clinical studies with both goals in mind can avoid some of the otherwise frustrating realizations that clinical studies designed for FDA purposes do not automatically provide assurances for coverage and reimbursement purposes.

This chapter is designed as a reference tool for device manufacturers and to facilitate an understanding of some of the admittedly arcane elements of Medicare coverage and payment methodologies as they apply to medical devices. It addresses how Medicare coverage decisions are made and provides considerations and examples of limitations on such coverage. Medicare payment systems are discussed generally, particularly as a framework for predicting issues related to payments for new medical devices. Finally, because coverage and payment issues are integral to the coding system used to identify products and services, some of the fundamentals of coding are discussed briefly.

2. Statutory Authority for Medicare Coverage

There are four parts to the Medicare program: Part A is the Hospital Insurance program, covering primarily hospital inpatient services, skilled nursing facility services, home health agency, and hospice services; Part B is the Supplementary Medical Insurance program, covering outpatient and physician services, as well as laboratory services and products provided for home use; Part C is the Medicare Advantage program,[3] which is a managed care program; and Part D is the new Prescription Drug Benefit program.[4] In addressing coverage and payment, we focus on Parts A and B as they relate to the longstanding Medicare fee-for-service program.

The Secretary of Health and Human Services (HHS), who administers the Medicare program, has delegated that authority to a component agency, the Centers for Medicare and Medicaid Services (CMS). Through its central and field offices, CMS contracts with private organizations to assist with this administration. Medicare contractors generally are insurance companies, who process claims, interact on a day-to-day basis with health care providers, and typically provide these services for prescribed geographic regions.

The Medicare program is a defined benefits program (i.e., Congress has expressly prescribed how beneficiaries become eligible to receive payment for products and services and defined the scope of those products and services). The Medicare statute does not specifically identify all covered items, services, treatments, procedures, or technologies but rather describes the benefit categories in which items and services are classified. For coverage eligibility, a medical device must fit into one of these recognized benefit categories and not be otherwise excluded. The classification of the device into a benefit category also determines the Medicare payment methodologies for the device.

2.1. Benefit Categories and Statutory Limitations

Statutorily defined benefit categories are designated at various parts of the Medicare statute and include, among others, hospital inpatient services,[5] hospital outpatient services,[6] physician services,[7] diagnostic laboratory tests,[8] durable medical equipment,[9] orthotic devices,[10] casts and splints,[11] and prosthetic devices.[12] Within these defined categories, CMS and its Medicare contractors have discretion to determine coverage.

Certain categories of items and services are expressly excluded from Medicare coverage.[13] For example, personal comfort items, custodial care, and cosmetic surgery are categorically excluded from coverage. Services provided by certain health care practitioners, such as dentists, are excluded.[14] Preventive services are not covered unless a separate benefit category is provided. Preventive service benefit categories include screening mammography services,[15] prostate-specific antigen testing,[16] and colorectal screening tests.[17]

Statutory provisions also set forth conditions of eligibility for those entities and individuals providing items and services to beneficiaries, including quality standards for the types of institutions that may provide the services and licensing and credentialing requirements for practitioners such as physicians. For example, providers (e.g., hospitals, skilled nursing facilities, comprehensive outpatient rehabilitation facilities, home health agencies, and hospice programs) must meet conditions of participation or conditions for coverage in order to participate in the Medicare program and provide services to program beneficiaries. Other entities and practitioners, referred to as suppliers, must meet licensing standards and other supplier-specific criteria. Practitioners participating in the Medicare program are also limited to performing services within the scope of their state's licensure provisions. In light of such standards, device manufacturers formulating a coverage and reimbursement strategy must evaluate not only whether their products fit into defined benefit categories but also who will use them and where they will be used.

2.2. Coverage Interpretation by the Centers for Medicare and Medicaid Services and Its Contractors

Because there is no statutory list of specific items or services to be covered, CMS and its contractors have discretion to determine whether a specific device meets the definition of a benefit category and, in turn, whether the device and/

or service using the device may be covered under the program. Decisions regarding coverage of specific devices not otherwise specified or expressly excluded by Congress are based on the statutory reasonable and necessary standard. The statutory standard is an exclusionary standard, which precludes coverage for those items and services "*not* reasonable and necessary for the diagnosis or treatment of an illness or injury or to improve the functioning of a malformed body member."[18] The authors refer to this standard as the "Medicare mantra" because it is the standard relied on for all coverage decisions rendered by CMS and its contractors.

What does this "Medicare mantra" mean? To date, CMS has not published definitions or criteria, despite a number of attempts to do so. Yet, there is a considerable amount of information available from the coverage decisions that CMS has made over the years. We know that the term is interpreted to exclude preventive services, because they are not designed to "treat or diagnose" and are considered covered only when Congress articulates specific benefit categories. For devices, CMS has historically interpreted *reasonable and necessary* to mean that the product must, at minimum, be safe and effective (i.e., unless exempt, it must have been approved or cleared for marketing by the FDA and must not be experimental).[19] CMS has also looked to authoritative objective evidence from published clinical studies and gives the most weight to those prospective, randomized studies published in peer-reviewed journals, as well as data regarding whether the item or service has been generally accepted in the medical community. More recently, CMS has adopted and employed principles of evidence-based medicine.[20] The agency seeks to identify quality data and whether the item or service has been generally accepted both for the condition for which it is used and for Medicare beneficiaries in general. Indeed, showing how the Medicare population would benefit from the technology should be a key feature for any strategy for Medicare coverage.

It has been more than 15 years since CMS first attempted to publish criteria for making coverage decisions.[21] It has withdrawn its most recent proposal, a notice of intent published to solicit public comments.[22] Much of the controversy surrounding proposed coverage criteria stemmed from the considerations of costs in the face of statutory language that did not require a cost–benefit analysis. Critics have argued that cost considerations should be factored into the payment methodology for a covered benefit. In other words, coverage should be evaluated separately from pricing.

In the Medicare Prescription Drug, Improvement, and Modernization Act of 2003 (Medicare Modernization Act), Congress directed CMS for the first time to publish the factors considered in making national coverage determinations on whether an item or service is reasonable and necessary. The factors will be published using guidance documents similar to those developed by the FDA.[23]

Although CMS on several occasions has indicated its intent to issue guidance documents for categories of items and services, it had not developed such documents.

2.3. Coverage Policies for Devices and Related Covered Services

There is no substitute for a comprehensive review of existing policies as one of the initial steps in evaluating coverage. Determinations regarding which specific items and services are to be included or excluded from coverage under the reasonable and necessary standard are made at both the national and local levels. CMS makes national determinations and relies on its Medicare contractors to issue the majority of coverage decisions at the local levels. Congress has only recently directed CMS to develop a plan to promote consistency in coverage decisions issued by contractors and consider which determinations should be adopted nationally.[24] Although it is not uncommon for Medicare contractors in different parts of the country to have different coverage policies, greater uniformity can be expected in the future.

2.3.1. National Coverage Determinations

National coverage determinations (NCDs) are CMS's national policy statements granting, limiting, or excluding Medicare coverage for a specific medical product or service. The policy statements not only interpret the reasonable and necessary standard, but they also include CMS's decisions regarding the benefit category under which the product or service is covered.[25] All Medicare contractors and Medicare managed care plans must abide by NCDs and may not disregard, set aside, or otherwise review them.[26]

CMS initiates the NCD review process through informal contacts or after receiving a formal request from an outside party[27] and may internally generate a request that it believes to be in the general health and safety interests of Medicare beneficiaries.[28] Requests are processed in one of two ways. First, the request may be processed by CMS medical staff based on evidence available. Second, the request may require additional external input in the form of a health technology assessment. CMS also may obtain expert advice and assistance from the Medicare Coverage Advisory Committee (MCAC),[29] a group of nationally recognized experts in a broad range of medical, scientific, and professional disciplines. Consumer and industry groups are also represented on the MCAC. For those NCDs for which no MCAC recommendation is sought, appropriate outside clinical experts are to be consulted.[30]

The MCAC holds open meetings and provides an opportunity for public participation in its consideration of evidence on technologies. In its advisory role, the MCAC makes recommendations to CMS on issues impacting whether the device or service can be considered reasonable and necessary.[31] CMS may adopt, adopt with modifications, or reject the recommendation. In a published decision memorandum, CMS explains its reasoning and how the MCAC recommendation factored into its decision.[32]

Under new timeframes prescribed by Congress for decisions on requests for new NCDs, CMS must render a decision no later than 6 months after the date of the request, unless a technology assessment is required by an outside entity or deliberation is required by the MCAC. Under the latter circumstances, the decision on the request must be made within 9 months.[33] CMS must make a proposed decision available to provide the public a 30-day comment period. After the close of the comment period, CMS makes a final decision, which includes summaries of comments received. This new procedure represents a positive outlook for openness in the coverage process. Further, new requirements direct CMS to assign a temporary billing code when the final decision grants a request for an NCD.[34]

At the present time, CMS's NCDs are initially published through decision memorandums, which offer the rationale for the decisions. These NCDs are implemented using program transmittals and manuals that can be found on the CMS Web site.[35] In formulating any coverage and reimbursement strategy, all NCDs should be reviewed to determine whether existing decisions offer coverage or noncoverage for devices and related services.

2.3.2. Local Coverage Determinations

CMS has delegated authority to Medicare contractors to develop and issue contractor-specific policies that identify the circumstances under which particular items or services will be covered in a geographic area. Local coverage determinations (LCDs) are coverage decisions interpreting the reasonable and necessary provision of the Medicare statute.[36] Each policy applies only in the jurisdiction of the individual contractor. Local policies may not conflict with any of CMS's national coverage decisions but may fill the gaps left by NCDs. LCDs comprise the vast majority of Medicare coverage determinations, and obtaining coverage at a local level traditionally has been a preferred strategy to pursuing coverage at the national level, where an unfavorable decision might have a devastating result.

Medicare contractors typically have developed policies in the form of local medical review policies (LMRPs), which include four different types of provisions—coding, benefit category, statutory exclusions, and medical necessity provisions (i.e., those provisions interpreting the reasonable and necessary provision of the Medicare statute).[37] CMS intends to work with contractors to divide existing LMRPs into documents that include only the medical necessity provisions (i.e., LCDs) and those with other, non-medical necessity provisions.

Contractors develop local policies for multiple reasons, including to address perceived inappropriate utilization or improper billing for a particular procedure or device. Local policies may set forth conditions on which to automatically deny inappropriate claims and may specify acceptable diagnoses, documentation requirements, and other guidelines. LCDs generally provide greater detail on the criteria for coverage of items or services that are the subject of an NCD. Most Medicare contractors are permitted to develop their own LCDs without regard to local policies in other regions of the country.[38] The policies are used in the processing of claims, but unlike NCDs, are not binding on adjudicators at all levels of the administrative appeals process.[39]

CMS requires local contractors to publish their policies in bulletins and manuals, which can be found on local contractors' or CMS's Web sites. In addition, under new law, as mentioned above, CMS must develop a plan to evaluate whether new local coverage determinations should be adopted on a national level and to what extent greater consistency can be achieved among LCDs.[40]

2.3.3. Individualized Determinations

If there are no specific coverage determinations at either the national or local level, a Medicare contractor will make an individualized or claim-by-claim coverage determination. In so doing, the contractor will consider both whether the device or related service was reasonable and necessary and whether other Medicare requirements are satisfied. The contractor considers whether the service was appropriate for the particular patient in terms of amount and frequency and in light of the medical condition. Without a coverage policy, there may be great variation on how individual claims reviewers decide coverage for a particular device or related service. There is not only a potential for variation among contractors but also within the same contractor if the claims are reviewed by different individuals.

2.3.4. Coverage of Investigational Devices

Because threshold coverage for a medical device generally depends on required FDA approvals or clearances, devices deemed experimental by the FDA are not considered reasonable and necessary under the statutory standard. Since 1995, special considerations were given to devices; the categories of devices that may be covered under Medicare now include:

- Devices approved by the FDA through the premarket approval (PMA) process.
- Devices cleared by the FDA through the 510(k) process.
- FDA-approved investigational device exemption (IDE) category B devices.
- Hospital Institutional Review Board-approved IDE devices.[41]

IDE devices are refinements or replications of existing technologies, and FDA places them into this category to gather scientific evidence and establish that the particular device is safe and effective.[42] To assist CMS in its coverage decision process, FDA assigns each device with an FDA-approved IDE to either category A or B.[43] Category A is for experimental/investigational devices; category B covers nonexperimental/noninvestigational devices. Category B devices are eligible for Medicare coverage; category A devices are not.

Category A devices are innovative Class III devices[44] for which absolute risk has not been established. In other words, the initial questions of safety and effectiveness have not been resolved and FDA is unsure whether the device can be deemed safe and effective.[45] Category A includes Class III devices that have not undergone the PMA process for any indication for use or have undergone significant modification for a new indication for use.

Category B devices are devices believed to be Class I or Class II as well as Class III devices for which the incremental risk is the primary risk in question (i.e., questions of safety and effectiveness have been resolved).[46] CMS approves category B IDE devices for coverage if they are used in the context of an FDA-approved clinical trial and according to the trial's approved patient protocols.[47] Payment for a category B device will be based on—and may not exceed—the amount that would have been paid for an existing device that serves the same medical purpose and that has been approved or cleared by FDA.[48]

2.3.5. Coverage of Clinical Trials

In the past, Medicare did not cover the services provided during clinical trials, which are designed to evaluate the safety and effectiveness of medical care. Effective September 19, 2000, certain qualifying clinical trials became eligible for Medicare coverage of routine costs.[49] Although these routine costs include costs for conventional services typically provided absent a clinical trial, they do not include costs for the investigational device or items or for services not used in the direct management of the patient.

Importantly, the policies for clinical trials do not withdraw Medicare coverage policies on IDEs. In addition, the Medicare statute has been revised to ensure payment of associated routine costs for category A investigational devices intended to be used for immediately life-threatening diseases or conditions.[50] This coverage, which became effective January 1, 2005, offers expanded payment opportunities for providers engaged in category A trials.[51]

2.3.6. Limitations on Coverage

In considering whether a device is reasonable and necessary, CMS and its contractors impose numerous limitations, including those relating to the site of service, length of use, patient diagnosis, and who can order or provide the device or related service. Such considerations are critical to any coverage strategy. For example, some devices and associated services might be covered only when provided in a particular setting, favoring inpatient settings for certain surgical procedures. Coverage also may be limited to the types of medical conditions for which a device is used. Contractors incorporate specific diagnosis codes (typically, International Classification of Diseases, 9th Revision, Clinical Modification [ICD-9-CM] codes[52]) in coverage decisions to identify further coverage parameters. Some coverage policies may set limits on the length of time during which a device may be used. Such parameters are common with durable medical equipment or equipment provided on a rental basis.

Significant to any coverage or reimbursement strategy is the recognition that the Medicare program, like most other payors, uniformly requires that covered items or services be ordered by a physician. In other words, the treating physician must communicate a request that an item or service be provided or performed for a beneficiary.[53] In addition, some services must be ordered or performed by designated specialists.

Importantly, there can be no coverage for services related to the use of a noncovered device (other than in qualifying clinical trials). Noncovered services include all services furnished contemporaneously with and necessary to the use of a noncovered device, and services furnished as necessary aftercare incident to recovery from the use of a noncovered device or related noncovered services.[54] Coverage is available for services furnished to address medical complications arising from the use of noncovered devices and services that are not incident to normal recovery.[55]

Any coverage strategy should anticipate and propose limitations that are appropriate for the technology. This would go a long way to helping develop a coverage policy that is suited for the particular device.

3. Overview of Medicare Reimbursement

3.1. Payment Methodologies—Generally

Since the inception of the Medicare program four decades ago, reimbursement methodologies have evolved significantly, largely owing to lawmakers' repeated efforts to contain Medicare spending. The volume of payment provisions in the Medicare statute has increased exponentially as Congress has refined statutory mandates for payment methodologies for specific services and items. Just as benefit categories define eligibility for coverage, statutorily prescribed payment methodologies define applicable reimbursement amounts. The significant breadth of payment systems makes understanding Medicare reimbursement for devices a challenge, because payment systems vary dramatically and depend on the site of service and the method of delivery. As Congress continues to revisit payment issues in its effort to modernize the program, it is likely that these various methodologies will undergo further adjustments or be supplanted by new approaches.

In this section, we provide an overview of the general payment methodologies for devices. These methods can be grouped into three primary categories: prospective payment systems (PPSs), fee schedules, and reasonable cost payments.

3.2. Prospective Payment Systems

As its name indicates, a PPS is forward-looking. Predetermined payment amounts are established before the actual provision of the item or service and are intended to reflect the typical costs of treating or diagnosing a beneficiary with a particular condition. The predetermined payment amount generally is adjusted annually using existing prior years' actual Medicare claims data and statutorily prescribed annual update factors intended to reflect variables, such as inflation.

PPS methodologies were designed to encourage efficiency and control costs by requiring providers to bear cost overruns. Perceived successes have led Congress to expand the use of PPS to a number of provider services. Originally prescribed for certain hospital inpatient services in 1982, Congress in the last several years has authorized PPS methodologies for services and items provided in several other settings: hospital outpatient departments, inpatient rehabilitation facilities, home health agencies, skilled nursing facilities, and long-term care facilities.

3.3. Fee Schedules

Many health care items and services are paid under nationally prescribed fee schedules, which set forth allowable payment amounts for specific items and services. Among these items and services are physician services, durable medical equipment, orthotics, prosthetics, supplies, and clinical diagnostic laboratory tests. Like PPS, fee schedules establish predetermined payment amounts—in this case, the maximum allowable amount that Medicare will pay for certain items and services. Medicare payment typically is based on the lesser of the fee-schedule amount or the actual charge for the particular item or service.[56]

3.4. Reasonable Costs

A reasonable cost methodology originally was used for most provider payments under the Medicare program but has largely been supplanted by PPS methodologies. The methodology continues to be employed for certain PPS-exempt hospitals' inpatient services (such as cancer hospitals' inpatient services and children's hospitals[57]) and is the basis of certain "pass-through" payments for a number of items, including eligible new technology used in hospital outpatient departments.

The term reasonable cost is defined as "the cost actually incurred, excluding therefrom any part of incurred cost found to be unnecessary in the efficient delivery of needed health services."[58] Under this methodology, intricate rules were issued to address which costs are related to patient care and how to calculate those costs. Reasonable costs are different for each provider, but payment will not be made for costs that differ substantially from those of other institutions with similar features (e.g., location, size, scope of services, utilization).[59] Periodic interim payments are made to providers during the year[60] and, after the close of the provider's fiscal year, the provider's Medicare cost report is reviewed for final settlement.

3.5. Reimbursement for Devices in Specific Settings

To determine the appropriate reimbursement for a device, one must first look at the setting in which it is furnished and evaluate the associated payment methodology. Within the associated payment methodology, there can be significant exceptions and adjustments, depending on the type of device. Thus, each payment methodology should be analyzed carefully for the specific device. Some of the key settings are hospital inpatient and outpatient departments, physicians' offices, and the patient's home.

3.5.1. Hospital Inpatient Departments: Acute Care Hospitals [61]

Under the inpatient PPS methodology for acute care hospitals, Medicare pays the hospital a predetermined amount based on the patient's discharge diagnosis for all hospital services furnished during a patient's stay. These services include operating costs for routine and ancillary services, intensive-care unit services, malpractice insurance costs, and certain preadmission services.[62] Capital-related costs also are paid under the PPS methodology.[63] With limited exception, separate payment is not made for medical devices. Instead, payment for devices is part of the bundle of costs incurred by a hospital treating a beneficiary. Hospitals thus lack incentive to incur the added costs of new technologies.

As to the specific payment for treating a beneficiary, the PPS amount is based on diagnosis-related groups (DRGs), which are classifications of cases that are similar clinically in terms of conditions and treatment, as well as use of hospital resources.[64] The DRG classification system groups cases involving similar resource uses, so that payment rates can be established for a limited number of cases. The DRG payment amount is based on the average cost of treating a patient with the condition described by the DRG. Actual costs for individual patients within a DRG vary; however, in the aggregate, the PPS payments are intended to compensate a hospital adequately.

There are more than 500 DRGs in 25 major diagnostic categories.[65] Hospitals report a series of ICD-9-CM codes reflecting the patient's principal and secondary diagnoses and procedures performed on the patient during the hospital stay. These entries, together with information on age, sex, and mortality, are the factors used for classifying cases into particular DRGs. By statute, CMS is required to make the necessary adjustments to DRG classifications at least annually.[66]

CMS assigns each DRG a numerical weighting factor representing the relative hospital resources necessary to treat the average case of that condition.[67] If a case using the average amount of resources is assigned a value of 1, a case using only half the resources as the average case would have a weighting factor of 0.5. Once the appropriate DRG is determined, the basic PPS payment to the hospital is calculated by multiplying the DRG weight for the condition involved by a standardized amount, which converts the weighting value to a dollar amount. The standardized amount will vary depending on a hospital's geographical location and wage rates in its area. For most purposes, two standardized amounts are determined—one for large urban areas (with populations greater than 1 million, or in New England, greater than 970,000) and another for all other areas.[68] These values must be adjusted by CMS annually.[69] The standardized amounts also are updated for each fiscal year by a factor that is based on the hospital market basket percentage increase—an indicator of inflation experienced by hospitals when purchasing goods and services to provide inpatient services.[70]

The inpatient base payment amount may be increased (i.e., an add-on payment is made) for unusually expensive cases.[71] This add-on payment for outlier cases is designed to protect hospitals from large financial losses. In addition, hospitals that treat a high percentage of low-income patients receive an additional payment to take into account the higher costs of low-income clientele in poorer health and in need of special services (e.g., social work, translation).[72] Hospitals with approved teaching programs receive payments in addition to the base DRG payment amount for costs associated with training medical residents.[73]

For new technologies, Medicare also provides separate add-on payment for cases that involve new and innovative treatments, including medical devices.[74] To qualify, the new technology must be a substantial improvement relative to previously available technologies and a showing also must be made that, absent an add-on payment, the new technology would be inadequately paid under the regular DRG payment.[75] As part of the Medicare Modernization Act changes to the Medicare statute, an opportunity for public input is now available for those technologies for which an application for additional payment is pending.[76] Also, effective starting fiscal year 2005, there is new funding for hospital inpatient technology, which means that the add-on payments no longer need to be budget neutral.[77] Before establishing an add-on payment amount, CMS identifies one or more DRGs associated with such technology, based on similar clinical or anatomical characteristics and the cost of the technology, and selects a DRG in which the average costs of care most closely approximates the costs of care using the new technology.[78] These changes provide a boost to hospitals' use of new technologies.

3.5.2. Hospital Outpatient Departments

For most services furnished to beneficiaries in the hospital outpatient setting, Medicare implements a PPS-based methodology.[79] Unlike the methodology used for the inpatient setting, outpatient PPS is based on the actual services provided during a visit. Payments are made under ambulatory payment classifications (APCs), which group together similar services comparable both clinically and with respect to use of resources. CMS has created more than 500 procedure-related APC groups encompassing thousands of procedures and products.

To calculate the applicable payment amount, each APC is assigned a relative weight, which is then multiplied by a standardized amount (called a conversion factor).[80] The conversion factor varies to reflect regional differences in labor costs.[81] Further, the conversion factor is adjusted annually to account for inflation, using an amount based on the hospital market basket percentage increase.[82] There is also an additional amount provided for outlier cases.[83]

Temporary additional payments, referred to as transitional pass-through payments, for certain drugs, biologicals, and devices are also made.[84] Congress authorized two categories of transitional pass-through payments. The first category is for certain products (e.g., cancer therapy drugs, orphan drugs, radiopharmaceutical drugs, brachytherapy, temperature-monitored cryoablation devices) that were in use when PPS went into effect on August 1, 2000.[85] The second category is for new and innovative drugs and medical devices, if the cost of the product is not insignificant compared to the PPS payment amount that would otherwise apply.[86]

For medical devices, the additional temporary payments, which are limited to a period of 2 to 3 years for each product,[87] equal the amount of the hospital's charge for the device, adjusted to the actual cost, minus the APC payment amount for the device.[88] To be eligible for pass-through payment, if required by the FDA, a medical device must have received FDA approval or clearance or classification as an IDE category B device.[89] The device must meet the reasonable and necessary standard,[90] be an integral and subordinate part of the service furnished, used for only one patient, come into contact with human tissue, and be implanted surgically or inserted, regardless of whether it remains with the patient when he or she is released.[91] In addition, the device may not be depreciable equipment or an item furnished incidental to a service (e.g., suture, surgical kit) or be used to replace human skin.[92] Pass-through devices now are classified by categories maintained by CMS and no longer by brand.[93]

CMS has also created a series of temporary APCs for certain new technologies that do not qualify under the above definition for pass-through payments. These APCs may be used where there is insufficient cost data to make an assignment to a clinical APC. The new technology APCs are based strictly on the estimated cost of the item and not on clinical use. New technologies may be classified under one of these APCs pending collection of accurate cost information, with possible eventual reassignment to a unique APC. The lowest-cost APC for new technologies is estimated at $0 to $50, with the most expensive new technology APC estimated at $9500 to $10,000. Payment rates are equal to the midway point between the cost parameters for each APC.[94]

3.5.3. Physician Offices and Clinics

Since 1992, physician services have been paid under the Medicare Physician Fee Schedule, derived from a resource-based relative value scale (RBRVS).[95] The fee schedule applies not only to services furnished by physicians but also to most office supplies and certain nonphysician practitioners' services, and services that are incident to a physician's service.[96]

Under the RBRVS fee schedule, each physician service is given a weight that measures its cost relative to other physician services.[97] The basic approach for calculating the relative weight is based on three components: a work component (reflecting the physician's time and intensity), a practice expense component (reflecting overhead costs, such as rent), and a malpractice component (reflecting malpractice expenses associated with the service).[98] Medical devices used to provide the service, with the exception of separately paid prosthetics, orthotics, and supplies, are paid under the practice expense component. A relative weight value is assigned to each of the three components and is converted to a payment amount by multiplying the value by an annual conversion factor and a geographic adjustment factor.[99] During the past few years, the payment levels for the physician fee schedule increased but only as a result of congressional interventions that avoided significant reductions.

3.5.4. Products for the Home-Care Setting

In the home-care settings, Medicare payments for medical devices generally are established under the fee schedule derived nationally for durable medical equipment, prosthetics, orthotics, and supplies (DMEPOS). Medicare payment is based on the lower of either the actual charge or the fee-schedule amount calculated for the item.[100] The fee-schedule amounts are adjusted geographically and are subject to nationwide floors and ceilings.[101] In accordance with the Medicare statute, the fee-schedule amounts are updated annually to account for inflation using a factor based on the consumer price index.[102]

To calculate the fee schedule for DMEPOS items, CMS reviews charge and purchase data and determines the reasonable charge for each state using prescribed formulas.[103] The nationwide floors and ceilings are then established based on the average or median state payment amounts.[104] If inadequate charge data exists, gap-filling methods are used to determine the payment amount.

CMS has the authority to adjust fee-schedule payment amounts if the statutory formula results in amounts determined not to be inherently reasonable.[105] CMS and its contractors can use a variety of factors to determine whether the existing payment amount is not inherently reasonable—defined as grossly excessive or grossly deficient—and, if so, can adjust such payments to a realistic and equitable amount. In addition, the Medicare program will be phasing in a competitive bidding program, designed to limit increases in expenditures through selection of DMEPOS suppliers in geographic areas, beginning in 2007.[106]

4. Overview of Coding Issues

To receive Medicare payment, health care providers, suppliers, and practitioners must submit claims to the appropriate Medicare contractor (i.e., a fiscal intermediary for services covered by Part A,[107] a carrier for services and items covered by Part B, and the Medicare Advantage risk plan for services covered by Part C). In addition to meeting administrative requirements, claims must describe the items and services for which payment is sought. One method of accomplishing this is the use of specific codes. Although a code does not represent a coverage or a payment decision, the existence of a code does have an integral role in ensuring whether the appropriate payment amount is received.

Uniform systems of codes, which depend on the service and type of provider, are prescribed for claims. The Administrative Simplification provisions of the Health Insurance Portability and Accountability Act of 1996 (HIPAA), directed the Secretary of HHS to prescribe medical data code sets for all electronic claims to third-party payors, including Medicare. In implementing these provisions, CMS prescribed the combination of the Healthcare Common Procedure Coding System (HCPCS) and the American Medical Association's (AMA's) Current Procedural Terminology, 4th Ed. (CPT-4) for professional services and products. HCPCS is an alphanumeric system consisting of three levels of codes and incorporating the CPT-4 codes. These codes are used for all claims for medical services, including physician services, and devices. For hospital reporting of inpatient services and diagnoses, the ICD-9-CM coding system was adopted.[108]

4.1. HCPCS and CPT-4 Codes

HCPCS Level I contains the CPT-4 codes.[109] CPT-4 codes are designed to describe physician and other services (e.g., radiologic procedures, clinical laboratory tests) and are maintained and annually updated by the AMA's CPT Editorial Panel with input from the CPT Advisory Committee. These codes are the basis of the Medicare physician fee schedule and are also used for reporting facility services in other settings. The AMA also has developed emerging-technology CPT codes to facilitate data collection for and assessment of new services and procedures. Because Medicare generally does not assign national payment amounts for these codes, assignment to one of these codes may not trigger payments. Requests for assignments to this code category should be approached cautiously.

HCPCS Level II codes are alphanumeric codes, primarily designed for items and services other than those covered by the CPT (e.g., ambulance services, dental services, DMEPOS).[110] The codes serve the important function of providing a standardized coding system. Under new procedures, CMS maintains the codes and the HCPCS Workgroup, in consultation with private payor organizations (e.g., the Blue Cross/Blue Shield Association, America's Health Insurance Plans), handles decision making. The HCPCS Workgroup coordinates an annual review cycle for making decisions about additions, revisions, and deletions to the permanent national alphanumeric codes. The HCPCS Workgroup meets periodically in nonpublic proceedings to consider applications for modifications to the coding system. Public meetings are held after the HCPCS Workgroup publishes its preliminary decisions. Permanent code changes adopted by the HCPCS Workgroup are effective January 1 of each year.

HCPCS Level II also includes a series of codes that are independent of the permanent national codes. These codes are called temporary codes and allow insurers, including Medicare, to establish codes that are needed before the next annual update for permanent national codes becomes effective or until consensus can be achieved on a permanent national code. About one-third of the HCPCS Level II codes are temporary codes. This is particularly important because CMS is now required to assign a temporary or permanent code whenever a final decision is made to grant the request for a national coverage determination.[111]

Level III codes are the subsystem of codes that have been developed by state Medicaid agencies, Medicare contractors, and private insurers for their specific programs or local areas of jurisdiction. For Medicare purposes, Level III codes are also referred to as local codes and were established when an insurer preferred that suppliers use a local code to identify a service for which there was no Level I or Level II code. These were alternatives to using a miscellaneous code or one describing an otherwise unclassified item or service. This allowed insurers to electronically process claims for new types of services for which a Level I or Level II code had not yet been established; however, Level III codes have been phased out under HIPAA.

4.2. ICD-9-CM Codes

Another important system of codes, ICD-9-CM codes, describe a patient's medical condition and are also required by HIPAA. ICD-9-CM codes, which were based on the World Health Organization disease classification system, are the official method of coding diagnoses and procedures associated with hospital utilization in the United States.[112] ICD-9-CM codes, Volumes 1 and 2, are used to report a patient's diagnosis or condition and are used by third-party payors to determine whether the service or product is warranted based on the patient's diagnosis or symptoms. For Medicare billing purposes, ICD-9-CM Volume 3 codes classify hospital inpatient procedures. These codes are also included on Medicare claims and drive the payment methodology for acute care hospital inpatient services. The National Center for Health Statistics and CMS maintain and annually update the ICD-9-CM procedure codes.

Under the Medicare Modernization Act, CMS is required to provide new diagnosis and procedure codes each April, although payments are not adjusted until each October 1.[113] This new timeframe facilitates introduction of new technologies and appropriate payments in a more timely manner. The addition of new codes helps ensure that cost data are available and can facilitate creation of a necessary DRG code, when appropriate.

5. Conclusion

It is never too early to develop a coverage and reimbursement strategy for devices being prepared for marketing, and this chapter has underscored some of the issues relevant to such a strategy. Considerations of Medicare coverage, payment, and coding issues will go a long way to ensuring that necessary steps can be integrated early in the development process and facilitate adequate and timely payment when the product is launched.

References

1. *See* 42 U.S.C. §§ 1395, *et seq.*(2005).
2. U.S. Department of Health and Human Services, 2004 CMS Statistics, *available at* http://www.cms.gov/researchers/pubs/CMSstatistics/2004CMSstat.pdf.
3. This program, previously called "Medicare + Choice," was created by Congress under the Balanced Budget Act of 1997. Pub. L. No. 105-33. The Medicare pro-

gram began as a fee-for-service payor, but through statutory amendments, CMS has the authority to contract with managed care companies to provide services under risk plans. The primary goal was to offer beneficiaries entitled to Part A and enrolled in Part B a wider range of plan choices. The plans represent a small percentage of benefit outlays. They are paid on a capitated payment system, with new payment options available after passage of the Medicare Prescription Drug, Improvement, and Modernization Act of 2003 (MMA). Pub. L. No. 108-173.

4. Part D offers the new outpatient prescription drug benefit enacted under MMA.
5. 42 U.S.C. § 1395x(b) (2005).
6. *Id.* § 1395x(s)(2)(B).
7. *Id.* § 1395x(q).
8. *Id.* § 1395x(s)(3).
9. *Id.* § 1395x(s)(6).
10. *Id.* 1395x(s)(9).
11. *Id.* § 1395 (s)(5).
12. *Id.* § 1395x(s)(8).
13. *Id.* § 1395y(a); *see also* 42 C.F.R. § 411.15 (2004).
14. Dental services may be covered in connection with an inpatient's underlying medical condition or if hospitalization is required for the procedure. 42 U.S.C. § 1395y(a)(12) (2005).
15. *Id.* § 1395x(jj).
16. *Id.* § 1395x(oo).
17. *Id.* § 1395x(pp).
18. *Id.* § 1395y(a)(1)(A) (also referred to by the non-codified provision—Section 1862(a)(1)(A) of the Social Security Act) (emphasis added).
19. *See, e.g.,* 54 Fed. Reg. 4302, 4304 (Jan. 30, 1989); 60 Fed. Reg. 48417, 48418 (Sept. 19, 1995).
20. Current literature offers a number of definitions for evidence based medicine. *See e.g.,* D.L. Sackett, et al., *Evidenced-Based Medicine: What It Is and What It Isn't,* British Medical Journal, 312: 71-72 (Jan. 13, 1996) ("Evidence-based medicine is the conscientious, explicit and judicious use of current best evidence in making decisions about the care of individual patients. The practice of evidence-based medicine means integrating individual clinical expertise with the best available external clinical evidence from systematic research.").
21. In 1987 and 1989, CMS (then "HCFA," the Health Care Financing Administration), first issued proposed regulations on coverage criteria. 52 Fed. Reg. 1557 (April 29, 1987) and 54 Fed. Reg. 5302 (Jan. 30, 1989).
22. *See* 65 Fed. Reg. 31124 (May 16, 2000).
23. *See* 42 U.S.C. § 1395y(*l*)(1) (2005) (added by MMA § 731).
24. *Id.* § 1395y(*l*)(5). This provision is effective for local coverage determinations made on or after July 1, 2004.
25. *See* 68 Fed. Reg. 55634, 55635 (Sept. 26, 2003). The definition for NCD was expanded by Section 522(b) of the Medicare, Medicaid and SCHIP Benefits Improvement and Protection Act of 2000 (Pub. L. No. 106-554), which also pro-

vided for appeal rights to challenge NCDs. *See* 42 U.S.C. § 1395ff(f)(1)(B) (2005). CMS has interpreted this newly expanded statutory definition to include the decision as to benefit category under which an item or service falls and statutory exclusion determinations. *See* 67 Fed. Reg. 54534, 54536 (Aug. 22, 2003). The same definition for the development of NCDs has been incorporated under Section 731 of MMA. *See* 42 U.S.C. § 1395y(*l*)(6)(A) (2005).

26. 68 Fed. Reg. at 55635; 42 C.F.R. §§ 405.732 & 405.860. Under new appeals procedures, NCDs may be challenged by certain beneficiaries who file complaints with the Department of HHS's Departmental Appeals Board. 68 Fed Reg. 63092 (Nov. 7, 2003). The Board may determine the validity of the NCD under prescribed appeal standards and the Board's decision is subject to judicial review.

27. *See* 68 Fed. Reg. at 55636.

28. *See id.* at 55638.

29. *See id.*

30. *See* 42 U.S.C. § 1395y(*l*)(4) (2005).

31. *See* 68 Fed. Reg. at 55640.

32. *See id.*

33. 42 U.S.C. § 1395y(*l*)(2) (2005). The nine-month timeframe does not apply to situations where a clinical trial is requested.

34. *See id.* § 1395y(*l*)(3)(C)(iv) (2005). The significance of billing codes is discussed below.

35. The longstanding manual compiling NCDs is the Medicare Coverage Issues Manual. *See* http://www.cms.hhs.gov/manuals/cmstoc.asp. The manual is broken down into the following categories: (1) Clinical Trials; (2) Medical Procedures; (3) Supplies and Drugs; (4) Diagnostic Services; (5) Dialysis Equipment; (6) DME; (7) Prosthetic Devices; (8) Braces, Trusses, Artificial Limbs and Eyes; (9) Patient Education Programs; and (10) Nursing Services. The new manual replacing the Medicare Coverage issues Manual is the Medicare National Coverage Determinations Manual, Publication 100-03. The new manual is an Internet-only (no hard copies) manual and may be accessed at the CMS web site: http://www.cms.hhs.gov/manuals. It contains more detailed classifications of NCDs: Chapter 1 describes the NCDs and Chapter 2, when available, will include billing codes related to each NCD.

36. 42 U.S.C. §§ 1395y(*l*)(6)(B), 1395ff(f)(2)(B) (2005).

37. 68 Fed. Reg. 63692, 63693 (Nov. 7, 2003).

38. The four Durable Medical Equipment Regional Carriers ("DMERCs"), which specialize in claims processing for durable medical equipment, prosthetics, orthotics and supplies, and drugs used with this equipment, on the other hand, are required to jointly develop and adhere to one set of policies, ensuring that DMERC coverage policies on prosthetics, orthotics, and supplies are uniform across the nation.

39. The decisions are not binding on administrative law judges or the Departmental Appeals Board. In addition, through new appeals procedures, a beneficiary may challenge the facial validity of an LCD by bringing a separate appeal to an administrative law judge. *See* 68 Fed. Reg. 63692 (Nov. 7, 2003).

40. Also, contractors within the same area are to consult on all new LCDs within the area. *See* 42 U.S.C. § 1395y(*l*)(5)(B) (2005); *see also supra* note 24.
41. Medicare Benefit Policy Manual, Chapter 14 § 10 (2005).
42. 60 Fed. Reg. at 48418.
43. 42 C.F.R. § 405.203 (2005); *see also* 60 Fed. Reg. at 48417.
44. Class III refers to devices that cannot be classified into Class I or Class II because insufficient information exists to determine that either special or general controls would provide reasonable assurance of safety and effectiveness. 42 C.F.R. § 405.201(b) (2005). Class III devices require pre-market approval.
45. *Id.*
46. *Id.*
47. Medicare Benefit Policy Manual, Chapter 14 § 50 (2005).
48. 42 C.F.R. § 405.209 (2005).
49. Medicare National Coverage Determination Manual, Chapter 1, § 310.1 (2005).
50. *See* 42 U.S.C. § 1395y(m) (2004) (added by MMA §731).
51. Details on Medicare coverage for the routine costs of clinical trials are addressed in another chapter of this publication.
52. The current standard for diagnosis codes includes the ICD-9-CM. *See,* e.g., 45 C.F.R. §162.1002(a)(l) (2004) (code set adopted under the Administrative Simplification provisions of the Health Insurance Portability and Accountability Act of 1996 ("HIPAA"), Pub. L. No. 104-191). These codes are also discussed further in the text.
53. *See, e.g.,* Medicare Claims Processing Manual, Chapter 23 § 10.1.2 (2005).
54. *See* 42 C.F.R. § 405.207 (2005).
55. *See id.*
56. *See, e.g.,* 42 U.S.C. § 1395w-4(a)(1) (2005) (physician fee-schedule payments).
57. *See* 42 C.F.R. § 412.23 (2005).
58. 42 U.S.C. § 1395x(v)(1)(A) (2005).
59. *See id.* § 1395x(v); 42 C.F.R. § 413.9(c)(2) (2005).
60. *See* 42. U.S.C. § 1395g(e) (2005); 42 C.F.R. §§ 413.60, 413.64 (2005).
61. CMS has recently implemented PPS for inpatient rehabilitation facilities and long term care hospitals. A PPS methodology for psychiatric hospitals also has been promulgated. For illustrative purposes, we focus here only on acute care hospital payment systems, which is the first setting transitioned to PPS and the model for those that followed.
62. *See* 42 U.S.C. § 1395ww(a)(4) (2005); 42 C.F.R. § 412.2(c) (2005).
63. *See* 42 U.S.C. § 1395ww(g) (2005); 42 C.F.R. § 412.300 (2005).
64. *See* 42 U.S.C. § 1395ww(d)(3) (2005); 42 C.F.R. § 412.60 (2005).
65. *See* 69 Fed. Reg. 48915, 48925 & 49592-49611 (Aug. 11, 2005).
66. *See* 42 U.S.C. § 1395ww(d)(4)(C)(i) (2005).
67. *See* 42 U.S.C. § 1395ww(d)(4)(B) (2005); 42 C.F.R. § 412.60(b) (2005).
68. *See* 42 U.S.C. § 1395ww(d)(3)(A)(iv) (2005); 42 C.F.R. § 412.63(c) (2005).
69. *See* 42 U.S.C. § 1395ww(d)(4)(C)(i) (2005).
70. *See id.* § 1395ww(b)(3)(B).

71. *See* 42 U.S.C. § 1395ww(d)(5)(A) (2005); 42 C.F.R. §§ 412.80, 412.84 (2005).
72. *See* 42 U.S.C. § 1395ww(d)(5)(F)(i)(I), (d)(5)(F)(v) (2005); 42 C.F.R. § 412.106 (2005).
73. *See* 42 U.S.C. § 1395ww(d)(5)(B) (2004); 42 C.F.R. § 412.105 (2005).
74. *See* 42 U.S.C. § 1395ww(d)(5)(K)-(L) (2005); 42 C.F.R. §§ 412.87, 412.88 (2005).
75. *See* 42 U.S.C. § 1395ww(d)(5)(K)(vi) (2005); 42 C.F.R. § 412.87(b)(1) (2005).
76. *See* 42 U.S.C. § 1395ww(d)(5)(K)(viii) (2005).
77. *See id.* § 1395ww(d)(5)(K)(ii)(III). The change was made under MMA § 503(d).
78. *See* 42 U.S.C. § 1395ww(d)(5)(K)(ix) (2005).
79. *See* 42 U.S.C. 1395*l*(t) (2005); 42 C.F.R. § 419.21 (2005). Critical access hospitals are exempt from outpatient PPS. *See* 42 U.S.C. § 1395m(g) (2005); 42 C.F.R. § 413.70(b) (2005).
80. *See* 42 U.S.C. § 1395*l*(t)(3)(C) (2005); 42 C.F.R. §§ 419.31, 419.32 (2005).
81. *See* 42 U.S.C. § 1395*l*(t)(4)(A) (2005); 42 C.F.R. § 419.31(c)(2) (2005).
82. *See* 42 U.S.C. § 1395*l*(t)(3)(C)(iv) (2005); 42 C.F.R. § 419.32(b) (2005).
83. *See* 42 U.S.C. § 1395*l*(t)(5) (2005); 42 C.F.R. § 419.43(d) (2005).
84. *See* 42 U.S.C. § 1395*l*(t)(6) (2005); 42 C.F.R. §§ 419.64, 419.66 (2005). The additional payments were authorized under section 201(b) of the Medicare, Medicaid, and SCHIP Balanced Budget Refinement Act of 1999, Pub. L. No. 106-113.
85. *See* 42 U.S.C. § 1395*l*(t)(6)(A)(i)-(iii) (2005); 42 C.F.R. § 419.64(a)(1-3) (2005).
86. *See* 42 U.S.C. § 1395*l*(t)(6)(A)(iv) (2005); 42 C.F.R. §§ 419.64(a)(4) & (b), 419.66 (2005).
87. *See* 42 U.S.C. § 1395*l*(t)(6)(B)(iii) (2005); 42 C.F.R. § 419.66(g) (2005).
88. *See* 42 U.S.C. § 1395*l*(t)(6)(D)(ii) (2005); 42 C.F.R. § 419.66(h) (2005).
89. *See* 42 C.F.R. § 419.66(b)(1) (2005).
90. *See id.* at § 419.66(b)(2).
91. *See id.* at § 419.66(b)(3).
92. *See id.* at § 419.66(b)(4).
93. *See* 42 U.S.C. §1395*l*(t)(6)(B)(i) (2005).
94. These new bands went into effect in 2005. *See* 68 Fed. Reg. 63398, 63485-86 (Nov. 7, 2003). CMS narrowed cost bands so that a greater number of APC codes were instituted to more precisely reflect costs.
95. *See* 42 U.S.C. § 1395w-4 (2005).
96. *See id.* § 1395w-4(j)(3).
97. *See id.* § 1395w-4(c).
98. *See id.* § 1395w-4(c)(1).
99. *See id.* § 1395w-4(b)(1).
100. *See id.* § 1395m(a)(1)(A)-(B).
101. *See id.* § 1395m(a)(2)-(9). Fee-schedule amounts for Alaska, Hawaii, and Puerto Rico are not subject to ceilings and floors. *See id.* § 1395m(a)(10)(A).
102. *See id.* § 1395m(a)(14)(F).
103. *See id.* § 1395m(a)(2)-(9).

104. *See id.*
105. *See id.* § 1395u(b)(8)-(9); 42 C.F.R. § 405.502(g)-(h) (2005).
106. *See* 42 U.S.C. § 1395w-3 (2005).
107. Fiscal intermediaries also process Part B Claims. For example, hospitals submit their inpatient Part A *and* outpatient Part B claims to the fiscal intermediary.
108. *See* 45 C.F.R. § 162.1002 (2005).
109. The AMA holds a copyright to the codes. *See* http:www.ama-assn.org/ama/pub/category/3882.html.
110. *See* http://www.cms.hhs.gov/medicare/hcpcs/.
111. *See* 42 U.S.C. § 1395y(*l*)(3)(C)(iv). In addition, revised procedures are required for the issuance of temporary codes under Medicare Part B. *See* MMA § 731(c).
112. *See* http://www.cdc.gov/nchs/about/otheract/icd9/abticd9.htm. Although released, the codes in ICD-10-CM are not currently valid for any purpose or uses. Prior to implementation of ICD-10-CM and adoption under the HIPAA, there will be a two year window, to occur after final notice has been published in the Federal Register. *See* http://www.cdc.gov/nchs/about/otheract/icd9/abticd10.htm.
113. *See* 42 U.S.C. § 1395ww(d)(5)(K)(vii) (2005). Procedures for requests for new ICD-9-CM procedures are found at http://www.cms.hhs.gov/paymentsystems/icd9/.

4

Postmarket Requirements for Significant Risk Devices

Suzan Onel

1. Background and Overview

The Food and Drug Administration (FDA) has developed multiple interrelated mechanisms for monitoring, evaluating, and taking remedial action (when necessary) against medical devices after they have gone to market. This chapter provides an overview of the various postmarket requirements that can be imposed on manufacturers, importers, distributors, and device-user facilities.[1]

The regulatory framework for medical devices that exists today was originally established in 1976 with the Medical Device Amendments[2] to the federal Food, Drug, and Cosmetic (FD&C) Act. Under this framework, medical devices are classified into one of three classes, depending on the level of risk of the device. Class I devices have the lowest level of risk, requiring only general controls, such as establishment registration and device listing. Class II devices pose additional concerns but are controlled by the use of general controls and certain special controls to ensure safety and effectiveness, including 510(k) premarket notifications and sometimes performance standards. Class III devices have the highest risk and are subject to the greatest premarket controls, including the submission and approval of premarket approval applications (PMAs).[3]

1.1. The Role of Postmarket Controls

Congress and the FDA established postmarket controls to help compensate for the perceived insufficiency of the premarket approval process to ensure safety and effectiveness. They reflect an effort to increase the likelihood that serious device problems are detected early, thereby minimizing the risk of widespread injury or death.

From: *Clinical Evaluation of Medical Devices: Principles and Case Studies, Second Edition*
Edited by: K. M. Becker and J. J. Whyte © Humana Press Inc., Totowa, NJ

This chapter focuses on the following postmarket controls that are exerted on Class III significant risk (SR) devices: mandatory reporting, device tracking, postmarket surveillance, and administrative remedies. Although some of the same controls apply equally to non-SR devices (i.e., Class I and Class II), others apply only to Class III devices. Because of the different levels of regulatory oversight necessary in the retail chain and the fact-specific nature of many of the postmarket controls, these controls represent a spectrum of mandatory and discretionary activities.

1.2. Complaint Files

All medical device manufacturers are subject to the complaint-filing requirements in 21 C.F.R. Part 820, the Quality System Regulation (QSR). The QSR broadly defines a *complaint* as "any written, electronic, or oral communication that alleges deficiencies related to the identity, quality, durability, reliability, safety, effectiveness, or performance of a device after it is released for distribution."[4]

Manufacturers are required to review, evaluate, and when appropriate, investigate complaints; establish and maintain written procedures describing the process used to perform these activities; and designate a responsible individual or entity to perform these tasks.[5] Complaints concerning death, serious injury, or malfunctions, as defined in the medical device reporting (MDR) regulation, must be reported to the FDA pursuant to the MDR regulations (*see* Section 2).

The FDA is authorized to review and copy a company's complaint files. In fact, the FDA often begins inspections this way because these records allow the agency to quickly assess whether a company's quality system and corrective actions are adequate. The FDA does not specify a standard complaint-handling system, but the QSR requirements specify certain minimum requirements that must be included in any manufacturing system. They are as follows:

- Document, review, evaluate, and file all complaints.
- Formally designate a unit or individual to perform these activities.
- Determine if an investigation is necessary.
- Record the reason if no investigation is made.
- Assign responsibility for deciding when not to investigate.
- Determine if the complaint requires an MDR report.
- If an investigation is necessary, prepare a written record of the investigation, any corrective action taken, and any reply to the complainant.[6]

1.3. Adverse Reaction and Device Defect Reporting

To provide continued reasonable assurance of the safety and effectiveness of a device, the FDA requires manufacturers of PMA devices to submit an adverse reaction or device defect report within 10 days of receiving information that meets one of the following conditions:

1. A mix-up of the device or its labeling with another device.
2. Any adverse reaction, side effect, injury, toxicity, or sensitivity reaction that is attributable to the device.
   a. Has not been addressed by the device labeling.
   b. Has been addressed by the device's labeling but is occurring with unexpected severity or frequency.
3. Any significant chemical, physical, or other change or deterioration in the device or any failure of the device to meet the specifications established in the approved PMA that could not cause or contribute to death or serious injury but is not correctable by adjustments or other maintenance procedures described in the labeling.[7]

Adverse reaction and device defect reports involve incidents that qualify as less than MDR-reportable events (*see* Section 2) but represent information the FDA deems to be important to track as part of a device's safety record.

2. Medical Device Reporting

Reporting requirements are the most basic form of postmarket control. The FD&C Act authorizes the FDA to implement MDR requirements for all medical devices, whether they are Class I, II, or III.[8] The information included in an MDR provides the FDA with data that can be used to determine whether and when remedial action must be taken.

2.1. General Terms and Definitions

The FDA's MDR regulation is concerned with adverse events, namely, deaths or serious injuries. Because not all deaths and serious injuries are device-related, the regulation seeks to identify the specific circumstances in which there is a duty to report an adverse event related to the functioning of a medical device. The FDA imposes reporting and record-keeping obligations on medical device manufacturers,[9] importers,[10] and user facilities.[11] Although the specific reporting and recordkeeping requirements vary according to each entity, there are a number of common concepts.

2.1.1. MDR-Reportable Event

There are two categories of MDR-reportable events. The first category, which applies to manufacturers, importers, and user facilities, covers situations in which an entity has become aware of information "that reasonably suggests that a device has or may have caused or contributed to a death or serious injury."[12] The second category, which applies only to manufacturers and importers, covers situations in which an entity receives or becomes aware of information that reasonably suggests that one of its marketed devices has malfunctioned and that the device or a similar device would be likely to cause or contribute to death or serious injury if the malfunction were to recur. *Malfunction*, in this context, is defined as the failure to meet performance specifications or to perform as intended.[13] The FDA assumes a similar malfunction is likely to occur.

2.1.2. Become Aware

The duty to report is triggered when any employee of a manufacturer, importer, or device-user facility *becomes aware* of information "reasonably suggesting that a reportable adverse event has occurred."[14] For all entities, this means information that reasonably suggests that a device has or may have caused or contributed to a death or serious injury. For manufacturers, this also refers to information that reasonably suggests that a marketed device has malfunctioned and that, if the device were to malfunction again, the device would be likely to cause or contribute to a death or serious injury.[15]

There is an additional element of awareness for manufacturers. The MDR regulation states that a manufacturer also becomes aware when an employee with a supervisory or management responsibility over "persons with regulatory, scientific, or technical responsibilities, or a person whose duties relate to the collection and reporting of adverse events" becomes aware of reportable MDR events that "necessitate remedial action to prevent an unreasonable risk of substantial harm to the public health."[16] In these situations, as noted below, the FDA requires expedited reporting.

2.1.3. Caused or Contributed

The FDA's reporting requirements focus on those events in which a medical device has or may have caused or contributed to a serious injury or death. The regulations define *caused or contributed* expansively.[17] This phrase applies not only to the failure or malfunction of medical devices but also to events occurring as a result of improper or inadequate design, manufacture, or labeling or user error. A report is not required for situations in which an individual who is qualified to make a medical judgment reaches "a reasonable conclusion that a device did not cause or contribute to a death or serious injury, or that a malfunction would not be likely to cause or contribute to a death or serious injury, if it were to recur."[18]

2.1.4. Serious Injury

The MDR regulation is concerned not only with deaths but also with situations in which a medical device may have caused or contributed to a serious injury. The FDA defines a *serious injury* as an injury or illness that is life-threatening, "results in permanent impairment of a body function or permanent damage to body structure,"[19] or requires intervention to prevent such permanent impairment or damage.[20]

2.2. Written MDR Procedures and Record-Keeping

Although device reporting regulations focus on the reports that various entities must submit to manufacturers and the FDA, the MDR regulation also includes requirements for developing MDR procedures and maintaining files on MDR-reportable events.

This internal system must include provisions for the timely identification and evaluation of events, a standardized review process, and a timely transmission of reports. These documents are subject to the FDA's inspectional authority. MDR event files should also include information about the deliberations on whether a serious injury, death, or malfunction is reportable and justification of the decision not to report an event to the FDA when appropriate.

2.3. Adverse Event Reporting

2.3.1. Mandatory Requirements

Manufacturers have the most extensive reporting requirements. They are responsible for submitting individual reports regarding reportable malfunctions, as well as actual deaths and serious injuries. These reports must be submitted to the FDA through the MedWatch system on FDA Form 3500A within 30 days of becoming aware of the MDR-reportable event. Manufacturers are also obligated to provide incomplete or missing information submitted by user facilities, importers, or other initial reporters and to investigate the cause of an MDR-reportable event.

When a manufacturer becomes aware that a reportable event, based on any information (including trend analysis), "necessitates remedial action to prevent an unreasonable risk of substantial harm to the public health," the manufacturer must submit a report within 5 business days.[21]

For user facilities, the reporting requirement varies for deaths and serious injuries. In the event of an MDR-reportable death, a user facility must report the event both to the device manufacturer and the FDA. In the case of MDR-reportable serious injuries, a user facility must report the event only to the manufacturer (or to the FDA if the manufacturer is unknown). In both cases, the reports must be submitted on FDA Form 3500A, within 10 working days of obtaining information about the event.

Reporting requirements for importers are similar to those of manufacturers. Importers must use the same FDA Form 3500A for reportable deaths, serious injuries, and malfunctions. Importers must send reports of such events to FDA,

Table 1
Summary of Reporting Requirements

Reporter	What to report	FDA form no.	To whom	When
Manufacturer	Death, serious injury, and malfunction	3500A	FDA	Within 30 days of becoming aware of an event
Manufacturer	Events requiring remedial action to prevent an unreasonable risk of substantial harm to the public health and other types of events designated by the FDA	3500A	FDA	Within 5 business days of becoming aware of an event or within 5 days of FDA request
Manufacturer	Baseline reports to identify and provide basic data on each device that is subject of an MDR report. At this time, FDA has stayed the requirement for denominator data requested in Part II, Items 15 and 16 on Form 3417	3417	FDA	With the initial 30 days report (Form 3500A)
Manufacturer	Corrections/removals that reduce health risk or remedy FD&C Act violation	N/A	FDA	Within 10 business days of correction/removal
Importer	Death, serious injury, and malfunctions	3500A	FDA	Within 30 days
Importer	Corrections/removals that reduce health risk or remedy FD&C Act violation	N/A	FDA	Within 10 business days of correction/removal
User facility	Death	3500A	FDA and manufacturer	Within 10 business days
User facility	Serious injury	3500A	Manufacturer (or FDA if manufacturer unknown)	Within 10 business days
User facility	Annual report	3419	FDA	January 1 (if there were individual reports in prior year)
Consumer/health professional	Death, serious injury, and malfunction	3500	FDA	N/A; voluntary

FDA, Food and Drug Administration; MDR, medical device reporting; FD&C, federal Food, Drug, and Cosmetic Act; N/A, not applicable.

along with a copy to the manufacturer, within 30 days of becoming aware of information from any source that reasonably suggests an imported device caused or contributed to a death, serious injury, or malfunction.[22] Table 1 summarizes reporting obligations.

2.3.2. Voluntary Requirements

In addition to the obligatory reporting requirements placed on manufacturers, importers, and user facilities, the FDA also invites health care professionals and consumers to submit reports about FDA-regulated products using FDA's MedWatch Form 3500.

2.4. Baseline and Annual Reports

Manufacturers are required to submit a baseline report (FDA Form 3417) to the FDA with the initial event report (Form 3500A). These reports help track the frequency of reportable events tied to specific devices and product families.[23] Manufacturers usually report a summary of complaints in annual PMA reports, although they are not required to by regulation.[24] If a manufacturer obtains information that was not provided in a report to the FDA because it was not known or available at the time of the initial report, the manufacturer must submit a supplemental report within 1 month of the receipt of this information.[25]

User facilities are required to submit annual reports on FDA Form 3419, including information for each reportable event that occurred during the annual reporting period.[26]

2.5. Reports and Records of Correction and Removals

The FD&C Act imposes an additional burden on manufacturers and importers to report certain actions concerning device corrections and removals and maintain records of all corrections and removals regardless of whether they must be reported to the FDA.[27] This requirement does not apply to actions taken by manufacturers or importers merely to improve the performance or quality of a device.

2.5.1. Definitions

A *correction* is "the repair, modification, adjustment, relabeling, destruction, or inspection (including patient monitoring) of a device without its physical removal from its point of use to some other location."[28]

A *removal* is defined as "the physical removal of a device from its point of use to some other location for repair, modification, adjustment, relabeling, destruction, or inspection."[29]

Risk to health means "(1) a reasonable probability that use of, or exposure to, the product will cause serious adverse health consequences or death; or (2) that use of, or exposure to, the product may cause temporary or medically reversible adverse health consequences, or an outcome where the probability of serious adverse health consequences is remote."[30]

Reports of corrections and removals do not apply to market withdrawals, stock recovery, or rotation.[31]

2.5.2. Reports of Corrections and Removals

Manufacturers and importers must report to the FDA within 10 business days any correction or removal if it was initiated to reduce a risk to health posed by the device or remedy a violation of the FD&C Act that may pose a risk to health, unless it was already reported pursuant to an MDR (*see* Table 1).

2.6. Public Disclosure

The FD&C Act requires manufacturers and importers to maintain records about corrections and removals and make such records available for FDA inspection, even if the corrections and removals do not qualify as reportable events.[32] In such cases, manufacturers must record information about the device and the events leading up to the correction or removal, along with a justification for not reporting the correction or removal to the FDA. These records must be maintained for 2 years beyond the expected life of the device.

All reports submitted to the FDA pursuant to the MDR and corrections- and removals-reporting requirements are available for public disclosure.

3. Medical Device Tracking

The FD&C Act allows the FDA to impose tracking obligations on manufacturers and distributors of types of devices that pose the greatest likelihood of serious adverse health consequences.[33] The tracking requirements are designed to facilitate notification to users and potentially product recall, in the event that a device presents a serious risk to health that requires prompt attention. Advanced tracking enables an efficient process to facilitate communication to the affected parties.

3.1. Significant Historical Changes

The FDA's tracking requirements were first added to the FD&C Act in 1990 by the Safe Medical Devices Act[34] and amended in 1997 by the Food and Drug Administration Modernization Act.[35] Under the amended law, the FDA requires manufacturers to track devices only when the agency issues an order to a manufacturer specifying that tracking is necessary. There is no statutory requirement to track, unless the FDA has issued an order.

3.2. Tracking Criteria

The FDA may require tracking for any Class II or III device that meets one of the following three criteria:

1. The failure of the device "would be reasonably likely to have serious adverse health consequences."
2. The device is a permanently implantable device.
3. The device is a life-sustaining or life-supporting device used outside a device-user facility.

Serious adverse health consequences are device-related experiences that are life-threatening or involve permanent or long-term injury or illness.[36] A *permanently implantable device* is intended for placement into a body cavity for more than 1 year to assist, restore, or replace a body function or structure.[37] A *life-sustaining* or *life-supporting device* either is essential or provides information that is essential to the continuation of human life.[38] When assessing whether to issue a tracking order, the FDA may consider the likelihood of sudden, catastrophic failure; the likelihood of significant adverse clinical outcome; and the need for prompt professional intervention.[39]

3.3. Devices That Require Tracking

The FDA has issued tracking orders to manufacturers of the following implantable devices:

- Glenoid fossa prostheses.
- Mandibular condyle prostheses.
- Temporomandibular joint prostheses.
- Abdominal aortic aneurysm stent grafts.
- Automatic implantable cardioverters/defibrillators.
- Cardiovascular permanent implantable pacemaker electrodes.
- Implantable pacemaker pulse generators.
- Mechanical replacement heart valves.
- Implantable cerebellar stimulators.
- Implantable diaphragmatic/phrenic nerve stimulators.
- Implantable infusion pumps.
- Dura mater.

FDA also has required tracking for the following devices used outside a device-user facility: breathing frequency monitors, continuous ventilators, DC defibrillators and paddles, and ventricular bypass (assist) devices.

The FDA may also issue subsequent orders terminating tracking requirements.

3.4. Tracking Methods and Requirements: Manufacturers

Although the FDA does not require a specific tracking method, manufacturers of tracked devices must develop written standard operating procedures to collect, maintain, and report data regarding devices not yet distributed to patients,[40] devices distributed to individual patients for use outside a device-user facility,[41] and devices distributed to a multiple distributor[42] for use by more than one patient.[43] Manufacturers must also maintain current records for each tracked device for as long as the device is in use or available for distribution.[44]

3.5. Tracking Methods and Requirements: Distributors and User Facilities

The tracking regulation distinguishes between three types of distributors—distributors, final distributors, and multiple distributors—all of which have a duty to report tracking information to the manufacturer. Both final distributors and multiple distributors are entities that make a delivery or sale to the ultimate user; the distinction between them is whether they distribute to a single patient (final distributor) or with the intention of distributing a device to more than one patient over the life of the device (multiple distributor). For purposes of tracking requirements, device-user facilities are considered either final distributors or multiple distributors (depending on whether they distribute devices to single or multiple patients). The federal government may also be considered a distributor, final distributor, or multiple distributor.

3.6. Patient Issues

Although the FDA does not require that patients give written consent to have a device tracked or to release their identity to the manufacturer, patients may refuse to have their device(s) tracked.[45] Patient refusal does not relieve manufacturers of their duty to track the device itself. In such cases, the FDA may need to assist in identifying patient information by working with the distributor or final distributor.[46] For situations in which the health or safety of the patient necessitates release of patient information to a manufacturer, physician, or other entity, such patient information will not be disclosed to the general public.[47]

4. Postmarket Surveillance

The FDA's postmarket surveillance procedures are the least refined of the various postmarket mechanisms in terms of formal regulations. Under FD&C Act §522, as amended in 1997, the FDA may require a manufacturer to conduct postmarket surveillance for any Class II or Class III device "the failure of which would be reasonably likely to have serious adverse health consequences or which is intended to be (1) implanted in the human body for more than one year, or (2) a life-sustaining or life-supporting device used outside a device-user facility."[48] Although these criteria are virtually identical to the primary criteria for tracking requirements, their purpose is much different. Whereas the purpose of device tracking is to facilitate notification and recall by having reliable information about device location and patients using certain devices, the purpose of postmarket surveillance is to conduct clinical studies that might reveal "unforeseen adverse events or other information necessary to protect the public health."[49] Postmarket surveillance can take the form of mandated postapproval studies and postmarket surveillance studies voluntarily undertaken by a company to accomplish a particular goal.

4.1. Mandated Postapproval Studies

It has become increasingly common for the FDA to require sponsors of Class III devices to conduct postapproval clinical studies as a mandatory condition of PMA approval. These studies are usually narrow in scope and focus on generating additional data to expand on results of pivotal trials in support of product approval. The objective is typically longer-term follow-up. The trend toward mandated postmarket surveillance is positive in that it represents an agency shift toward allowing a device to enter into the marketplace with certain postmarket constraints in place rather than postponing approval until further data is generated. For example, the FDA-approved Johnson & Johnson/ Independence Technology's iBOT™ stair-climbing wheelchair with the condition that the company file reports "summarizing usage information obtained from data logs, reported device failures, and reported adverse events" for the first 2 years of the iBOT's postmarket use.[50] The FDA further required that physicians and other health professionals undergo special training to prescribe the iBOT.

To date, the FDA has not issued formal regulations regarding its mandatory postmarket surveillance policy, but the agency has developed several nonbinding guidance documents. Within the broad category of devices for which postmarket surveillance may be required, the FDA has identified internal criteria to determine when to order surveillance.[51] The principal criterion is whether there are important unanswered questions about marketed devices.

4.2. Voluntary Postmarket Surveillance Studies

Most voluntary postmarket surveillance studies are initiated by the manufacturer, sometimes at the request of the FDA, to develop systematic data on either long-term failure modes and/or the potential for adverse events occurring in a small number of patients. It is worth noting that compliance with medical device Good Manufacturing Practices/QSRs also compels sponsors to engage in postmarket surveillance of marketed devices. This includes requirements to evaluate and, if appropriate, act on complaints, product failures, and adverse events associated with product use. This process also requires a manufacturer to maintain procedures for a systematic evaluation of the postmarket experience.

Other postmarket research can include performing comparative studies with alternative or competitive treatments or devices with the goal of providing support for comparative effectiveness claims, pharmacoeconomic claims, expanded label claims, or reimbursement. Vigilance in the evaluation of adverse events, returned goods, and user complaints are also important sources of information on clinical experience. Although anecdotal, these data can provide a useful foundation for further product research into design deficiencies and strengths and consequently lead to better products.

5. Administrative Remedies: Notification and Recalls

Class III devices are potentially subject to a variety of other postmarket mechanisms designed to control their use and marketing.

5.1. Mandatory and Voluntary Notification

The FD&C Act provides that the FDA may require a company to notify users, user facilities, and other appropriate entities when a device "presents an unreasonable risk of substantial harm to the public health" and there is no more practical means to eliminate that risk.[52] In the event that notifying the user directly would present a greater harm than providing no notification at all, the FD&C Act provides that prescribers of the device be given the notification, so that they can determine whether any action by the user or on behalf of the user should be taken to eliminate or reduce the risk. Typically, manufacturers, on their own or after dialogue with the FDA, voluntarily issue a safety alert notifying prescribing doctors, hospitals, and clinics of the potential for injury.

5.2. Repair, Replacement, or Refund

Notification is the lowest action level for instances in which the FDA determines that a device presents an unreasonable risk of substantial harm to the public health. If notification is considered inadequate in eliminating the risk, the FDA can take additional steps to require that a device be repaired, replaced, or refunded;[53] however, the agency must first provide an opportunity for an informal hearing before issuing such an order and find that the following four conditions apply:

1. The device "presents an unreasonable risk of substantial harm to the public health."

2. There are reasonable grounds to believe "that the device was not properly designed or manufactured with reference to the state of the art as it existed at the time of its design or manufacture."
3. There are reasonable grounds to believe that the unreasonable risk was not posed by the device user.
4. Notification is inadequate.

5.3. Recalls

Whereas notification, repair, replacement, and refund are associated with determinations that a device presents an unreasonable risk of substantial harm to the public health, the FDA can require that a device be immediately recalled when it has a reasonable probability of causing "serious, adverse health consequences or death."[54]

5.3.1. Voluntary Recalls

The regulatory requirements for voluntary recall of medical devices are part of the FDA's general regulations for all FDA-regulated products.[55] The FDA regulations define *recall* as a removal or correction of a marketed product that the FDA considers in violation of the laws it administers and against which the FDA would initiate legal action.[56]

A manufacturer or importer "may decide of its own volition and under any circumstances to remove or correct a distributed product."[57] If the firm believes that the device in question is in violation of device laws, it must notify the FDA immediately with detailed information about the device, the level of risk, the scope of distribution, and the strategy for conducting the recall. Based on a health-hazard analysis, the FDA then classifies the recall according to the level of risk posed by the device into the following categories:

Class I: A situation in which there is a reasonable probability of "serious adverse health consequences or death."
Class II: A situation in which the use of or exposure to a device may cause "temporary or medically reversible adverse health consequences or where the probability of serious adverse health consequences is remote."
Class III: A situation in which adverse heath consequences are unlikely.

After advising the manufacturer or importer of the recall classification, the FDA may suggest changes in the recall strategy and will place the recall in its weekly FDA Enforcement Report.

FDA regulations for voluntary recalls of products specify that a recall may be initiated either by an entity's own volition or at the request of the FDA. The FDA has two paths of recourse for entities that do not comply with its request for voluntary recalls. First, the FDA may initiate a seizure of the device in question pursuant to the voluntary-recall provisions. Second, the FDA can demand that the product be recalled under its mandatory-recall authority, as described in the next section.

5.3.2. Mandatory Recalls

Although rarely used, the FDA's mandatory-recall authority is a potent threat to help drive a voluntary recall. Unlike a voluntary recall in which the FDA requests that a manufacturer conduct a recall, a mandatory recall is legally binding.

The FDA's mandatory-recall authority is limited to situations in which there is a reasonable probability of death or serious adverse health consequences. In other words, the FDA's mandatory-recall authority corresponds to voluntary Class I recalls; however, before taking any action, the FDA must first conduct an informal hearing.

Both voluntary and mandatory recalls can be terminated on request of the manufacturer or importer. Alternatively, the FDA may terminate a recall when the device has been removed or corrected to such an extent that the device no longer poses serious, adverse health consequences or death. (In the case of voluntary recalls, the standard varies according to the recall classification.) The FDA may also terminate "cease distribution" and "notification orders" when all reasonable efforts have been taken to ensure that all appropriate individuals and entities have been notified and instructed to cease use.

6. Conclusion

In sum, the FDA has a variety of mechanisms to monitor SR medical devices in the postmarket period to ensure their safety and effectiveness. These mechanisms include mandatory reporting under the MDR regulation, device tracking, postmarket surveillance, and administrative remedies such as notification and recall.

That said, the FDA's postmarket controls tend to be weaker than its premarket controls owing to the lack of systematic data in the postmarket environment. This is caused by several factors: (1) rapid evolution of technology making postmarket studies obsolete; (2) lack of incentives for industry to conduct or complete studies; (3) lack of interest in the clinical community and inadequate postmarket focus by the FDA; (4) lack of clearly specified public health questions; (5) under-recognition/under-reporting of device errors or adverse events; and (6) uneven enforcement of the postmarket requirements.

In recent months, the FDA has refocused its attention on the need to strike a balance between pre- and postmarket requirements. From a practical standpoint, this results in greater enforcement of the postmarket requirements, improved flow of real-world information on a device from clinical settings, increased focus on the lifecycle of a device, and ultimately improved device safety and effectiveness.

References

1. Entities covered by these requirements should consult the statute, regulations, and guidance materials directly for further detail.
2. Medical Device Regulation Act of 1976, Pub. L. No. 94-295, 90 Stat. 539 (1976).
3. Federal Food, Drug, and Cosmetic Act of 1938, Pub. L. No. 75-717, §515, 52 Stat. 1040 (1938), 21 U.S.C. §360e.
4. 21 C.F.R. §820.3(b). Service and repair requests may also meet this definition.
5. 21 C.F.R. §820.198.
6. Food and Drug Administration, Center for Devices and Radiological Health, 1997. MEDICAL DEVICE QUALITY SYSTEMS MANUAL, §15.
7. 21 C.F.R. §814.82(a)(9).
8. Federal Food, Drug, and Cosmetic Act of 1938, Pub. L. No. 75-717, §519, 52 Stat. 1040 (1938), 21 U.S.C. §360i.
9. A manufacturer is defined as "any person who manufactures, prepares, propagates, compounds, assembles, or processes a device by chemical, physical, biological, or other procedure," in Medical Device Reporting, Definitions. 21 C.F.R. §803.3(o) (2003).
10. 21 C.F.R. §803.3(m) defines an importer as "any person who imports a device into the United States and who furthers the marketing of a device from the original place of manufacture to the person who makes final delivery or sale to the ultimate user, but who does not repackage or otherwise change the container, wrapper, or labeling of the device or device package."
11. 21 C.F.R. §803.3(f) defines a device-user facility as "a hospital, ambulatory surgical facility, nursing home, outpatient diagnostic facility, or outpatient treatment facility" (each of which are individually defined in paragraphs (b), (l), (t), (u), and (v)). The definition of device-user facility does not apply to facilities that meet the definition of a "physician's office" in §803.1(x), school nurse offices, and employee health units.
12. 21 C.F.R. §803.3(r). With respect to user facilities, the regulation specifies becoming aware of information. With respect to manufacturers and importers, the regulation specifies either becoming aware of information (i.e., internally) or receiving information (i.e., externally).
13. 21 C.F.R. §803.3(n).
14. 21 C.F.R. §803.3(c).
15. 21 C.F.R. §803.3(r)(2).
16. 21 C.F.R. §803.3(c)(2).
17. 21 C.F.R. §803.3(d).
18. 21 C.F.R. §803.20(c)(2).
19. 21 C.F.R. §803.3(bb)(1)(ii).

20. "Permanent," in this context, is defined as "irreversible impairment or damage to a body structure or function, excluding trivial impairment or damage." 21 C.F.R. §803.3(bb)(1)(iii)(2).
21. 21 C.F.R. §803.53(a).
22. 21 C.F.R. §803.40.
23. 21 C.F.R. §803.55.
24. 21 C.F.R. §814.84.
25. 21 C.F.R. §803.56.
26. 21 C.F.R. §803.33.
27. Federal Food, Drug, and Cosmetic Act of 1938, Pub. L. 75-717, §519(f), 52 Stat. 1040 (1938), 21 U.S.C. §360i(f).
28. 21 C.FR. §806.2(d).
29. 21 C.FR. §806.2(i).
30. 21 C.FR. §806.2(j).
31. *Market Withdrawal* means "a correction or removal of a distributed device that involves a minor violation of the act that would not be subject to legal action by FDA or that involves no violation of the act, e.g., normal stock rotation practices." 21 C.F.R. §806.2(h). *Stock Recovery* means the correction or removal of a device that has not actually been marketed or released for sale. 21 C.F.R. §806.2(l).
32. 21 C.F.R. §806.20(b).
33. FD&C Act §519(e); 21 U.S.C. §360i(e).
34. Safe Medical Devices Act of 1990, Pub. L. No. 101-629, 104 Stat. 4511 (1990).
35. Food and Drug Administration Modernization Act of 1997, Pub. L. No. 105-115 (1997), 111 Stat. 2295.
36. 21 C.F.R. §821.3(e).
37. 21 C.F.R. §821.3(f).
38. 21 C.F.R. §821.3(g).
39. Food and Drug Administration, Center for Devices and Radiological Health, 2003. Guidance on Medical Device Tracking. Rockville, MD, p. 9.
40. 21 C.F.R. §821.25(a)(1).
41. 21 C.F.R. §821.25(a)(2).
42. 21 C.F.R. §821.3(k). "A multiple distributor is a device-user facility, rental company, or any other entity such as a home health care agency that distributes a life-sustaining or life-supporting device intended for use by more than one patient over the useful life of the device."
43. 21 C.F.R. §821.25(a)(3).
44. With respect to multiple-use devices, the current records need only include identifying information about the device, the date of shipment, and contact information for the multiple distributor. §21 C.F.R. 821.25(b).
45. Food and Drug Administration, Center for Devices and Radiological Health, 2003. Guidance on Medical Device Tracking. Rockville, MD, p. 19.
46. Id.
47. 21 C.F.R. §821.55(b).

48. 21 U.S.C. §360*l*(a).
49. 21 U.S.C. §360*l*(b).
50. Center for Devices and Radiological Health, INDEPENDENCE iBOT 3000 Mobility System (PO20033) Premarket Application Approval Letter (August 13, 2003), *available at* http://www.fda.gov/cdrh/PDF2/P020033a.pdf.
51. Food and Drug Administration, Center for Devices and Radiological Health, 1998. GUIDANCE ON CRITERIA AND APPROACHES FOR POSTMARKET SURVEILLANCE. ROCKVILLE, MD.
52. 21 U.S.C. §360h(a)(1).
53. U.S. Congress. Federal Food, Drug, and Cosmetic Act of 1938, Pub. L. 75-717, §518(b), 52 Stat. 1040 (1938), 21 U.S.C. 360h(b).
54. 21 U.S.C. §360h(e)(1).
55. 21 C.F.R. Part 7. While the Part 7 regulations use the language of "product" to encompass all products, this chapter references only devices.
56. 21 C.F.R. §7.3(g).
57. 21 C.F.R. §7.46(a).

5

Applications of Bayesian Methods to Medical Device Trials

Telba Z. Irony and Richard Simon

1. The Bayesian Framework
1.1. Introduction

The fundamental idea in the Bayesian approach is that one's uncertainty about an unknown quantity of interest is represented by probabilities for possible values of that quantity. For instance, unknown quantities of interest in device trials are the parameters of the clinical safety and effectiveness distribution for the treated and control groups.

Before an experiment is performed and data are obtained, the investigator assigns prior probabilities to the possible values of the unknown quantity, which is known as the *prior distribution*. The prior distribution reflects the investigator's knowledge about the quantity and is usually based on familiarity with relevant previous trials. If absolutely nothing is known about that quantity, a noninformative prior distribution can be specified. In the case of medical devices, prior information resulting from bench testing, animal studies, or previous clinical trials is often available and may be considered formally within the Bayesian framework.

After data are gathered and information becomes available, the prior probabilities are updated, according to Bayes' theorem describing mathematically how probabilities should be updated as information is accrued. These updated probabilities, known as the *posterior distribution*, represent probabilities for values of the unknown quantity after data are observed. When valid prior information is available, Bayesian methods may lead to faster conclusions, with smaller and shorter pivotal trials.

From: *Clinical Evaluation of Medical Devices: Principles and Case Studies, Second Edition*
Edited by: K. M. Becker and J. J. Whyte © Humana Press Inc., Totowa, NJ

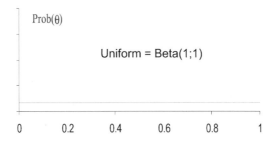

Fig. 1. Noninformative prior distribution.

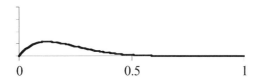

Fig. 2. Informative prior distribution.

The Bayesian approach encompasses a number of key concepts, some of which are not part of the frequentist approach. We briefly discuss these concepts in the subsections that follow. Section 1.9 contrasts the Bayesian and frequentist approaches.

1.2. Initial Information About Endpoints: Prior Distributions

Suppose that θ is an endpoint (parameter) of interest in a medical device clinical trial. The initial uncertainty about θ should be described by a probability distribution for θ, also called *prior distribution* and denoted by $p(\theta)$. If θ is the chance of an adverse event, its possible values lie between 0 and 1. The prior distribution of θ could be a uniform distribution indicating no preference for any value of θ. Fig. 1 illustrates the uniform distribution over the interval from 0 to 1. The flat line indicates that intervals of same length have the same probabilities. In other words, Fig. 1 shows that the probability that θ lies between 0.1 and 0.2 is the same as the probability that θ lies between 0.4 and 0.5 or between 0.65 and 0.75.

Alternatively, another prior distribution can give preference to lower values of θ. For instance, in Fig. 2, it is more likely that θ lies between 0.2 and 0.3 than between 0.7 and 0.8.

Good informative priors may decrease the required sample size in a trial; consequently, it is in the investigator's best interest to identify all relevant sources of prior information. Possible sources of prior information include clinical trials conducted previously, earlier studies of previous generations of the device, bench testing, data registries, clinical data on similar products, pilot studies, and literature controls. Quantitative priors (i.e., priors based on data from pilots, overseas studies, or similar devices) may be the easiest to use. Prior distributions that originated from studies that were similar to the current study in terms of protocol, endpoints, target population, sites, physicians, and timeframe are easiest to use. Priors that are not based on data but on other information, such as expert opinion, can be controversial (refer to ref. *1* for choosing an appropriate prior distribution for Bayesian medical device trials in the regulatory setting). In such a setting, the authors encourage data-dependent rather than expert opinion priors.

Bayesian medical device clinical trials may use prior distributions for the device under study, for the control device, or for both. When it is difficult to recruit the required number of control patients in a trial, prior information based on historical controls may be used, providing there is a basis for believing that there would be little inter-trial variation in outcome for control patients. Such borrowing of information can significantly decrease the required sample size for the concurrent control group. Consequently, investigators are able to allocate a greater proportion of patients to the experimental device arm, resulting in a faster trial.

1.3. Bayes Theorem and Posterior Distributions

Data in a clinical trial are collected with the objective of obtaining the posterior probability of each possible value of the endpoint θ given the observed data, which is denoted by $p(\theta|\text{data})$. Exclusively using the laws of probability, Bayes' theorem combines the prior probability for θ, $p(\theta)$, with the likelihood provided by the data from the trial, $p(\text{data}|\theta)$, to obtain the posterior distribution for θ, $p(\theta|\text{data})$. The process is illustrated in Fig. 3.

As more data are obtained, more updating can be performed. Consequently, the posterior distribution that has been obtained today may serve as a prior distribution later, when more data are gathered. The uncertainty about θ usually decreases as more information is accrued and the posterior distribution becomes sharper (Fig. 4). If enough data are collected, the prior information may be washed away. When the prior information does not support the information provided by the current study, the posterior distribution may become initially wider before it sharpens up. In any case, all available information about θ is summarized by the posterior distribution $p(\theta|\text{data})$, and all inferences about

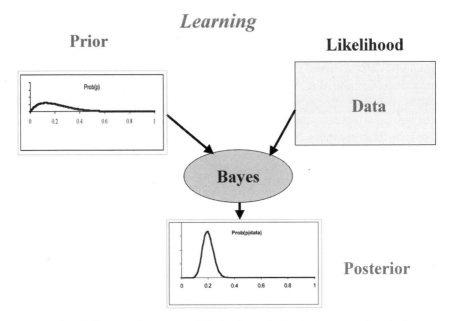

Fig. 3. Want to learn more? Today's posterior is tomorrow's prior!

the endpoint should be based on it (refer to refs. *2* and *3* for an introduction to Bayesian statistics and to ref. *4* for Bayesian estimation in the case of discrete distributions).

1.4. Exchangeability

Exchangeability is a key idea in Bayesian inference. Two observations are exchangeable if their roles can be switched. In other words, if two patients implanted with the same device are exchangeable, their chances of success are the same. If the patients in a sample are exchangeable with the patients in the rest of the population for which the device is intended, the sample can be used to estimate population parameters, such as the probability of success. Otherwise the sample is of little value for making inferences about the population. If patients in two samples are exchangeable, these patients can be combined or *pooled* for making inferences.

Exchangeability may be also thought of in terms of subpopulations (or centers). Several subpopulations are exchangeable if they are independent samples from the same hyperpopulation. Because the hyperpopulation may be hetero-

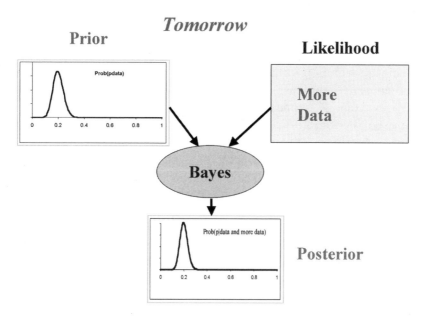

Fig. 4. The more information is gathered, the sharper the posterior distribution becomes.

geneous, patients in two subpopulations are *not* necessarily exchangeable or *poolable*, although the subpopulations can be combined using a Bayesian hierarchical model (*see* Section 1.5). That is, the subpopulations will not be completely pooled but may be pooled to a certain degree. The degree of poolability will depend on the ratio between the variability among the subpopulations and among the patients within each subpopulation.

For in-depth discussions on exchangeability *see* ref. 5 and Sections 4.2. and 4.3.

1.5. Borrowing Strength Using Bayesian Hierarchical Models

1.5.1. Borrowing Strength From Other Studies (6)

The direct use of Bayes' theorem incorporating all prior information from previous studies in order to make inferences about a parameter in a current study assumes that the parameter is the same across studies. Thus, patients across studies are considered exchangeable or poolable and can be treated as if they were in the same study. If that is the case, full strength is borrowed from previous studies.

Bayesian methods also allow for partial use of prior information through hierarchical models. They are indicated when previous and current studies may differ slightly and researchers are not completely convinced that the patients in the studies providing prior information are exchangeable with the patients in the current trial. In this case, strength can be partially borrowed from previous studies. For example, suppose that information from a registry of device trials will be used as prior information to strengthen a trial for a new device. Patient success is the primary outcome, and the parameter of interest is the proportion of successes in the study. In the hierarchical model framework, the patients in the registry and in the current study are not exchangeable and the proportion of successes may differ across studies; however, the proportions of success in the studies will be considered random samples from a hyperpopulation of proportions that follow a probability distribution with the same mean and variance (hyperparameters). The assumption is that the studies (parameters)—not the patients—are exchangeable.

The hierarchical model will automatically calibrate the similarity of the registry data with the new data and prior information will be used only partially. When the proportions in different studies are highly similar, there will be a considerable amount of pooling, resulting in a great deal of prior information being added to the new trial. The amount of pooling depends on the variation among patients across studies compared to the variation among patients within the same study. It also depends on the probability distributions used in the hierarchical models and on the sample sizes of the studies. If the sample size in the new trial is too small, compared with the registry sample size, too much pooling may occur and the new data will be swamped by the registry data. To avoid excessive pooling of data not similar to the data in the new study, simulations based on likely outcomes should be carried out at the planning stage using different sample sizes for the new trial. These simulations will assess the operating characteristics of the design (probabilities of type I and II errors) and should show that the proposed sample size will be large enough to avoid too much borrowing from the registry when the data sets are different. Because one cannot estimate variance among studies when there are only two studies, it is advisable to combine more than two studies when using hierarchical models.

1.5.2. Borrowing Strength Among Centers

In contrast to drug trials, outcomes for devices can vary substantially by site owing to differences in implantation technique, physician experience with the device, patient standard of care, and patient population. In those multicenter trials for which center results cannot be completely pooled, Bayesian hierarchical models can be used to combine data across centers. A hierarchical model on centers assumes that the parameters of interest vary among centers but come from a hyperpopulation of exchangeable centers. This model takes into account center-to-center variability when estimating parameters across centers.

For examples, refer to refs. 2, Chapter 8; 3; and 6, Chapter 5.

1.6. Predictive Distributions

The Bayesian approach allows for the derivation of the probability of future observation given outcomes that have already been observed (i.e., predictive probability). Collectively, the probabilities for all possible values of an unobserved outcome are encompassed by the predictive distribution. Although any predictive analysis assumes that unobserved patients are exchangeable with the observed patients, this assumption may not always be valid. For example, patients enrolled later in a trial could be more successful with a device than those enrolled earlier because physicians may have had to overcome a learning curve in implanting the device.

Early Termination of a Trial: By using Bayesian predictive probabilities, one may stop a trial early based on results obtained at an interim point. If the predictive probability that a trial will be successful is sufficiently high based on interim results, one may stop the trial and declare success early. On the other hand, if the predictive probability that the trial will be successful is small, one may stop the trial for futility and cut his or her losses. Exchangeability is a key issue in this case. Simulations to assess the operating characteristics of the design with interim analyses (probabilities of type I and II errors) are essential in the regulatory setting. For more details, *see* Section 2.4.

Predicting a Clinical Outcome From a Surrogate: Predictive probability has other applications. For example, when patients have two measurements, the first at an earlier follow-up visit and the second at a later follow-up visit, predictions for the later follow-up visit may be made (even before the follow-up time has elapsed) based on measures at the early follow-up visit, provided there are some patients who have results from both follow-up visits and there is a correlation between the early and the later measurement.

Labeling and Medical Decision Making: The predictive probability of the outcome of a future patient given the observed outcomes in a clinical trial can be calculated. In fact, that probability is the answer to the key questions: "Given the results of the clinical trial, what is the probability that a new patient receiving the experimental treatment will be successful?" "What would that probability be if the patient were treated in the control group?" These probabilities are essential in helping physicians and patients make decisions and should be included in the labeling.

Predicting Missing Data: Missing data can be predicted based on observed data, and trial results can be adjusted accordingly. The adjustment strongly depends on the assumption that missing patients follow the same statistical model as observed patients (i.e., the two groups of patients are exchangeable). Thus, one assumes that data are missing at random. When this assumption is not valid, missing outcomes may be predicted using a different model to determine whether conclusions change. For example, the model for patients with missing outcomes could assume that control patients are more successful and experimental device patients are less successful than under the model for patients with observed outcomes. A sensitivity analysis quantifies how discrepant the two models would need to be before conclusions would change. If the models have to be largely discrepant, conclusions are said to be robust to deviations of the missing outcomes from the statistical model. For patients with missing outcomes, these adjustments are an improvement over the common but often unrealistic approaches of considering the "worst-case scenario" or "carrying the last observation forward." *See* Section 2.6. for further discussion on missing data in clinical trials for medical devices.

For a complete review on Bayesian statistical prediction, refer to refs. *7* and *8*.

1.7. The Likelihood Principle

The likelihood principle is fundamental to the Bayesian methodology. It states that all information about the endpoint of interest, θ, obtained from a clinical trial is contained in the likelihood function. Consequently, the posterior distribution of θ, from which all inferences about θ are made, should include only the prior information and the information provided by the experiment through the likelihood function.

This principle provides a great deal of flexibility in conducting Bayesian clinical trials, particularly with respect to stopping rules and interim looks. If a trial that was conducted with interim looks provided exactly the same results as another trial conducted without interim looks, the likelihood functions generated by both trials will be the same. The posterior distributions will also be the same, provided the prior distributions were the same. Consequently, inferences based on the Bayesian paradigm will be the same, although the inferences based on the traditional frequentist approach will differ owing to the interim looks.

Accordingly, if two trials that differ only in the stopping rules provide likelihood functions that are proportional to each other, the posterior distributions generated by both trials will be exactly the same, provided the prior distributions are the same. Again, Bayesian inferences will be the same in both cases.

Section 2.4 elaborates on interim analyses in the regulatory setting. For more on the topic, *see* ref. *2*, Chapter 7.1. and refs. *9–11*.

1.8. Bayes Risk

The posterior distribution of the parameters of interest is the basis for inference about the parameters and functions of the parameters, such as the difference in success probabilities for two treatments. When decisions have to be made and actions taken, it is often important to take into account the costs of wrong decisions. Statistical decision theory, although originally developed in a non-Bayesian context, fits well into the Bayesian approach. Let D(data) denote a function that maps from the data to a set of possible decisions and let $L(D[\text{data}];\theta)$ denote the loss resulting from decision D when the true parameter vector has value θ. The Bayes risk is the expected value of the loss computed with respect to the posterior distribution of θ. The optimal decision is the one that minimizes the Bayes risk. The decision theoretic approach is limited in applications by the difficulty of defining realistic and broadly acceptable loss functions.

For more on Bayes risk, see ref. *2*, Chapter 7.5. and refs. *12–15*.

1.9. Comparison: Bayesian and Frequentist Approaches

Traditional frequentist statistical methodology does not allow the formal mathematical use of any prior information to strengthen a current clinical study. Alternatively, Bayesian methodology requires the use of prior information, either completely, through the direct use of Bayes' theorem, or partially, through the use of Bayesian hierarchical models. The absence of information *a priori* should be represented by a noninformative prior distribution (*see* refs. *5* and *14*).

The Bayesian approach bases all inferences about θ on the posterior distribution of θ, which is the conditional probability of θ given previous knowledge and data; it is denoted by $p(\theta|\text{data})$. Inferences are performed in the parameter space. Thus, the probability is evaluated for values of the parameter θ, which is the unknown quantity, conditionally on the data that have been observed.

The frequentist approach bases some inferences (but not all) on the likelihood function. Frequentist methods are generally based on the p value, which is the probability of obtaining data that is extreme or more extreme than the data actually obtained if the parameter θ were stated by the null hypothesis (i.e., if $\theta = \theta^*$): $p(\text{Data} > \text{data} \mid \theta = \theta^*)$. In general, this is different than the likelihood function $p(\text{data}|\theta = \theta^*)$. When the p value is small, the null hypothesis is rejected. Inferences are performed in the sample space. In other words, the probability is evaluated for possible values of the data, conditionally on the value of the parameter stated by the null hypothesis. It assumes that the data that were actually observed and known are random and that the parameter, which is actually unknown, is known. In certain cases, frequentist methods based on likelihood-ratio tests correspond to Bayesian methods using noninformative prior distributions. Table 1 summarizes these differences.

Table 1
Bayesian and Frequentist Inference

Approach	Bayesian	Frequentist
Measure	Posterior distribution $P(\theta|\text{data})$	p value $P(\text{Data} \geq \text{data}|\theta = \theta^*)$
Endpoint	Unknown: θ	Assumed known: $\theta = \theta^*$
Observations	Known: data	Assumed unknown: data

2. Clinical Trials: Bayesian Design and Analysis Plans

2.1. Hypotheses

Like frequentist trials, clinical trials designed within the Bayesian framework require careful statistical planning. Bayesian inference depends on specification of the likelihood function and prior distributions for all parameters. Consequently, all variables that will have a role in the analysis must be identified in advance. For example, suppose that the outcome of interest is binary and that the probability of response can be influenced by treatment and covariates $x_1,\ldots,x_m$. In a clinical trial comparing a device to a control, let t be an indicator of the treatment group ($+1/2$ for treatment, $-1/2$ for control). An additive logistic model for the probability p of response is:

$$\text{logit } p = \mu + \alpha t + \beta_1 x_1 + \ldots + \beta_m x_m \tag{1}$$

Bayesian inference about the device effect is based on the posterior distribution of α.

2.1.1. Superiority Claims

Often the objective of the trial is to test whether there is an effect of treatment. That may be viewed as testing the hypothesis that $\alpha = 0$. Cornfield *(16)* pointed out that if one is interested in testing whether a treatment effect is zero, one must have believed *a priori* that a zero treatment effect is a real possibility. This *a priori* belief should be reflected by a prior distribution for the treatment effect that places positive probability on $\alpha = 0$. With this approach, a superiority trial may be based on examining three hypotheses:

$$\begin{aligned} H_0 &: \alpha = 0 \\ H_1 &: \alpha > 0 \\ H_2 &: \alpha < 0 \end{aligned} \tag{2}$$

An alternative approach is to take the view that a treatment effect of zero is implausible and to use a continuous prior distribution for α that does not place positive probability on $\alpha = 0$. Then, H_1 remains the same but H_0 becomes the hypothesis $\alpha \leq 0$.

2.1.2. Noninferiority Claims

With the above formulations, rejection of the null hypothesis and acceptance of H_1 leads to a claim for the new device. These formulations are appropriate for superiority trials; however, for a noninferiority trial, one wishes to make a claim for the new device if one can conclude that $\alpha \geq 0$. Often, one considers the hypotheses:

$$H_0: \alpha \leq \varepsilon \qquad (3)$$
$$H_1: \alpha \geq 0$$

In this hypothesis, ε represents the smallest medically meaningful treatment effect. For H_0, the new treatment is inferior. If the true state is $\alpha = 0$, it is unlikely that the trial will result in a large posterior probability for H_1. With a continuous prior distribution and an infinite sample size, the posterior probability that $\alpha \geq 0$ is 0.5. Thus, the claim of noninferiority is based on having a small posterior probability for H_0.

2.1.3. Effectiveness Claims With Active Controls

In some trials, one is interested in a claim that the new device is effective relative to an older standard of care, but a randomized trial of the new treatment compared to the older standard is not possible because of the availability of a newer active-control treatment. The claim of effectiveness relative to the older standard is different than the claim of noninferiority compared to the active control. Simon *(17)* developed a Bayesian approach for designing active control trials based on a model with an additional treatment effect parameter, α_2, representing the effectiveness of the control relative to the previous standard. The prior distribution for α_2 is based on a random effects meta-analysis of previous relevant trials. The effectiveness of the new device relative to the previous standard is evaluated based on the posterior distribution of $\alpha + \alpha_2$. The active-control trial compares the new device to the active control and provides information about α.

2.2. Controls

Most claims of effectiveness are based on a comparison of outcomes for a treated group of patients relative to a control group. The control group often receives an established standard of care and should be prognostically equivalent to the treatment group in all respects. Bayesian analyses are based on like-lihood functions for models, such as that shown in Section 1. Unfortunately, specifying a model does not ensure that it is correct. For example, inherent in the model in Section 1 is the assumption that other covariates not represented in the model are ignorable with regard to estimation of treatment effect (*see* ref. *1*). Randomized treatment assignment is the most satisfactory way to ensure that the ignorability assumption is valid. Although eandomization is not always feasible, randomization has as key a role in Bayesian analyses as in non-Bayesian methods.

2.3. Sample Size

Sample sizes for Bayesian clinical trials could be determined to minimize an expected loss function that combines the cost of conducting the trial and the gain of a positive result. The expected value would be taken with regard to the sponsor's prior distribution on the efficacy and safety parameters for the treatment and control groups. This approach is rarely used, however, because of the difficulty of quantifying the value of a positive clinical trial.

For Bayesian regulatory trials, sponsors generally determine sample size based on limiting the probability of falsely rejecting the null hypothesis when it is true and of falsely failing to reject the null hypothesis when the treatment effect is of a specified medically important magnitude. This is similar in spirit to planning methods used for frequentist trials, but Bayesian trial base the decision of whether to reject the null hypothesis on the posterior probability of the null hypothesis rather than on the result of a frequentist test of statistical significance (*see* ref. *19*).

2.4. Interim Analyses and Midcourse Modifications of Trials

In the Bayesian approach, the decision-making process is based on posterior probabilities that typically do not depend on the experimental design (*see* ref. *11* for situations in which the experimental design may influence decisions). On the other hand, when the frequentist approach is used, decisions are based on *p* values that strongly depend on the experimental design. As a result, the Bayesian approach provides more flexibility in both design and analysis because experiments can be altered midcourse and interim analyses are easily implemented without interfering with posterior or predictive probabilities. For regulatory purposes, interim analyses and possible midcourse modifications should be planned in advance, however.

To assess the probabilities of type I and type II errors when interim analyses or midcourse modifications of trials are performed, simulations should be carried out at the planning stage. The simulations should assess the probabilities of making type I and type II errors when the proposed experimental design is adopted. For example, when a hypothesis to support a noninferiority claim is tested simulations considering the following factors should be performed: for a borderline noninferiority case, how often would one declare noninferiority, how often would one declare superiority, and what would be the average sample size? The simulations should be carried out for a few likely values of the parameters. Analogous simulations could be performed by assuming superiority cases (type II errors), equivalence but not superiority (type II errors), and inferiority (type I errors).

Bayesian interim analyses are generally based on the posterior distribution of the treatment effect or the predictive distribution of the treatment effect. The former provides information about the relative merits of the two treatments with regard to the primary endpoint available at the time of the interim analysis. For example, let α denote the treatment effect, with $\alpha = 0$ indicating no difference, $\alpha > 0$ indicating superiority of the new treatment, and $\alpha < 0$ indicating inferiority of the new treatment. At interim analysis, we compute posterior probability of these three quantities. If the posterior probability of α ≤ 0 is small (i.e., <1%), the protocol may have provided for early termination of accrual. Let D_i denote the data available at the interim analysis and D_f denote the data available at the planned end of the study. If the study continues to the planned end, one would be able to compute the final posterior probability $\Pr(\alpha \leq 0|D_f)$ at that time. The null hypothesis may be rejected if $\Pr(\alpha \leq 0|D_f)$ $\leq 5\%$. At the time of the interim analysis, the final data D_f is not available for patients not yet accrued and for those with incomplete follow-up, although some of the data may have been obtained. The components of D_f not available in D_i have a probability distribution that can be computed based on the posterior distribution available at the interim analysis. Hence, at the interim analysis, one can compute the probability of ultimately having $\Pr(\alpha \leq 0|D_f) \leq 5\%$. This is called a *predictive probability*. The interim analysis plan may specify that accrual is to be terminated early if this predictive probability is large (e.g., 99%). This is a more stringent requirement than basing early termination on the posterior distribution at an interim analysis.

2.5. Multiplicity

The frequentist approach to clinical trials is generally based on addressing a single question and controlling the error rates for answering that question. The frequentist paradigm does not adequately address the numerous multiplicity issues encountered in the analysis of major clinical trials. There are often multiple interim analyses, multiple efficacy and safety endpoints, and multiple covariates that can be used for adjusting treatment effects and for defining subsets, as well as multiple centers. Frequentist methods deal with many of the multiplicity issues by attempting to count the number of possible hypotheses that can be tested and reducing the error level for each test accordingly. Bayesian methods do not require ad-hoc multiplicity adjustments but depend on the use of appropriate prior distributions. With the Bayesian approach, however, one must define a model that incorporates all endpoints, covariates, and subsets in advance.

In the Bayesian framework, inferences about parameters of interest are not affected by the number of parameters that are being estimated; however, Bayesians may worry about the consequences of making multiple mistakes. Consequently, although multiplicity of parameters will not affect the marginal posterior distribution of each parameter, Bayesian decisions may differ depending on how many parameters are estimated.

For more information on multiplicity, *see* refs. *20–23.*

2.5.1. Patient Subsets

Suppose that there are many subsets of patients and that there is interest in examining whether the treatment effect varies among them. With binary response, the statistical model for the data might be a generalization of Eq. 1, which incorporates treatment by covariate interaction terms:

$$\text{logit } p = \mu + \alpha t + \beta_1 x_1 + \ldots + \beta_m x_m \, \gamma_1 t x_1 + \ldots + \gamma_m t x_m \tag{4}$$

In this example, we assume for simplicity that the covariates are binary and coded as 0 or 1. The treatment effect for a patient with covariates $i_1, i_2, \ldots, i_k$ at level 1 and the others at level 0 is thus $\alpha + \gamma_{i_1} + \ldots + \gamma_{i_i}$ *(23)* points out that we perform the clinical trial as a single trial, sized for evaluating the average treatment effect for the eligible patient population because we implicitly assume that qualitative interactions are unlikely. In a qualitative interaction, one treatment is preferred for one subset but not for others *(24)*. Whereas quantitative interactions are dependent on the scale of measurement, qualitative interactions are not. The assumption that qualitative interactions are unlikely can be used to specify a prior distribution for the interaction parameters γ_i as described by *(23)*. This same approach can be used to determine whether treatment effect varies substantially by center by including binary center variables among the covariates. For additional Bayesian approaches to subset analysis, *see* ref. *25.*

2.5.2. Multiple Endpoints

For many clinical trials, there are numerous safety endpoints that must be examined, although there may be no *a priori* evidence of a safety problem with any specific endpoint. There can be substantial opportunity for an unexpected result on a safety endpoint occurring by chance. The Bayesian approach is to interpret the results in light of prior distributions that properly reflect the context of *a priori* information on safety endpoints. Assume that there are K independent binary safety endpoints and that for each endpoint k, we have a logistic model such as:

$$\text{logit } p_k = \mu_k + \alpha_k t + \beta_{1k} x_1 + \dots + \beta_{mk} x_m \tag{5}$$

In this example, the α_k parameters represent the treatment effects on the safety endpoints and the x_i values are covariates for safety endpoints. For a trial considering safety endpoints and that has *a priori* evidence that the new treatment is more or less safe than the control, the α_k parameters may be assumed to be independent draws from a normal distribution with mean 0 and variance σ_α^2. It is easy to show that if the maximum likelihood estimates $\hat{\alpha}_k$ are independent with sample variances s_k^2, the posterior distributions of the α_k parameters are approximately independent normal with means $\hat{\alpha}_k / \left(1 + s_k^2 / \sigma_\alpha^2\right)$ and variances $s_k^2 / \left(1 + s_k^2 / \sigma_\alpha^2\right)$. Hence, the means are shrunken toward zero, the mean of the prior distributions, but the variances are reduced compared to the variances of the maximum likelihood estimates. As shown by Simon *(23)*, if δ^* represents the smallest treatment effect on a safety endpoint (on the odds-ratio scale) of medical importance and λ denotes the prior probability of such a treatment effect existing for a specified safety endpoint then $\sigma_\alpha = (-\delta^*/k_\lambda)$. Here k_λ is the 100λth percentile of the standard normal distribution. When the endpoints are not independent, similar approaches can be employed *(19)*. For other Bayesian approaches for multiple endpoints, *see* ref. *26*.

2.6. Missing Data

Missing data is an important issue in the analysis of medical device clinical trials. For randomized trials, the groups are made comparable by the randomization, but the data may not give an unbiased estimate of the effect of the device under study if data are missing or excluded in a biased manner. The intent-to-treat principle for randomized trials generally means that all patients randomized are included in the primary analysis, even those who did not receive the treatment assigned by the randomization or did not receive it in a protocol-specified manner. The objective of the principle is to assure that the probability of type I error does not exceed its nominal level (usually 5%). The intent-to-treat principle may result in reduced statistical power, but it preserves the probability of a type I error occuring.

Analysis consistent with the intent-to-treat principle is difficult when patients drop out of the trial and their key endpoints are not measured. Excluding such patients from the primary analysis may bias the results because the dropouts may be prognostically different than patients who remain in the trial, there may be more dropouts in one treatment group than in the other, and the reasons for dropping out may differ among treatment groups. Excluding the dropouts from the primary analysis is equivalent to assuming that the dropouts are selected completely at random, which is often not an appropriate assumption.

Bayesian multiple imputation is a method of handling missing data that is not based on the assumption that patients dropout completely at random. For example, assume that patient evaluations are planned at baseline (time = t_0) and at several follow-up points $t_1, \ldots, t_f$. and that a scalar patient status variable is defined at each visit, denoted $S_0, S_1, \ldots, S_f$. The primary endpoint is defined based on S_0 and S_f, and S_0 is measured for all patients. We write $E = g(S_0, S_f)$ to indicate that the endpoint is defined based on the initial and final status of the patient.

A model for the data vector $(S_0, S_1, \ldots, S_f)$ is defined for each treatment group, and prior distributions are selected for the parameters of the model. At any analysis point, there is likely to be some missing data. Because some patients will not have been followed long enough at interim analyses for all of their follow-up visits to have taken place, only partial follow-up information will be available. Some patients will miss follow-up visits or drop out of the study; only partial information will be available for them. At any analysis, one can compute the posterior distribution of the parameters of the model based on the available data and the prior distribution. Let P denote this multivariate posterior distribution of the parameters. In computing P, one assumes that missing data is missing at random, conditional on the observed data for those patients. Compared to assuming that the data is missing completely at random, this is a weaker assumption.

Consider a patient for whom partial data is available but not the final status (S_f). Let D denote the partial data available for this patient and suppose it consists of his or her status until dropping out. From the posterior distribution P, one can compute the predictive distribution of S_f given D. If one samples an S_f value from this predictive distribution for each patient missing final data, one then has an S_f value for all of the patients in the study. Thus, endpoints may be computed from the function $E = g(S_0, S_f)$. If this is done for each treatment group, there is complete endpoint data that can be used to compare the treatment groups. Suppose we summarize the treatment comparison by some statistic X, which may be an odds ratio, a difference in means, or some other statistic dependent on the nature of the endpoint E. With Bayesian multiple imputation, one repeats this sampling process many times. Each time an S_f value is sampled for each patient with missing final data, the endpoints $E = g(S_0, S_f)$ are computed, and then complete data are analyzed to obtain a treatment comparison statistic X. The final analysis consists of summarizing the distribution of X over the multiple imputations.

3. Conclusion

Bayesian methods provide a sound and consistent framework for the design and analysis of medical device clinical trials. Although facilitating the use of prior information, they require that the assumptions and uncertainties be explicit.

These prior distributions should reflect relevant previous evidence or lack of such data about the endpoints and hypotheses. They should be selected carefully and must be acceptable to the consumers of the results of the clinical trial. With properly defined prior distributions, Bayesian methods can be particularly effective in dealing with the many problems of multiplicity (e.g., multiple endpoints, multiple patient subsets, multiple analyses, multiple covariates) in a trial.

In trials, the Bayesian approach offers more flexibility than does the frequentist approach with respect to midcourse modifications, such as interim analyses and multiple looks, early stopping, and augmentation of sample size. For regulatory purposes, simulations to demonstrate that the probability of type I error is under control should be conducted at the planning stage.

Bayesian trials require careful advance planning because the prior distributions used in all analyses of all endpoints of interest must be specified in advance and flexible designs and modifications of trials in midcourse are often desirable.

Although Bayesian methods provide an effective framework for using information external to the trial, the Bayesian paradigm is not a justification for conducting nonrandomized clinical trials. The appropriate use of external control groups requires the same types of assumptions regardless of whether one is working in a Bayesian or frequentist framework.

Once prior distributions that have a strong empirical basis or an adequate community acceptance are established, the Bayesian approach provides a consistent and flexible framework for conducting and analyzing the clinical trial. In addition, recent breakthroughs in computational algorithms and increases in computing speed have enabled the calculation of the necessary quantities for any Bayesian analysis. As a result, Bayesian approaches are increasingly being used in medical device trials.

References

1. Irony, T.Z. and Pennello, G. A. 2001. Choosing an appropriate prior for Bayesian medical device trials in the regulatory setting. American Statistical Association 2001 Proceedings of the Biopharmaceutical Section. American Statistical Association, Alexandria, VA.
2. Lee, P. M. 1977. *Bayesian Statistics: an Introduction*. John Wiley and Sons, New York.
3. Berry, D. A. and Stangl, D. K. (eds). 1996. *Bayesian Biostatistics*. Marcel Dekker, New York.
4. Irony, T. Z. 1992. Bayesian estimation for discrete distributions. *J. Appl. Statist.* 19:533–549.
5. Bernardo, J. M. and Smith, A. F. M. 1994. *Bayesian Theory*. John Wiley and Sons, New York.
6. Gelman, A., Carlin, J. B., Stern, H. S., Rubin, D. B. 1996. *Bayesian Data Analysis*. Chapman and Hall, New York.

7. Aitchison, J. and Dunsmore, I. R. 1975. *Statistical Prediction Analysis*. University Press, Cambridge.

8. Geisser, S. 1993. *Predictive Inference: an Introduction*. Chapman and Hall, New York.

9. Berger, J. O. and Berry, D. A. 1988. The relevance of stopping rules in statistical inference. *Statistical Decision Theory and Related Topics IV* (Gupta, S. S. and J. O. Berger, J. O., eds.), Springer Verlag, New York.

10. Berger, J. O. and Wolpert, R. L. 1984; 1988. *The Likelihood Principle* (2nd ed. in 1988). IMS, Hayward, CA.

11. Irony, T. Z. 1993. Information in sampling rules. *J. Statist. Plan. Infer.* 36:27–38.

12. Lindley, D. V. 1985. *Making Decisions* (2nd ed.) John Wiley and Sons, New York.

13. Raiffa, H. and Schlaifer, R. 1961. *Applied Statistical Decision Theory*. Harvard University, Cambridge, MA.

14. Berger, J. O. 1985. *Statistical Decision Theory and Bayesian Analysis*. Springer, Berlin.

15. DeGroot, M. H. 1970. *Optimal Statistical Decisions*. McGraw-Hill, New York.

16. Cornfield, J. 1966. A Bayesian test of some classical hypotheses—with applications to sequential clinical trials. *J. Am. Statist. Assoc.* 61:577–594.

17. Simon, R. 1999. Bayesian design and analysis of active control clinical trials. *Biometrics*. 55:484–487.

18. Rubin D.B. 1978. Bayesian inference for causal effects: The role of randomization. *Ann. Statist.* 6:34–58.

19. Thall, P. F., Simon, R. M. and Shen, Y. 2000. Approximate Bayesian evaluation of multiple treatment effects. *Biometrics*. 56:213–219.

20. Carlin, B. P. and Louis, T. A. 2000. *Bayes and Empirical Bayes Methods for Data Analysis*. Chapman and Hall, New York.

21. Simon, R. 1994. Problems of multiplicity in clinical trials. *J. Statist. Plan. Infer.* 42:209–221.

22. Simon, R. and Freedman, L.S. 1997. Bayesian design and analysis of 2 by 2 factorial clinical trials. *Biometrics*. 53:456–464.

23. Simon, R. 2002. Bayesian subset analysis: application to studying treatment by gender interactions. *Statist. Med.* 21:2909–2916.

24. Gail, M. and Simon, R. 1985. Testing for qualitative interactions between treatment effects and patient subsets. *Biometrics*. 41:361–372.

25. Dixon, D.O. and Simon R. 1991. Bayesian subset analysis. *Biometrics*. 47:871–882.

26. Simon, R. 1995. Discovering the truth about tamoxifen: problems of multiplicity in the statistical evaluation of biomedical data. *JNCI*. 87:627–629.

6

Intellectual Property Protection for Medical Devices

Fariborz Moazzam and Michael D. Bednarek

1. Introduction to Types of Intellectual Property

Intellectual property (IP) broadly encompasses the concepts derived from the intellectual and artistic achievement of innovators. More specifically, IP includes discoveries, inventions, technological advancement or development, literary or artistic works, and unique names for trade or business. A concise summary of the available IP tools appears in Table 1.

1.1. Patents

A patent is a grant of a property right by the government to the inventor for an invention that is deemed by the government to be novel, useful, and unobvious. The government agency that reviews and grants US patents is the US Patent and Trademark Office (PTO). A patent gives its owner the right to exclude others from making, using, selling, offering for sale, or importing the claimed invention without first obtaining a license from the patent holder. The scope of the right to exclude is defined by the patent claims. In the United States, there are three kinds of patents: utility patents, design patents, and plant patents.

The question of what is patentable is discussed in greater detail below, but simply, an invention must pass a three-part test to be patentable. The first inquiry is whether the invention is useful. One aspect of the utility test is that the invention cannot be a mere theoretical phenomenon. Thus, it must have at least one useful purpose. There are limits to the usefulness requirement, including exclusion of inventions that are used for the sole purpose of causing harm (e.g., bombs).

From: *Clinical Evaluation of Medical Devices: Principles and Case Studies, Second Edition*
Edited by: K. M. Becker and J. J. Whyte © Humana Press Inc., Totowa, NJ

Table 1
Summary of US Intellectual Property

Type (marking)	Term (nature of right)	How obtained	Test for infringement	Limitations	Recommended steps
Utility Patent Patent Pending, US Patent No., or Patented	**20 years from filing** Prevents others from making, using, or selling products incorporating the claimed feature, method, or process. Scope of rights is determined by claim language rather than by the description of the specific product or method.	Utility patent, application filing, examination process, and issuance. This process is often initiated by filing a provisional application before the utility application.	Manufacture, use, sale, or offer to sell the claimed invention. Does the claimed invention "read" on accused product or process?	Rights are obtained only after 2–3 years examination process and significant upfront expense.	Keep good records of innovation. File at least a provisional application within 1 year of: (1) offer for sale; (2) public use; or (3) publication. File before any public disclosures or sales to preserve foreign rights.
Trademark or Service Mark (®, TM, or SM, an unregistered marks ®, federally registered marks	**As long as in use; registrations are renewable every 10 years** Prevents others from using a mark that is likely to cause confusion among customers.	State "common law" rights created by using unique mark in commerce. Expanded rights available through federal registration.	Likelihood of confusion, mistake, or deception as to source or sponsorship.	Strength of rights depends on distinctiveness. Marks including common or descriptive terms are subject to encroachment.	Select unique, nondescriptive marks. Use appropriate marking (®, TM, SM) and use trademarks as adjective immediately preceding the generic noun.
Design Patent Patent Pending or US Patent No.	**14 years after grant** Prevents others from making, using, or selling product with similar ornamental appearance.	Design patent application filing, examination process, and issuance.	Designs look alike to the ordinary observer.	Protects only ornamental (nonfunctional) aspects of product (e.g., product housing).	File within 1 year of: (1) offer for sale; (2) public use; or (3) publication. File before any public disclosures or sales to preserve foreign rights.
Copyright Copyright [year] (e.g., © 2001)	**Generally author's life plus 70 years** Prevents direct copying of text, software code, etc. Registration gives additional rights.	Create and fix in tangible form an original work of authorship. Expanded rights through federal registration process. Valid copyright notice: 1. © or Copyright 2. Year 3. Owner's name 4. All Rights Reserved	Unauthorized copying or use.	Prevents only direct copying. Does not protect underlying facts, ideas, or concepts. No protection against independent development. Does not protect short phrases and slogans.	1. Mark all copies with valid copyright notice. 2. Register commercially significant copyrights at the Copyright Office to expand enforcement rights.
Trade Secret "Confidential," "Proprietary," or other markings as determined by nondisclosure agreement	**As long as information remains secret** State and federal laws protect against misappropriation of commercially sensitive information that is not generally known.	Take reasonable steps to maintain confidentiality of proprietary information.	Acquisition of information by improper means or unauthorized disclosure of information.	No protection against reverse engineering or independent development of the same information. Protection ends when secret becomes public.	1. Nondisclosure agreements. 2. Employment and noncompete agreements. 3. Mark sensitive documents: • Proprietary • Confidential 4. Physical security

The second inquiry is whether the invention is novel. In other words, the invention must be something that no one has developed before. The test for novelty has a relatively low standard, in that an invention will be deemed novel even if it only slightly differs in way, function, or form from an existing product.

The final inquiry is whether the invention is obvious in the view of someone in the field of the invention in light of everything that is already known in the field. Most issues of patentability, as well as patentability disputes occurring after patents are issued, hinge on this requirement. Even if an invention appears to be a slight improvement over what is already known in the field, it may still be granted a patent, as long as the invention is unobvious in the field. The level of unobviousness required for patentability depends on the particular scientific area encompassing the invention. Frequently, the unobvious part of an invention is simply identifying the problem, even if the solution is obvious once the problem has been identified.

1.2. Trademarks

A trademark is a word, name, symbol, or device that is used in trade with goods to indicate the source of the goods and distinguish it from the goods of others. A servicemark is the same as a trademark except that it identifies and distinguishes the source of a service rather than goods. The terms *trademark* and *mark* are commonly used to refer to both trademarks and servicemarks.

Trademark rights may be used to prevent others from using a confusingly similar mark, but not to prevent others from making the same goods or from selling the same goods or services under a clearly different mark. Trademarks that are used in interstate or international commerce may be registered.

Similar to patent applications, trademark applications are filed at the PTO; however, anyone who claims rights in a mark may use the TM (trademark) or SM (servicemark) designation to alert the public to the claim, regardless of whether they have filed an application with the PTO. The federal registration symbol "®" may be used only after the PTO actually registers a mark and not while an application is pending. Also, the registration symbol may be used only with the mark on or in connection with the goods and/or services listed in the federal trademark registration.

1.3. Copyrights

A copyright protects the *works* (i.e., the particular form of expression of an idea or concept) of an author or artist against copying, performance, display, or use as an underlying work. Copyright protects the form of expression rather than the subject matter or idea of the writing. Materials within the scope of copyright protection include books, articles, musical and artistic works, as well as computer software. Registration of copyrights is handled by the Copyright

Office and is not associated with the PTO. Such requests for copyright registration are typically not reviewed by a designated copyright examiner in the same manner as patent and trademark applications are reviewed by patent examiners and trademark attorneys, respectively. There is no set procedure for a copyright examiner to make a comprehensive search of databases for all other similar published material before registering the particular copyright being considered. Thus, the copyright process is deemed to be more of a registration process rather than a review and examining process.

1.4. Trade Secrets

A proprietary technology or trade secret may be defined as secret information that gives one having such knowledge a competitive advantage in business because the general public does not know it. Examples of trade secrets include an unpatented invention, formula, blueprint, machine, process, customer list, or tool. Know-how, including the expertise and knowledge of employees and staff, is a body of knowledge outside the public domain that has commercial value. Typically, patent applications filed by companies do not contain information the company wishes to keep as a trade secret. Regulation and laws relating to trade secrets and know-how are set by the particular states and relate to the state's fair trade and business practice policies, although such regulation and laws have limited effect against legitimate reverse engineering by competitors.

1.5. Applicability of Intellectual Property Tools in the Context of Medical Devices

Although each of these subject areas can be relevant and important in the field of medical devices, utility patents are of paramount importance as demonstrated by the following summary of each of the aforementioned tools:

- *Utility patents* can be used to prevent others from making, using, or selling products as well as from incorporating the claimed feature, method, or process. Thus, in the context of medical devices, utility patents can provide exclusive rights to devices, processes for making devices, and methods of using devices.
- *Design patents* can be used to prevent others from making, using, or selling products with similar ornamental appearance. Thus, in the context of medical devices, a design patent might be used to protect the nonfunctional (i.e., ornamental) appearance of a device. The requirement of nonfunctionality limits the usefulness of design patents in the medical device field.
- *Trademarks* are used to prevent others from using a mark (e.g, a product or service name or logo or symbol) that is likely to cause confusion among customers. Thus, in the context of medical devices, trademarks are useful for protecting the goodwill associated with commercial products or services but do not provide exclusivity with regard to the products or services, *per se*.

- *Copyrights* can be used to prevent direct copying of works such as text, art, sound recordings, and software code, although it does not give the owner exclusive rights to the underlying idea. Thus, in the context of medical devices, copyright is useful for protecting against copying of software code and published works.
- *Trade secret protection* refers to an array of state and federal laws that protect against misappropriation of commercially sensitive information that is not generally known. Thus, in the context of medical devices, trade secret protection might protect internal information from being used by former employees and others; however, trade secret protection is not useful for protecting information or designs that enter the public domain.

Because utility patents are most relevant to the field of medical devices, the remainder of this chapter focuses on the IP protection tool of patents. The reader is encouraged to seek the guidance of an experienced IP attorney for further information beyond what is generally provided here.

2. How to Read a US Patent

Patents granted in the United States typically have a general format that provides for consistency and readability. To be issued a patent, US patent law provides general requirements that a US patent application should have, although there is no strict requirement for a specific format. The format described in this section is a conventional format for a US patent and most patents will follow such a format or a variation of it; however, other formats are also possible. Patents issued in foreign countries may have some features in common with those described here.

2.1. Sample US Patent

Although each patent has its own style and substance, there are some general similarities between most patents. Presented below is a sample US patent in its entirety that we have examined (Fig. 1). It was selected only for its short length and simple and effective style. The sample patent is shown here only to provide the reader with an idea on what is contained in a typical patent. We have not assessed the quality or validity of the patent.

2.2. General Parts of a US Patent

A typical patent issued in the United States has several sections, each with a specific purpose and function. The major components of a patent include the written description, claims, and drawings, if any exist.

2.2.1. Written Description

The written description includes any part of the patent other than the claims section and the drawings. The more commonly used name for this section is

US005827487A

United States Patent [19]

Holmes

[11] **Patent Number:** **5,827,487**

[45] **Date of Patent:** **Oct. 27, 1998**

[54] **MEDICAL INSTRUMENT FIXATION METHOD AND MEANS**

[75] Inventor: **Russell P. Holmes**, Boston, Mass.

[73] Assignee: **Riley Medical, Inc.**, Auburn, Me.

[21] Appl. No.: **807,812**

[22] Filed: **Feb. 26, 1997**

[51] Int. Cl.⁶ **A61L 2/00**
[52] U.S. Cl. **422/297**; 422/297; 422/300;
248/424; 248/172; 248/176.1; 248/298.01;
211/70.6; 206/483; 206/370
[58] **Field of Search** 422/300, 297;
248/424, 172, 176.1, 298.1, 346.07; 211/59.1,
70.6; 206/474, 480, 483, 370

[56]   **References Cited**

U.S. PATENT DOCUMENTS

337,888	3/1886	Swan .
635,284	10/1899	Adair .
4,135,868	1/1979	Schainholz 422/310
4,262,799	4/1981	Perrett 206/363
4,317,416	3/1982	Baum et al. 108/157
4,573,569	3/1986	Parker 206/1.7
4,635,801	1/1987	Oren 211/70.6
4,643,303	2/1987	Arp et al. 206/370
4,859,423	8/1989	Perlman 422/102
5,384,103	1/1995	Miller 422/310
5,424,048	6/1995	Riley 422/300
5,433,930	7/1995	Taschner 422/300

5,441,709	8/1995	Berry, Jr. 422/297
5,492,671	2/1996	Krafft 422/26
5,525,314	6/1996	Hurson 422/300

OTHER PUBLICATIONS

Sterilization Systems Corporation, The Advanced Laparoscopy/Pelviscopy Sterilization Basket, Protect Your Instruments, 3 pages, no date available.

Primary Examiner—Robert J. Warden
Assistant Examiner—Fariborz Moazzam
Attorney, Agent, or Firm—Cesari and McKenna, LLP

[57]   **ABSTRACT**

A medical instrument fixation system for a sterilization tray having a bottom wall and an array of ventilation holes in the bottom wall, the holes being arranged in columns and rows with a selected spacing between the columns and rows includes a plurality of elongated rails. Each rail has a bottom surface, a top surface and a plurality of projections extending from the bottom surface and spaced apart along each rail a distance equal to, or an intregal multiple of, the selected spacing, the projections being sized to plug into the holes so that the rail extends along a column or row of holes. The system also includes a plurality of posts, each post having opposite ends and means for keying one end of each post to a different one of the rails so that so that each post can slide along the corresponding rail to a selected position thereon. A method of fixating the instruments in the tray using special retention devices is also disclosed.

7 Claims, 2 Drawing Sheets

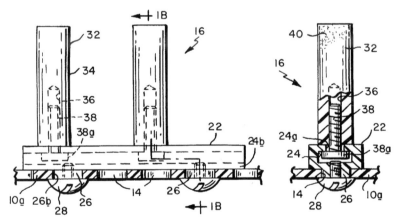

Fig. 1. Sample US patent.

U.S. Patent   Oct. 27, 1998   Sheet 1 of 2   **5,827,487**

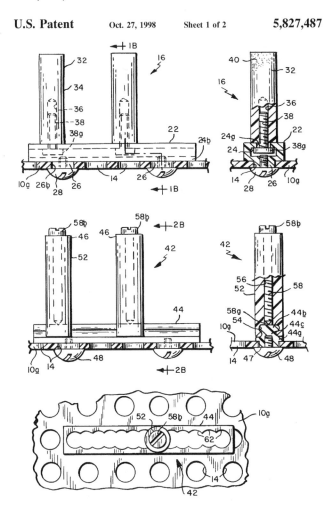

Fig. 1. *(Continued)*

the *specification*, although technically, this term refers to all parts of the application other than the drawings. The written description describes the features of the invention. Often, the written description is further divided into more specific parts to further simplify the readability of the patent.

Among the major divisions in the written description are the background, brief summary of the invention, brief description of the drawings, detailed description of the invention, and abstract. The background section may be subdivided into *field of the invention* and *background of the invention*. Additional sections are also possible in the written description and often include examples, such as commonly used for pharmaceutical or chemical inventions.

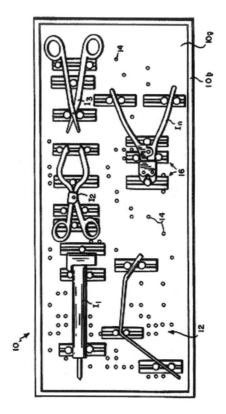

Fig. 1. *(Continued)*

The front page of a patent has relevant historical information relating to the patent and includes the title, the inventor(s), the assignee, the date of filing and issue of the patent, the examiner(s) who reviewed and issued the patent, the law firm or attorney who conducted at least part of the prosecution of the patent application, and any background information and identification of related US and foreign patents. An abstract of the invention is also placed on the front page.

Following the technical information is a list of patents and other documents that the patent examiner reviewed as part of the patent prosecution process. Such patents and documents are listed on the issued patent as a sign that the specific patent was issued after the listed documents were taken into consideration by the patent examiner.

After the list of patents and documents considered usually is a section labeled *background of the invention*. This section may include an additional section titled *field of the invention*, which presents the overall scientific field of

5,827,487

| 1 | 2 |

MEDICAL INSTRUMENT FIXATION METHOD AND MEANS

This invention relates to the field of sterilization trays. It relates more particularly to improved instrument fixation devices for holding surgical instruments at fixed positions in the tray.

BACKGROUND OF THE INVENTION

Surgical instruments are often transported in trays prior to use. The instruments are usually laid out in a certain way in the tray and subjected to sterilization in a steam autoclave or similar sterilization apparatus. In order to maintain separations between the various instruments in the tray, the instruments are supported or retained by brackets positioned in the tray. Following sterilization, the tray full of instruments may be transported to an operating room and placed close to the surgical team whose members may withdraw the instruments from the tray as needed for a particular surgical procedure. Many times, the instruments are selectively arranged in the tray so that they can be picked from the tray in the general order that they are needed for the particular procedure. Examples of such trays are found in U.S. Pat. Nos. 4,643,303; 5,424,048 and 5,492,671.

As seen from the above patents, the known devices for organizing and fixating medical instruments in a tray include various types of slotted brackets, clips and posts which project up from the bottom of the tray, the instruments being held in place within the slots and clip openings and between the posts. As shown there, a plurality of such fixation devices can be spaced parallel or perpendicular in the tray so that they engage and support the opposite sides or ends of various different length instruments.

Most such fixation devices are able to effectively locate and hold instruments which are more or less straight and regular. However, they are not particularly suitable for fixating oddly shaped and irregular instruments such as retractors and other longer instruments that have ring handles. This is because there is insufficient flexibility in the placement of the various fixation devices within the tray as to enable the devices to closely engage the instruments while still organizing the instruments in an efficient layout within the tray. This results from the fact that the fixation devices are often plugged into the ventilation holes usually present in the bottom of the tray such that a fixation device can only be placed where there are holes in the bottom of the tray.

As the number of such holes is limited by manufacturing cost, required tray bottom strength and the need to prevent the instruments from projecting through the holes, so also are the positions of the various fixation devices. Consequently, either the tray contains too few properly fixated instruments or a larger number of instruments some of which may not be properly fixated. Thus, if the tray is shaken or tilted, those loose instruments may become disengaged from the fixation devices and assume skewed positions in the tray so that they may become damaged and difficult to remove without upsetting other instruments in the tray. In extreme case, these loose instruments may even fall out of the tray and become contaminated. Since a tray may contain a complete set of instruments needed for a particular surgical procedure, this may require that another full tray of sterilized instruments be made available to the surgical team.

Another consideration is that the instruments required to perform a specific surgical procedure may vary greatly between hospitals. Therefore, it is practically impossible to design a standard tray configuration that will be acceptable

in more than one hospital. Thus, an optimum instrument fixation arrangement is one which is enormously flexible so that it can be applied to each individual hospital, because the numbers and types of instruments being presented in the trays change constantly. Moreover, the instruments vary greatly in shape and size. Therefore, if these instruments are to be efficiently laid out and fixated within the trays, the fixation devices must be placed at the best holding locations and preferably be spaced no more than $\frac{1}{8}$ inch from the instrument's surface. Fixation devices positioned in this manner effectively locate the instruments within the tray, yet allow their quick removal during surgery. Such an arrangement may also facilitate the daily tray assembly by clearly depicting the outlines of the instruments to be presented in the tray.

SUMMARY OF THE INVENTION

Accordingly, the present invention aims to provide a medical instrument fixation device which can be positioned at almost any location in a sterilization tray.

Another object of the invention is to provide such a fixation device which effectively holds the instrument, yet allows its ready removal when needed.

Yet another object of the invention is to provide an instrument fixation device which allows an efficient and effective layout of different instruments in a sterilization tray.

Another object of the invention is to provide a medical instrument fixation system which is particularly suitable for retaining and presenting oddly shaped instruments.

Another object of the invention is to allow the tray configuration to be changed easily without effort at minimal cost.

Other objects will, in part, be obvious and will, in part, appear hereinafter. The invention accordingly comprises the sequence of steps and the apparatus embodying the features of construction, combination of elements and arrangement of parts which will be exemplified in the following detailed description, and the scope of the invention will be indicated in the claims.

Briefly, the fixation device of this invention comprises a rail of optional length and having at least two pegs projecting from its underside which are sized and is spaced apart so as to be able to plug into at least two of the ventilation holes present in the bottom wall of a conventional sterilization tray. These holes are usually arranged in a rectangular array of columns and rows so that the rail can be positioned at any location in that hole array within the resolution of the hole array. The rail may be releasably fixed in position by fastener means secured to the ends of the pegs from the underside of the tray.

The fixation device also includes one or more upstanding posts of optional length which inter-fits with the rail and may be positioned at various locations along the length of the rail. Preferably, means are provided for releasably fixing the post position along the rail.

Resultantly, while the position of each rail within the sterilization tray may be limited to some extent by the spacings of the columns and rows in the hole array at the bottom of the tray, the position of each post relative to the hole array is not so limited. In other words, each post can be positioned at substantially any location on the tray bottom.

The invention thus allows the person who is setting up the tray to lay out the instruments in the tray as specified by the particular hospital or surgeon. Rails of various lengths may

Fig. 1. *(Continued)*

5,827,487

3

then be inserted under the instruments and loosely fastened to the tray. Next, posts may be engaged to the rails and slid down the rails until they are positioned next to the instruments. When the tray is complete and the posts and instruments are positioned correctly, the posts and rails may be tightened down so that the instruments are fixated at their assigned positions within the tray.

Thus, a fixation system incorporating my instrument retention devices is extremely flexible and may be accommodated to many instrument layouts desired within sterilization tray, as well as to irregularly shaped instruments within those layouts. Accordingly, the invention should find wide application in clinics and hospitals which perform surgical procedures and whose routines require a variety of different presentations of the surgical instruments.

BRIEF DESCRIPTION OF THE DRAWINGS

For a fuller understanding of the nature and the objects of the invention, reference should be had to the following detailed description taken in connection with the accompanying drawings, in which:

FIG. 1A is a side elevational view on a larger scale illustrating one of the instrument retention devices comprising an instrument fixation system according to the invention,

FIG. 1B is a sectional view taken along line 1B—1B of FIG. 1A;

FIG. 2A is a view similar to FIG. 1A showing another retention device embodiment for use in the fixation system;

FIG. 2B is a sectional view taken along line 2B—2B of FIG. 2A,

FIG. 2C is a plan view of the FIG. 2A retention device,

FIG. 3 is a plan view of a tray incorporating an instrumentation fixation system according to the invention.

DESCRIPTION OF THE PREFERRED EMBODIMENTS

A more or less conventional sterilization tray is an integral part of my medical instrument fixation system. The tray has a generally rectangular bottom wall $10a$ surrounded by upstanding side walls. Formed in the bottom wall $10a$ is a rectangular array of holes 14 arranged in columns and rows. Typically, the holes are spaced apart $1\frac{1}{2}$ inch on center. These holes are utilized to set the configuration of the fixation system as will be described.

The fixation system also includes a plurality of instrument retention or fixation devices indicated generally at 16 which are positioned and oriented within the tray and secured to the tray bottom wall $10a$, in such a way as to fixate or hold a variety of different medical instruments that have been arranged in a selected layout within the tray.

Refer to FIG. 3 which shows a typical tray 10 having a bottom wall $10a$ and side wall $10b$ and with an instrumentation fixation system 12 composed of devices 16 retaining various medical instruments $I_1, I_2, I_3, \ldots I_n$.

As shown in FIGS. 1A and 1B, each fixation device 16 comprises an elongated generally rectangular rail 22 having a generally C-shaped cross section so that the rail defines a internal keyway 24 which includes an longitudinal slot $24a$ at the top of the rail. The rails 22 comprising the different fixation devices 16 may come in a variety of lengths, e.g., two, three and five inches.

The bottom wall $24b$ of each rail 22 is formed with a plurality of depending pegs or collars 26 which are sized to fit in the holes 14 in the tray bottom wall $10a$. That is, the

4

outside diameter of each peg 26 is slightly smaller than the diameter of holes 14 and the length of each peg is comparable to the thickness of the tray wall $10a$. Furthermore, the centerline distance between adjacent pegs 26 is equal to, or an integral multiple of, the centerline spacing of the holes 14. Thus, the pegs 26 of a three inch long rail 22 depicted in FIG. 1B may be spaced apart a distance such that the pegs can be plugged into holes 14 spaced two holes apart. The rail 22 may be releasably secured to the tray wall $10a$ by threaded fasteners 28 screwed into the ends of the pegs 26 at the underside of the tray. As shown in FIGS. 1A and 1B, the fasteners 28 are provided with heads which are larger than the tray holes 14 so that the edge margins of the heads engage the underside of tray wall $10a$ around the perimeters of holes 14.

Still referring to FIGS. 1A and 1B, each fixation device 16 also includes one or more upstanding posts 32 movably mounted to rail 22. The device illustrated in those figures has two such posts 32. Each post 32 includes generally cylindrical sleeve-like nut 34 which may come in a variety of lengths, e.g., 0.5, 1.0 and 1.3 inches. The diameters of the nuts are usually the same, e.g., 0.25 inch. A threaded passage 36 extends in from the lower end of each nut 34 for receiving a threaded fastener 38. Each fastener 38 has a head $38a$ whose sides are flat as best seen in FIG. 1B so that the head can key into the keyway 24 formed in rail 22 with the shank of the fastener extending out through the slot $24a$ in the top of the rail. Thus, when the fastener 38 is keyed to the rail as shown in FIGS. 1A and 1B, it cannot rotate relative to the rail; however, it can slide along the keyway 24 in the rail. Each post 32 may be releasably positioned at a selected location along the rail by rotating the nut 34 so that the associated fastener 38 screws into the nut passage 36, thereby gripping the slotted top wall of the rail between the fastener head $38a$ and the lower end of nut 34.

In order to install the fixation system 12 to fulfill the objectives of the invention, the various instruments should be arranged in the tray at the desired locations for the particular surgical procedure to be performed using those instruments. For example, the instruments may be positioned in the order in which they will be used beginning from one side or end of the tray. Alternatively, the instruments may be arranged so that a maximum number of instruments can be placed in the tray while maintaining adequate spacings of the instruments from each other so that they will not become entangled and so that they can be removed easily when needed.

Once the desired instrument layout has been set, the rails 22 can be inserted under the instruments and loosely attached to the tray bottom wall $10a$ using fasteners 28 at the positions and orientations to best fixate the overlying instrument particularly at the instrument extremities. Then, the posts 32 may be slid onto the rails 22 from one or both ends such that the post fasteners 38 key into the rail keyways 24. Each post may be moved along the associated rail until it nearly engages the adjacent surface or critical holding area of the associated medical instrument. At that point, the nut 34 of each post may be tightened as needed to fix the position of the post. In this way, a selected instrument layout can be fixed so that even if the tray is tipped or jostled during handling, the instruments will remain in their proper positions within the tray.

It is important to note also that the rails 22 space the instruments above the tray bottom wall $10a$ assuring good circulation of steam under the instruments during the sterilization process.

Since the arrangement of the fixation devices 16 is customized for each instrument layout, the locations of the

Fig. 1. *(Continued)*

5,827,487

5

posts **32** actually provide a rough outline of the layout. Thus, if particular instruments have to be removed from the tray temporarily, their proper locations in the tray will be obvious from the outlines provided by the associated posts. Actually, the sleeve-like nuts **34** and/or posts **32** that retain the various instruments may be color coded as shown at **40** in FIG. 1B so that the posts for each instrument have a unique color. This also helps to identify the correct position for each instrument in the tray.

Refer now to FIGS. 2A—2C, which show another fixation device **42** for use in the sterilizer tray. The device **42** comprises a rail **44** which supports one or more upstanding posts **46** which may be movably positioned along the rail. In this case, however, the rail **44** is more or less like a railroad track in that it has a relatively wide base **44**a and a relatively wide head portion **44**b connected by a narrower neck portion **44**c as best seen in FIG. 2B. As we shall see, rail **44** functions as a key. A plurality of pegs or collars **47** project from the underside of base **44**a which pegs are arranged and adapted to plug into one or another of the holes **14** in the tray bottom wall **10**a. The rail **44** may be releasably secured to the tray bottom wall by threaded fasteners **48** in the same manner described above for rail **22**.

Each post **46** comprises a tubular sleeve **52** whose lower end is slotted to provide a keyway **54** for the associated rail **44**. In other words, the keyway **54** has essentially the same profile as the rail **44** cross section as best seen in FIG. 2B. Also, the sleeve **52** has an axial passage **56** which extends from the top of the sleeve down to keyway **54**. Screwed down into passage **56** is a long threaded fastener **58**. Fastener **58** is longer than passage **56** so that when the fastener is turned all the way down, its tip **58**a will engage the top of the underlying rail **44**.

The fastener tip **58**a may be pointed as illustrated in FIGS. 2A and 2B. In this event, when the post is slid to the desired position along the rail and the fastener **58** is screwed down, its tip **58**a will "bite" into the top of rail **44**, thereby fixing the position of the post on the rail.

Alternatively, the tip of the fastener can be made flat and the top of the rail **44** formed with overlapping discoid indentations **62** as illustrated in FIG. 2C. In that event, when the fastener is turned down, its tip will seat in one or another of the indentations **62** thereby fixing the location of the post along the rail. Preferably, the fastener is dimensioned so that the fastener head **58**b will seat against the top of sleeve **52** before the fastener tip bottoms in the indentation **62** so that the position of the post on the rail is fixed by the fastener tip engaging the side of the indentation rather than the bottom. This helps to minimize stresses on the post and rail.

It will be appreciated from the foregoing that the fixation system described herein should facilitate the efficient organization of medical instruments in a sterilization tray and effectively hold those instruments in position during the sterilization process and during the transportation of the tray to and from the sterilizing facility. Thus, it assures that the instruments will be protected against damage and presented properly to the medical personnel performing a surgical procedure using those instruments while Is allowing ready removal of the instruments from the tray when needed.

It will be seen that the objects set forth above among those made apparent from the preceding description are efficiently attained. Also, certain changes may be made in the above method and in the constructions set forth without departing from the scope of the invention. For example, the pegs can be designed to snap lit in the holes in the tray bottom wall. Therefore, it is intended that all matter contained in the

6

above description or shown in the accompanying drawings shall be interpreted as illustrative and not in a limiting sense.

It should also be understood that the following claims are intended to cover all of the generic and specific features of the invention described herein.

What is claimed is:

1. A medical instrument fixation system comprising

a sterilization tray having a bottom wall;

an array of ventilation holes in the bottom wall said holes being arranged in columns and rows with a selected spacing between the columns and rows;

a plurality of elongated rails, each rail having a bottom surface and a top surface;

a plurality of projections extending from the bottom surface and spaced apart along each rail a distance equal to, or an integral multiple of, said selected spacing, the projections being sized to plug into said holes so that each rail extends along a column or row of holes in position for engaging and supporting one or more medical instruments;

a plurality of posts, each post having
opposite ends,
means for keying one end of each post to a different one of said rails so that each post can slide along the corresponding rail to a selected position thereon adjacent to an instrument engaged by the corresponding rail, and
fixing means for releasably fixing the position of each post on the corresponding rail, said fixing means including a threaded fastener threaded axially through said post, said fastener having an end portion engaging said rail so that when the post and fastener are rotated relatively in one direction, the post is clamped to the rail and when the post and fastener are rotated relatively in the opposite direction the post is unclamped from the rail.

2. The system defined in claim **1** and further including a plurality of additional posts releasably fixed to selected ones of said plurality of rails.

3. The system defined in claim **1** and further including means for releasably fixing each rail to the tray bottom wall through the holes underlying that rail.

4. The system defined in claim **1** wherein

each rail defines a keyway having an internal wall;

said post one end defines a key configured to slide in said keyway, and

said fixing means include means for pressing the key against said keyway wall.

5. The system defined in claim **1** wherein

each rail defines a key;

said post one end defines a keyway configured to slide along said rail, and

said fixing means include means on the post for engaging the rail.

6. A retention device for fixating medical instruments in a sterilization tray of the type having a bottom wall with a rectangular array of ventilation holes, said device comprising

an elongated generally rectangular rail having opposite walls;

a plurality of projections extending from one of said walls and spaced apart along said rail a distance equal to, or an integral multiple of, the hole spacing, said projections being sized to plug into the holes;

a plurality of posts, each post having opposite ends;

Fig. 1. *(Continued)*

5,827,487

7

means for keying one end of each post to the other wall of said rail so that the post can slide along the rail to a selected position, and

fixing means on each post for releasably fixing the position of the post on the rail, said fixing means including a threaded fastener threaded axially through said post, said fastener having an end portion engaging said rail so that when the post and fastener are rotated relatively

8

in one direction, the post is clamped to the rail and when the post and fastener are rotated relatively in the opposite direction the post is unclamped from the rail.

7. The device defined in claim **6** and further including securing means attached to the posts for releasably securing the rail to a tray bottom wall through the holes therein.

* * * * *

Fig. 1. *(Continued)*

which the invention is part. This field statement is typically no longer than one or two sentences and helps the classification expert at the PTO classify the patent application in a particular scientific field on receipt of the filed patent application. The background section typically includes a background of the invention in terms of the state of the art in the field and explanations of why the current application is needed or desired in the field.

A summary of the invention section follows the background of the invention. The summary presents a more general description of the invention and some of its unique features, objectives, or capabilities. Sometimes, the summary of the invention presents the features of the invention that specifically address the shortcomings described in the background of the invention.

A brief description of the drawings is presented only in patent applications that have drawings and typically provides a one-sentence description of each of the drawings. This brief description is not intended to be the only description of the drawing and merely provides a general idea of what each drawing is intending to show.

The detailed description of the invention is typically the lengthiest portion of the patent. It is here that the invention and each of the drawings are described in full detail. The level of description of the invention does not have to be completely exhaustive but should be sufficient to allow one having an ordinary level of knowledge in the scientific field to make and use the invention. Often, this section is labeled *detailed description of preferred embodiments*. Under US patent law, if there is a best way to practice the invention, that method, conventionally labeled as *best mode*, must be described or the patent may be deemed invalid in later litigation. The detailed description may contain a number of examples, each describing how to use the invention in a different way.

Finally, the end of the patent application contains an abstract that describes the invention in brief and gives a quick explanation of it. Because this abstract will be placed on the front page of a future patent, applicants should take care to present an accurate description of the invention.

2.2.2. Claims

The least understood and most scrutinized section of a patent is the claims section. The claims are the legally binding words that set forth the metes and bounds of the invention. In other words, the language of the claims is what determines the actual invention for which the patent has been issued. Thus, the language of the claims and the meaning of each word and phrase within the claims are important parameters to consider when determining the scope of a patent and whether an offending device may infringe on the patent.

Drafting proper claims is likened to an art form and should be done by an experienced patent attorney. Ideally, claims are broad enough to encompass the particular inventions described in the patent application and variations thereof but not so broad as to pierce the property rights of other patents (i.e., prior art).

2.2.3. Drawings

Many patents have drawings that help describe the invention or the state of the art that lead to the invention. Although not a requirement for patent applications that have only method claims, drawings are a requirement for any patent application that includes device or system claims. Thus, in the medical device field, drawings will almost always be required for a patent.

3. Obtaining Patents in the United States

The procedure used to obtain patents in the United States is mechanical and relatively simple in theory. An inventor must file an application for the invention at the PTO either before disclosing the invention to others or within 1 year of the earliest date of disclosure to others. The actual rules are more complicated and require detailed analysis regarding the level and manner of disclosure.

To be safe, one should file a application for a US patent before commencing any activities that may be construed later as invoking certain historic barring activities. Among these activities are disclosure in the public, through either publishing or oral statements, sales of the invention or offers to sell the invention, and any use of the invention in the public domain. Any such activities may be potentially detrimental to the preservation of patent rights and may be used to invalidate a patent in a future legal proceeding. It is imperative that an inventor be cognizant that such activities, once done, typically cannot be "undone." Thus, a comprehensive patent strategy plan should be considered, preferably with the guidance of an experienced patent attorney, to prevent any loss of rights in a potential future patent.

3.1. US Patent Office

The US PTO is one of the best known and least understood government agencies. One of the biggest myths is that the PTO is a registration center for inventions and that it will "rubber stamp" any application it receives. To appreciate the procedure required for obtaining a patent, one should closely consider the process that takes place between the time when a patent application is filed until a patent is granted, as well as who is involved in that process.

The official date of filing of a patent application is the date that the PTO receives a complete patent application. This important date determines the life of the patent, because patents issued typically have a 20-year lifetime from the date of the filing.

Soon after the patent application is filed, a classification expert and a security expert will briefly review the patent application. The classification expert determines the particular scientific field that covers the patent application. For example, patent applications may be classified as automobile accessories, protein catalysts, or orthopedic medical devices. The classification expert classifies each application using a designated classification number that is needed to transfer the application to the appropriate expert(s) in the PTO that are charged with reviewing all patent applications within that field. Such experts are called *patent examiners.*

Before the patent application is transferred to the patent examiner, a security expert considers whether the patent application contains any material that may be detrimental to national interests if made public. Typically, inventions in the national defense industry are designated as security risks, as well as other inventions that may have potential for public misuse, such as radiation devices or certain types of chemicals or compounds. Patent applications designated as having security risks will not be published or made available to the public, although a patent could be granted. These patents are secret and are known only to the particular patent examiner, the patent attorney, and the inventor.

Patent examiners are the government's counterparts to the patent attorney. They typically are charged with reviewing patent application in specific scientific areas to gain experience and knowledge in such fields, even if their background education does not relate exactly to such fields. All patent examiners have at least a bachelor's degree in a scientific field. Many patent examiners have advanced degrees, such as PhDs, MDs, DDSs, or MBAs. Thus, very experienced patent examiners are considered by some as foremost experts in their particular fields of review because they are the first to see the latest developments in a particular field, outside of the actual inventors and their attorneys. Thus, patent examiners are not rubber stamp bureaucrats who do not have the capacity to understand the merits of a patent application.

Once a patent examiner receives a patent application and places it on his or her docket, a considerable amount of time—sometimes several months to several years—may pass before the examiner actually reviews the application. This delay is caused in part by the relatively small number of examiners (a few thousand) compared to the number of patent applications received per year (several hundred thousand). The patent examiner must review these applications in the order that they were received. Thus, it may take years for a patent examiner to review a patent application, even though it may have been on his or her docket for several years. This elapsed time is typically considered part of the 20-year period of the patent and may not be "reimbursed" once the patent has been issued.

3.2. Patent Prosecution

The process of applying for, arguing for, and receiving a patent is traditionally referred to as *patent prosecution*, although most of the activities that occur in patent prosecution occur after the patent examiner reviews the claims, which may be years after the patent application is filed. As the patent examiner reviews the patent application, he or she primarily considers the language of the claims in light of prior art. Under US law, a patent applicant who has properly filed a patent application is entitled to a patent unless certain barring circumstances exist or barring activities have occurred. Barring circumstances mostly relate to prior published patents (i.e., prior art) and may also include other publishings, statements, or activities that may bar the patent application from being granted. Often, the prior art applied include only other patents or publishings.

The patent examiner will reject one or more of the patent claims by deeming them unpatentable in light of the prior art. The reasons for deeming such claims as unpatentable must be set forth in the "office action," the actual written document prepared by the patent examiner and sent to the patent attorney. The office action typically sets a 3-month period for a response. Thus, the inventor and patent attorney should carefully consider the reasons for the rejections and plan a response to be filed within that 3-month period. The rejected claims may be canceled, amended, or maintained along with a written explanation of why they should not have to be changed in order for the patent to be granted.

After a response is filed in response to the patent examiner's office action, the patent examiner must consider the merits of the response, including any new or amended claims, and make a final decision on the status of the claims. The patent examiner may reject all of the claims, allow all of the claims, or allow some of the claims and reject the rest. The patent examiner must make a decision typically no more than 2 months from receiving the patent attorney's response. Thus, patent attorneys and inventors should expect to hear back from patent examiners no more than about 3 months after they have filed the response. Final decisions are forwarded to the patent attorney. If the patent examiner does not allow all of the claims, the patent attorney then may advise the inventor to either continue the prosecution process, take any claims that are allowed, appeal the patent examiner's decision, abandon the patent application, or some combination of the above.

4. Relevant Inquiries Before Seeking to Patent

It is good practice to consider whether an invention or inventive concept could be patentable before taking the steps needed to procure a patent. Furthermore, it is also important to consider the value of such a patent. Common inquiries regarding the patenting process and preapplication considerations are presented in this section.

4.1. What Can Be Patented?

A patent can be issued for any novel, useful, and unobvious invention. Congress intended the scope of possible inventions to "include anything under the sun that is made by man." Courts have affirmed that software, business methods, sales techniques, and even artificially created life forms can be patented if they satisfy the three-part test for patentability.

Medical devices are inherently useful and will typically not encounter difficulty when their usefulness is considered, although it remains informative to consider the test for patentability as a three-part inquiry. First, is it useful? Almost any level of utility is sufficient. Second, is it novel? The inventors must have been the first to develop the idea. One cannot patent something that others have previously developed and publicized, even if the others did not seek a patent. Finally, is it unobvious? The invention must not be so close to previously known devices that it would be obvious to someone having ordinary (not a high-level) skill in the relevant field.

The standard for getting a patent is lower than most people think. Talented technical people sometimes assume patents are only for remarkable inventions and thus dismiss patenting an improvement because it is obvious to them. One should not blindly accept assertions by medical researchers or engineers that every feature of a new product is obvious. Also, each feature of a new product may support a patent claim, and features can be considered separately or in combination to determiine patentability.

4.2. What Should Be Patented?

The decision on whether an invention should be patented requires consideration of whether a patent will provide competitive advantage.

To assess the competitive advantage that would be gained from a patent, it is useful to understand the purpose of a patent. A patent does not grant the patent owner the right to practice the claimed invention. Instead, a patent grants the owner the right to exclude others from making, using, importing, selling, or offering to sell the inventions during the life of the patent.

The right to exclude others from making, using, selling, or importing a product or process (invention) may be used for several competitive purposes. It may be used to protect market share by keeping competitors from offering unique features and functions or to generate income through licensing the invention to others and collecting royalties. Another purpose of patents is to cross-license, which reduces the royalty rate that might otherwise be paid for another's technology. Also, shareholder value may be affected via an increase in a company's market capitalization based on its ability to generate future revenues on patent-protected technology. Finally, patents may be used as a defense to claims that involve asserting patents against a plaintiff in a patent dispute. Alternatively, patents may be cross-licensed to settle (or avoid) a dispute or to reduce the amount the defendant has to pay to induce a settlement. With these factors in mind, consideration of whether a patent will (or could) provide competitive advantage should determine whether an application should be filed.

4.3. What To Do When a Patent's Competitive Advantage Is Not Clear?

Filing a provisional patent application while assessing the value of a full-utility patent application could help establish some rights to an invention. A provisional application is a simplified way of starting the utility patent application process. Costs are lower than for full utility applications because these applications do not require claim drafting and adherence to formatting rules.

A provisional application filing may provide a precisely identifiable piece of IP that may be licensed, sold, or used to gain credibility with potential investors; evidence of invention by an earlier filing date, which provides procedural advantages, especially if others file similar applications; the right to advertise the product as *patent pending*; and an entire year to seek capital, test markets, refine the product, and find out whether the idea has critical mass before incurring the substantial expense of filing a regular utility patent application.

Unlike a standard utility patent application, the provisional application is not examined and does not issue as an enforceable patent. The provisional application has a life of 1 year. Before the end of that year, if protection is still desired for the idea, a utility application and any desired foreign applications must be filed. Provisional filings preserve rights, stopping the 1-year disclosure clock under US law. Filing a provisional application before divulging the invention will preserve potential rights in most foreign countries.

Full-utility applications (i.e., nonprovisional applications) must follow more detailed formatting requirements than for provisional applications and include a set of claims. In deciding whether to file a provisional application or proceed directly with a full-utility application, one should balance the advantages of postponing expense and obtaining an early filing date (both favoring provisional filings) against the desire for early patent issuance (favoring an early full-utility filing).

4.4. What Activities May Bar US Patent Application Filing?

The occurrence of one or more of the following statutory bar activities may prohibit an inventor from filing a US patent application for an invention if the delay is greater than 1 year: publishing a document disclosing the invention; publicly using the invention; and selling or offering to sell the invention (usually a product incorporating the invention). Beginning on the earliest date any one of these activities occurs, US law imposes a 1-year period for filing a patent application.

Foreign countries also have rules regarding barring activities. The vast majority of these countries use different rules regarding when an application must be filed. Few foreign countries allow a grace period as does the United States, and those that do usually do so only in limited circumstances. To the contrary, most countries bar the filing of an application unless it is filed before any public disclosure of the invention, however, if the US application is filed before the public disclosure, the foreign application may be filed under an agreement that will treat the foreign application as if it were filed on the same day as the US application.

4.5. What If One Is About To Publish Information About the Invention?

It is imperative to ensure that a patent application is on file or that an affirmative decision not to file a patent application has been made. Even filing a provisional patent application quickly before such a publishing may be sufficient to preserve full rights in a future patent.

4.6. What Is the Difference Between Design and Utility Patents?

In general, a utility patent protects the way an article is used or the way it works, whereas a design patent protects the way an article looks. Both design and utility patents may be obtained on a medical device if the invention resides both in its utility and ornamental appearance. Although utility and design patents afford legally separate protection, the utility and ornamentality of an article are not easily separable. Articles of manufacture may possess both functional and ornamental characteristics.

Design patents cannot be used to protect a design for an article of manufacture that is dictated primarily by the function of the article. Specifically, if there was no unique or distinctive shape or appearance to the article at the time the design was created not dictated by the function that it performs, the design lacks ornamentality and would not be patentable. If given a choice between utility-patent protection and design-patent protection, it is generally better to seek utility-patent protection because of its stronger enforcement. An experienced patent attorney can prepare a utility-patent application and claim both the utility and the physical features of an invention. Factors such as actual shape of a medical device that is essential to its function, size, and materials used should be primarily considered in a utility-patent application.

Because design patents protect only the appearance of an article of manufacture, it is possible that minimal differences between similar designs can render each patentable. Therefore, even though an inventor may ultimately receive a design patent for a product, the protection afforded by such a patent may be somewhat limited.

4.7. What Is the Difference Between Utility Patents and Copyrights?

A utility patent differs from a copyright in several important ways. A copyright typically covers only the expression of a work and does not stop others from appropriating ideas that are embodied in that work. By contrast, a patent can be used to prevent a person other than the inventor from extracting ideas from a copyrighted work and creating a new system based on them.

Additionally, a US patent is obtained only after preparing a detailed patent application and then only after a patent examiner has concluded that the patent application is allowable. By contrast, copyright protection is automatic and a copyright registration is granted almost as matter of course on filing a relatively simple and inexpensive copyright registration application.

4.8. When Does a Proposal Infringe on a Patent?

For some medical device areas, the answer to this question is easy to find, whereas the answer may be difficult to ascertain in other areas. For example, if a product has been on the market for a long time, there is a chance that any patent that might have covered it would have long since expired, although the mere passage of a long period is not, by itself, complete assurance that a patent would not be infringed. A patent could issue from a patent application that had been pending before the PTO for a long time.

Depending on the particular business circumstances, it may be advisable to conduct a search for patents that may be infringed in certain areas. Such a search may take place through online public databases or an in-person physical search of patents at the PTO's public search room. One may also look at patent markings on products made by competitors.

Sometimes, one can be fairly certain that the only patents that would be of concern are those held by a particular competitor. In such a case, the first step typically is to identify and study all patents and published patent applications owned by that competitor.

4.9. What Is an Opinion Letter and What Is Its Value?

As the term is generally used, an *opinion letter* is a document representing a legal opinion given by counsel. The opinion might be on any of a wide range of topics. In patent law, an important category of opinion letters involves an opinion of patent counsel regarding a patent held by a potential adversary. For example, the losing party, or patent infringer, in a patent-infringement casemay be held liable for damages. In addition, if the court decides that the infringement was willful, the damages may be increased by a factor of up to three. This so-called *trebling* of damages occurs only if the infringement was found to be willful; if the infringer had a good-faith belief that the conduct was noninfringing or that the patent was invalid, the court may not find willfulness and damages will not be trebled.

Thus, it is prudent for a manufacturer to seek advice of patent counsel when it encounters a patent that it may have infringed on. If patent counsel determines that the patent is invalid or that its claims are not infringed, a decision may be made to memorialize this finding in an opinion letter. As described above, a well founded and well reasoned opinion letter may negate a subsequent finding of willfulness and thus may avoid trebling of damages.

4.10. What Do "Patent Pending" and "Patent Applied For" Mean?

Manufacturers or marketers of an article use the terms *patent pending* and *patent applied for* to inform the public that an application for a patent on that article is on file in the PTO. The law imposes a fine on those that use these terms falsely to deceive the public.

4.11. Could the PTO Give Others Information Contained in a Pending Application?

Most patent applications filed on or after November 29, 2000, will be published 18 months after the filing date of the application or any earlier filing date relied on under applicable US law. Otherwise, all patent applications are maintained in the strictest confidence until the patent is issued or the application is published, although a member of the public may request a copy of the application file after the it has been published. After the patent is issued, the PTO file containing the application and all correspondence leading up to its issuance is made available in the Files Information Unit for inspection by anyone, and copies of these files may be purchased from the PTO.

4.12. Who Will Receive the Patent if Two or More People Jointly Make an Invention?

If each has had a share in the ideas forming the invention as defined in the claims—even if only one claims—each is a joint inventor and a patent will be issued jointly on the basis of a proper patent application. On the other hand, if one of these persons has provided all of the ideas of the invention and the other has only followed instructions in making or developing the invention, the person who contributed the ideas is the sole inventor and the patent application and patent shall be only in his or her name.

Thus, an inventor is one who makes a contribution, no matter how small, to the overall concept of an invention, and whose contribution is in a claim; one who merely follows the inventor's instructions for building or testing an invention is generally not a coinventor.

4.13. Should Patents Be Filed Jointly by Inventors and Investors?

The application must be signed by the true inventor and filed in the PTO in the inventor's name, not in the name of the employer or an investor. The inventor (e.g., the first person in the above fact pattern) is the person who furnishes the ideas.

Because intentional misrepresentation of inventorship in a patent application, either by naming a noninventor as an inventor or omitting a true inventor, is grounds for completely invalidating a patent, inventors and others who have interest in a patent application must take inventorship issues seriously and must examine them closely before the application is filed.

5. Patent Marking

Patent marking is one of two approaches to maximizing damages that may be sought in an infringement suit. By enforcing a comprehensive patent-marking policy, including periodic reviews of the patent portfolio and making any necessary changes to the marking of patented articles, constructive notice is provided to potential infringers of an inventor's rights. This approach requires the analysis of an inventor's new products for coverage by one or more patents in the portfolio. A failure to mark appropriately may result in an inability to sue for damages prior to actual notice of infringement.

Under US law, a patentee who makes or sells patented articles or a person who does so for or under the patentee is required to mark the articles with the word *Patent* (or the abbreviation *Pat.*) together with the relevant patent number(s). The penalty for failure to mark is that the patentee may not recover damages from an infringer unless he or she was duly notified of the infringement and continued to infringe after the notice.

The marking of an article as patented when it is not yet patented is unlawful and subjects the offender to a penalty. Although the terms *Patent Applied For* or *Patent Pending* have no legal effect, they give information that an application for patent has been filed in the PTO. The protection afforded by a patent does not start until the patent has been granted. False use of these phrases or their equivalent is prohibited.

The marking statute is applicable to the patentee, as well as to the patentee's licensees who make, offer for sale, or sell a patented article in the United States or who import a patented article into the United States; however, many patentees have found it difficult to force a licensee to mark, even though the license agreement contains a marking provision. A patentee may still be able to recover damages based upon a constructive notice theory, even if the licensee does not consistently mark, as long as the governing license agreement contains a marking provision and the patentee exercises reasonable efforts in an attempt to force the licensee to comply with that provision. Such reasonable efforts may include letters to the licensee requesting compliance with the license terms and an acknowledgment of the duty to mark and an assurance that such duty is being fulfilled. The patentee should retain these letters in case it needs to establish that reasonable efforts were indeed employed.

When a patentee produces a family of products, with each product being covered by a different subset in a series of patents, it is appropriate to be overly inclusive, as long as the patentee has a reasonable belief that each of the listed patents covers at least one of the products in the family. In this case, courts have approved an indication that the product is covered by *one or more* of the listed patents. Likewise, listing an expired patent number is not likely to be a problem for a patentee, as long as at least one of the products was covered by the patent before it expired.

6. Nondisclosure Agreements

A nondisclosure agreement (NDA) is typically used by an inventor when relaying information about the invention to a third party that is not in a position to know about the invention. For example, there is no need for an NDA when disclosing the details of an invention to a coworker or other member of a company that is developing it because of the confidential relationship that is inherent in such a company. However, a contractor or other noncompany personnel may need to sign NDAs if they will receive information about an invention. Although an NDA cannot guarantee that a third party will not exploit the knowledge relating to an invention, it will serve as evidence in future court proceedings that the third party had received such proprietary information and had promised not to exploit or share such information with others.

7. Obtaining Patents in Foreign Countries

Most industrialized countries have a system for inventors to patent or at least register their inventions. Each system is governed by the rules of that particular country. Thus, a patent practitioner should inform an inventor of the rules and procedures in countries that the inventor is interested in pursuing a patent in order to ensure that the inventor has not violated any rules of that particular country. Many inventors prefer to obtain a patent only in the United States for various reasons including better enforceability, higher industrial value, lower cost, and greater flexibility in time rules. Thus, the processes provided in this section were mostly inherent with the United States patent. A US patent, no matter how strong or broad in scope, is only valid in the United States and is unenforceable in other countries. An inventor is encouraged to consult an experienced patent attorney when there is interest in pursuing patents in countries other than the United States.

7

HIPAA Compliance in Clinical Research

Peter M. Kazon and Sharon D. Brooks

1. Introduction

The passage of the Health Insurance Portability and Accountability Act of 1996 (HIPAA) has generated widespread concern that clinical research may be increasingly difficult to conduct. Certainly, those involved in clinical research may feel frustrated or confused by HIPAA's requirements, particularly when the media report that investigators have been denied access to patient records or that patient enrollment in studies has dramatically decreased. Innovative drugs and devices that improve and save lives would not be possible without the contributions of clinical research and human subject volunteers.

Because health data on patients are among the most sensitive types of personal information, protecting the confidentiality of personally identifiable health data is critical. Insufficient protections leading to unauthorized use and disclosures of data may subject individuals to adverse consequences such as loss of personal privacy and social stigma, as well as damaged financial standing, employability, and insurability.

Although a September 2004 US Government Accountability Office report found that organizations believe that HIPAA implementation went more smoothly than expected, the report also found that organizations continue to face many challenges involving access to health information for public health monitoring, research, and patient advocacy.[1] By understanding what is required of those involved in clinical research, HIPAA compliance may become less challenging. This chapter highlights the HIPAA requirements for those involved in clinical research and illustrates how and why HIPAA's Privacy Rule protects identifiable health information.

From: *Clinical Evaluation of Medical Devices: Principles and Case Studies, Second Edition*
Edited by: K. M. Becker and J. J. Whyte © Humana Press Inc., Totowa, NJ

2. What is HIPAA?

HIPAA mandated that the US Department of Health and Human Services (HHS) issue regulations regarding standards for privacy of individually identifiable health information.[2] These regulations are known as the Privacy Rule and, as of April 14, 2003, compliance has been required for most covered entities.[3] The Privacy Rule is a response to public concern over the unauthorized use and disclosure of individually identifiable health information. The rule applies to all past, present, and future protected health information (PHI; i.e., any oral, written, or electronic information that can be identified to a specific individual).[4] PHI must be created or maintained by a covered entity, which is any health plan, health care clearinghouse, and health care provider that engages in certain electronic transactions. Researchers are not considered covered entities unless they also are health care providers that transmit PHI as specified under the rule,[5] although the Privacy Rule's protections of PHI will have a wide-ranging impact on how research is conducted.

As noted, the Privacy Rule's purpose is to define when a covered entity can use or disclose PHI. Under the Privacy Rule, disclosures involving payment, treatment, or health care operations—all of which are defined terms in the regulations—do not require a specific permission from the patient. The rationale for this position is that these types of disclosures are so common that it would interfere with a provider's ability to furnish care if a specific consent was required for each such disclosure. Under the rule, the covered entity needs to take reasonable steps to ensure that each patient receives a Notice of Privacy Practices, which describes how the covered entity may use and disclose PHI, at the first patient encounter. Because research typically involves greater use and disclosure of the PHI, additional permission from the patient will usually be required.

2.1. Basic Rule

Under HIPAA's Privacy Rule, covered research entities—and organizations that use or disclose individually identifiable health information from or on behalf of a covered entity—may use or disclose PHI only when this use or disclosure has been authorized by the individual, obtained under a waiver of authorization, or provided through the use of a limited data set.[6] If the researcher is not a covered entity, he or she may still be affected by the Privacy Rule because he or she may not be able to obtain information from a covered entity unless these conditions are met. Furthermore, the Privacy Rule imposes a minimum necessary requirement on most uses and disclosures of PHI by a covered entity—that is, instead of using or disclosing all possible PHI, a covered entity may use or disclose only the PHI that is reasonably necessary to accomplish the purpose for which it is being used.[7]

Covered entities that fail to comply with the Privacy Rule may be subject to civil monetary penalties, criminal monetary penalties, and/or imprisonment. Notably, individually identifiable health information that is held by anyone other than a covered entity is not considered to be PHI and may be used or disclosed without regard to the Privacy Rule, although other applicable federal or state laws and regulations may limit the use or disclosure of such information.[8]

2.2. Relationship to HHS "Common Rule" and US Food and Drug Administration Human Subjects Regulations

In fact, much of the clinical research conducted in the United States is already subject to HHS "Common Rule" regulations"[9] and/or US Food and Drug Administration (FDA) regulations regarding the protection of human subjects.[10]

Specifically, HHS's Common Rule is intended to safeguard the rights and welfare of human subjects taking part in federally funded research by requiring an Institutional Review Board (IRB) to ensure that (a) risks to research subjects are minimized, (b) risks to research subjects are reasonable (in relation to anticipated benefits), (c) there is an equitable selection of subjects, (d) informed consent has been obtained from each subject and has been appropriately documented, (e) the research plan has adequate provisions for monitoring data, and (f) there are adequate provisions to protect the privacy of subjects and maintain the confidentiality of data.[11] These regulations apply to 16 federal agencies, including HHS, FDA, the Department of Agriculture, the Social Security Administration, the Department of Energy, and the Environmental Protection Agency (hence, the term *Common Rule*).

The FDA is responsible for protecting the rights of human subjects enrolled in certain studies that investigate the use of drugs, biologics, and medical devices, as well as those that investigate foods and dietary supplements. The relevant FDA regulations require researchers to obtain informed consent from all research participants and for all research (unless otherwise exempted) to be reviewed and approved by an IRB.[12] Researchers who fail to follow these requirements risk having the agency terminate the study or withhold the approval of new studies, which may ultimately result in the agency not approving the relevant drug, device, or food application.[13] Moreover, an IRB or institution that fails to follow these regulations may be subject to civil or criminal proceedings and any other appropriate regulatory action.[14] HIPAA's Privacy Rule does not replace the HHS or FDA regulations—that is, the Common Rule applies to certain *federally funded research* involving living persons. The FDA's regulations apply to certain *clinical investigations of drugs, biologics, and medical devices* involving living persons, and HIPAA's Privacy Rule applies to *all protected health information* regarding living or deceased persons. Thus, some researchers may be required to comply with all of these regulations under certain circumstances.

Sometimes, a person or organization may perform certain activities for a covered entity that involve the disclosure of PHI. Such persons or organizations are known as *business associates.* Before the covered entity discloses the PHI to the business associate, a contract ensuring that the business associate will appropriately safeguard the information is required.[15] Notably, the disclosure of PHI for research purposes does not require a business associate contract; the Privacy Rule requires a covered entity only to enter into a written contract with persons or businesses performing certain covered functions on their behalf that involve PHI (e.g., legal, actuarial, accounting, consulting, data aggregation, management, administrative, accreditation, financial services). Research is not usually one of those functions.[16]

3. Authorization

As mentioned, one of the three conditions under which PHI may be disclosed is when the disclosure has been authorized by the individual. A valid authorization under the Privacy Rule is a document that the individual signs, permitting the covered entity to use or disclose the individual's PHI for specified purposes and to specified recipients. The authorization must also be written in plain language.[17]

The following are considered to be the core elements of a valid authorization under the Privacy Rule:[18]

- A specific identification of the PHI to be used or disclosed.
- The identification of all parties who are authorized to use or disclose the PHI.
- The identification of all parties who receive the PHI from the covered entity.
- A description of each purpose of the requested use or disclosure.
- The expiration date of the authorization or the expiration event. An authorization for research purposes does not need a fixed expiration date or to state a specific expiration event. For research purposes, the document may state "none" or "the end of the research project."
- Signature of the individual and date. If a legally authorized representative signs the authorization, a description of the representative's authority to act for the individual must also be provided.

In addition, the authorization is required to provide a statement of the individual's right to revoke the authorization, instructions on how to do so, and, if applicable, the exceptions to the right to revoke the authorization. Furthermore, the authorization must provide a statement regarding whether treatment, payment, enrollment, or eligibility of benefits can be conditioned on authorization, as well as any consequences of refusing to sign the authorization (such conditional activities are typically permitted under HIPAA when associated with clinical research). For example, because HIPAA allows patients to access and review their medical records, this activity could compromise the results of a clinical investigation. Therefore, researchers may include language in the authorization that conditions treatment in an investigation upon subjects waiving their right to access their records during the ongoing investigation. Finally, the authorization is required to provide a statement regarding the potential risk that PHI may be redisclosed by the recipient, and if so, that the Privacy Rule may no longer protect that redisclosed information.[19]

For the use of a deceased individual's PHI in research, covered entities are not required to obtain an authorization, waiver, or a data-use agreement from the personal representative or next of kin. However, the covered entity must obtain the following from the researcher seeking access to this PHI: (1) an oral or written statement that the use and disclosure of the decedent's PHI is solely for research purposes, (2) an oral or written statement that the PHI is necessary, and (3) documentation of the death.[20]

Researchers typically are familiar with a document known as a *consent form* or *informed consent*. An informed consent differs from a Privacy Rule authorization in that the informed consent, among other things, provides research subjects with a general description of the study, the study's possible risks and benefits, and the manner in which the confidentiality of information will be protected. Even though an authorization also provides research subjects with a description of how, why, and to whom certain information will be used and disclosed for research, the authorization requires that the covered entity specifically name all persons who may view the PHI (the informed consent does not require such specificity). In addition, the authorization may contain a specific expiration date or event.

An authorization may be combined with an informed consent document or other permission to participate in research; however, to ensure that Privacy Rule requirements are met for such combined documents, the authorization information tends to be identifiably separate from other sections of the document. The signed authorization must be retained by the covered entity for 6years from the date of creation or the date it was last in effect, whichever is later.

4. Waiver or Alteration of Authorization Requirements

The second condition under which PHI may be used or disclosed is if the use or disclosure has been obtained under a waiver of authorization. In some instances, obtaining an authorization for research purposes may be difficult or unfeasible (e.g., contact information is unknown for several hundred research subjects). Under such circumstances, an IRB or a Privacy Board may approve a waiver or alteration of authorization for the use and disclosure of PHI.

As mentioned, an IRB is body that typically reviews research protocols, informed consent documents, and related materials to ensure protection of the rights and welfare of human subjects in research. A Privacy Board is a review body that is created under the Privacy Rule. It is established to act solely on requests for a waiver or an alteration of the authorization requirement under the Privacy Rule for a particular research study. Privacy Boards are not involved in creating authorization forms and do not monitor uses and disclosures of PHI made pursuant to an authorization.

A Privacy Board must meet certain membership requirements, such as (a) members must have varying backgrounds and appropriate professional competency to review the effect of the research protocol on the individual's privacy rights and related interests; (b) there must be at least one member who is not affiliated with the covered entity, at least one member who is not affiliated with any entity conducting or sponsoring the research, and at least one member who is not related to any persons affiliated with any of such entities; and (c) the board must not have any member participating in a review of any project in which the member has a conflict of interest.[21] These requirements are not necessarily the same as those for IRB membership under other relevant federal regulations.

A Privacy Board or IRB may waive or alter all or part of the authorization requirements. The waiver or alteration must meet the following criteria: (a) the use or disclosure of PHI involves no more than a minimal risk to individual privacy based on an adequate plan to protect the PHI from improper use and disclosure, there is an adequate plan to destroy the PHI at the earliest opportunity consistent with conduct of the research, and there are adequate written assurances that PHI will not be reused or disclosed to any other person or entity, except as otherwise permitted under the Privacy Rule; (b) the research could not practicably be conducted without the waiver or alteration; and (c) the research could not be conducted without access to and use of the PHI.[22]

A covered entity may use and disclose PHI once it receives proper documentation of waiver/alteration approval from the Privacy Board or IRB.[23] Documentation must be retained by the covered entity for 6years from the date of its creation or the date it was last in effect, whichever is later.[24] When IRBs and Privacy Boards coexist, the Privacy Rule does not require approval of a waiver or an alteration of authorization by both bodies; a covered entity may rely on a waiver or an alteration of authorization approved by any appropriate IRB or Privacy Board.[25]

5. Limited Data Sets and Data-Use Agreement

The third condition under which PHI may be used or disclosed by a covered entity is if certain PHI has been provided through the use of a limited data set.

A limited data set is information stripped of certain direct identifiers (e.g., name, address) but may include other data not considered to be direct identifiers (e.g., city, state, elements of a date). A covered entity may use or disclose only a limited data set for research, public health, or health care operations purposes.[26]

In addition, a limited data set may be used or disclosed by a covered entity only if the covered entity and the recipient of the data set enter into a data-use agreement.[27] A data-use agreement provides satisfactory assurances that the recipient of the limited data set only will use or disclose the enclosed PHI for the purposes mentioned in the document. Any person, including an employee or other member of a covered entity's workforce, requesting a limited data set from a covered entity must enter into a data-use agreement.[28]

If the covered entity providing the limited data set knows of recipient activities that violate the data-use agreement, the covered entity must take reasonable steps to correct the inappropriate activities. If the covered entity has been unsuccessful in correcting those activities, it must discontinue disclosure of PHI to the recipient and notify HHS.[29] Many researchers are already familiar with data-use agreements, as such documents are required to access federal, population-based health utilization, and morbidity and mortality data.

6. De-Identifying Protected Health Information

If a person has not received an authorization, waiver of authorization, or a limited data set with regard to PHI, the covered entity may use or disclose only health information that is de-identified. In other words, PHI that has been stripped of certain identifiers is no longer considered PHI. Covered entities intending to use or disclose this health information must ensure that the de-identification has occurred through one of two ways: removal of 18 specified pieces of information from each record, or statistical verification by a qualified expert.[30]

De-identifying data via the first method of information removal requires the elimination of 18 elements that may identify the subject or the subject's relatives, employers, or household members. Moreover, the remaining information cannot be used alone or in combination with other information to identify the individual who is the subject of the information. *All* of the following 18 elements must be removed in order to de-identify data under the Privacy Rule:[31]

- Names.
- All geographic subdivisions smaller than a state (e.g., street address, city, county, precinct, zip code, equivalent geographical codes under specified conditions).
- All elements of dates that directly relate to an individual (e.g., elements of a birth date, admission date, discharge date, date of death), and all ages over 89 and elements of dates indicative of such age, unless these ages and elements are aggregated into a single category of *age 90 or older*.

- Telephone numbers.
- Facsimile numbers.
- E-mail addresses.
- Social security numbers.
- Medical record numbers.
- Health plan beneficiary numbers.
- Account numbers.
- Certificate/license numbers.
- Vehicle identifiers and serial numbers, including license plate numbers.
- Device identifiers and serial numbers.
- Web site addresses.
- Internet protocol address numbers.
- Biometric identifiers, including fingerprints and voiceprints.
- Full-face photographic images and any comparable images.
- Any other unique identifying number, characteristic, or code, unless otherwise permitted by the Privacy Rule for re-identification.

De-identifying data through statistical methods requires a qualified statistician or a person with appropriate knowledge of and experience with generally accepted statistical and scientific principles and methods for rendering information not individually identifiable. The qualified expert must provide certification that there is a very small risk that the de-identified information could individually identify the subject of the information. The qualified expert must also document the methodology and analytic results used to assess the risk.[32] A covered entity is required to retain this certification, in written or electronic format, for at least 6years from the date of its creation or the date when it was last in effect, whichever is later.[33] The covered entity also may retain a code for re-identification of the de-identified information; however, the code by itself cannot be able to identify the individual, and the covered entity cannot generally disclose how to use the code for re-identification.[34]

Unfortunately, de-identified data sets provide information that many consider useless for various research purposes. For example, because age is a risk factor for many diseases, a study regarding possible treatment for a disease may need to account for the age of study subjects—either through study sample eligibility criteria or by using age as a variable in a statistical test. If components of birth dates or death dates are removed from data sets, it may be impossible to account for the effects of age on treatment outcomes. In addition, researchers may need to verify whether multiple or questionable entries are present in a data set. For example, if there are similar data regarding a "Jean Williams" and a "Jean M. Williams" or if there are data that do not make sense for a particular person (a hysterectomy procedure for Sam Jones, who, on further review, could be Samantha Jones), researchers need to determine whether these data correspond to the same person. One of the few ways to make such verifications is to run tests of the data through checks of individually identifiable information (e.g., date of birth, address, date of admission). Without the ability to do this, it is difficult to ensure the validity of study results.

7. The Effect of HIPAA on Clinical Research

7.1. FDA-Regulated Products and Activities

When clinical research is conducted in support of activities related to any FDA-regulated drugs and devices, the agency requires certain reporting and record-keeping activities (e.g., product tracking, postmarketing surveillance, adverse-event reporting) that help to ensure the quality, safety or effectiveness of the regulated product. Under the Privacy Rule, a covered entity may use or disclose PHI to a public health authority that is authorized by law to collect or receive the information for purposes of preventing or controlling disease, injury, or disability. Providing information to the FDA as required under agency regulations constitutes a permissible disclosure of PHI to a public health authority.[35]

7.2. Activities Preparatory to Research

The Privacy Rule allows covered entities to use and disclose PHI to researchers for certain activities preparatory to research. Such activities include drafting research protocols and identifying prospective research participants. Under these circumstances, PHI must be examined onsite.[36] Furthermore, under these circumstances, the researcher may identify, but not contact, potential study participants. Authorization or waiver of authorization is needed to contact the potential participants. Researchers who are covered entities may discuss various treatment options, which may include the option of enrolling in a clinical trial, with their patients without an authorization or waiver,[37] although any screening interviews or other preparatory activities that result in recording PHI before conducting the formal study must take place after securing an authorization or waiver for this activity.

7.3. Tracking Study Subjects Who Are Lost to Follow-Up

A covered entity may use or disclose PHI in order to determine the location of research participants during an ongoing investigation. For example, the entity may disclose information needed to locate those who are lost to follow-up, as long as the use or disclosure of the PHI is covered in each research participant's authorization.

7.4. Existing Data Banks or Repositories Maintained by Covered Entities

Access to a database of health information, including archived medical records and disease registries, may be an essential element of various types of research. Covered entities may collect and enter into a database any treatment, payment, and health care operations data that contain their patients' PHI without authorization.[38] However, any future research requiring use of the database will require an authorization or waiver.[39]

7.5. Reactions to the Privacy Rule

The Association of American Medical Colleges (AAMC) reported in July 2004 that researchers have been increasingly frustrated with Privacy Rule requirements.[40] Specifically, some researchers contend that the additional documentation required by HIPAA has resulted in an arduous process that makes it difficult to recruit subjects. In addition, according to the AAMC, some covered entities have exhibited overly cautious behavior, resulting in the protracted release—or no release—of PHI. Thus, particularly for those studying diseases that are rapidly fatal, some believe that the HIPAA Privacy Rule has irreparably harmed the research world.[41]

Indeed, the September 2004 US Government Accountability Office report regarding the Privacy Rule's implementation found that researchers believed it was increasingly difficult to obtain patient data; however, the report also stated that many of the difficulties cited by researchers can be attributed to misunderstandings or confusion about how to interpret the Privacy Rule in conjunction with other federal requirements.[42] Still, the report noted that researchers contend that ambiguity remains in determining whether a health survey activity is considered a health care operation or research and whether a public health entity's data request is part of its public health activities or is for research.[43] Although the report made no recommendations on how to alleviate these misunderstandings and ambiguities, understanding basic Privacy Rule requirements may help resolve some of the confusion and frustration. An education in the basics may be the passport to an even more efficient implementation.

8. Conclusion

The Privacy Rule provides criteria for protecting PHI held by a covered entity or its business associates. These criteria include certain minimum standards for the use and disclosure of PHI for research purposes. Although researchers who are not subject to the Privacy Rule (noncovered entities) may view the rule as an impediment to accessing information, the Privacy Rule was not intended to be such. Instead, the rule is intended to be a vehicle for ensuring additional privacy safeguards when accessing certain information for research. Because many clinical researchers must also comply with other federal and state regulations that protect human subjects, a properly implemented Privacy Rule should provide a familiar rubric from which to secure information privacy. In this manner, the protection of research subjects from foreseeable research risks, including risks to privacy, may be maintained.

Footnotes

1. US Government Accountability Office, Health Information: First-Year Experiences under the Federal Privacy Rule (2004).
2. 65 Fed. Reg. 82,462 (Dec. 28, 2000).
3. *Id.* Please also note that small health plans were required to be compliant as of April 14, 2004.
4. 45 C.F.R. § 160.103; *See also* M Frank-Stromborg, *They're Real and They're Here: The New Federally Regulated Privacy Rules Under HIPAA*, 16 DERMATOL NURS 13-14, 17-18, 22-25 (2004).
5. US Department of Health and Human Services, Clinical Research and the HIPAA Privacy Rule, 1-2 (2004)[hereinafter HHS Report on Clinical Research and HIPAA]; *See also* 45 C.F.R. § 160.102.
6. 45 C.F.R. §§ 164.508, 164.512, 164.514(e).
7. *Id.* § 164.502(b).
8. US Department of Health and Human Services, Protecting Personal Health Information in Research: Understanding the HIPAA Privacy Rule 8 (2004) [hereinafter HHS Report on Privacy Rule].
9. 45 C.F.R. Part 46, Subpart A.
10. 21 C.F.R. Parts 50 and 56.
11. 45 C.F.R. § 46.111.
12. 21 C.F.R. Parts 50 and 56.
13. *Id.*, Part 56.
14. *Id.* § 56.124.
15. 45 C.F.R. §§ 164.502(e), 164.504(e).
16. *Id.* § 160.103.
17. *Id.* § 164.508.
18. *Id.* § 164.508(c)(1).
19. *Id.* § 164.508(c)(2).
20. *Id.* § 164.512(i)(1)(iii).
21. *Id.* § 164.512(i)(1)(i)(B).
22. *Id.* § 164.512(i)(2)(ii).
23. *Id.* § 164.512(i)(2).
24. HHS Report on Privacy Rule, *supra* note 8, at 13.
25. 45 C.F.R. § 164.512(k)(1)(i); *see also* 65 Fed. Reg. 82,692 (Dec. 28, 2000); 67 Fed. Reg. 53,232 (Aug. 14, 2002).
26. 45 C.F.R. § 164.514(e)(3).
27. *Id.* § 164.514(e).
28. HHS Report on Privacy Rule, *supra* note 8, at 16.
29. 45 C.F.R. § 164.514(e)(4)(iii)
30. *Id.* § 164.514(b).
31. *Id.* § 164.514(b)(2).
32. *Id.* § 164.514(b)(1).
33. HHS Report on Privacy Rule, *supra* note 8, at 10.
34. 45 C.F.R. § 164.514(c).

35. *Id*. § 164.512(b)(1)(iii).
36. *Id*. § 164.512(ii).
37. HHS Report on Clinical Research and HIPAA, *supra* note 5, at 4.
38. US Department of Health and Human Services, Research Repositories, Databases, and the HIPAA Privacy Rule 2 (2004).
39. HHS Report on Privacy Rule, *supra* note 8, at 11.
40. S. Santana, *Researchers Criticize New HIPAA Regulations*, AAMC Reporter, July 2004.
41. Sabin Russell, *Medical Privacy Law Said To Be Chilling Cancer Studies; Scientists Fight For Fast Access To Patient Files*, San Francisco Chronicle, (September 26, 2004) available at: http://sfgate.com/cgi-bin/article.cgi?file=/c/a/2004/09/26/PRIVACY.TMP
42. US Government Accountability Office, Health Information: First-Year Experiences under the Federal Privacy Rule 1, 13, 16 (2004).
43. *Id*. at 17.

8

Overview of Medicare Coverage of Clinical Trials

Julie K. Taitsman

1. Introduction

The Social Security Act establishes several criteria that must be satisfied in order for an item or service to be eligible for reimbursement under the Medicare program. These threshold Medicare coverage criteria include the requirement that the item or service be *reasonable and necessary* for the diagnosis or treatment of an illness or injury or to improve the functioning of a malformed body member. Administrators of the Medicare program have generally held that experimental therapies are not reasonable and necessary. This interpretation of the reasonable and necessary criteria traditionally precluded Medicare coverage for services provided to beneficiaries as part of clinical investigations. Denying Medicare reimbursement for investigational interventions and costs related thereto created a barrier to Medicare beneficiaries' participation in clinical trials. This posed a dual problem. First, in the short term, it could deny Medicare beneficiaries access to potentially useful therapies available in the clinical trial context. Second, limited enrollment of Medicare beneficiaries in clinical trials hindered collection of sufficient evidence to demonstrate a new therapy's effectiveness for use in the Medicare population.

Removing barriers to beneficiaries' participation in clinical trials recently became a pronounced goal of the Medicare program. Two distinct policies were developed to allow the Medicare program to cover certain costs for Medicare beneficiaries' participation in select clinical trials. The first policy allowed for coverage of the full scope of items and services within a narrow class of clinical trials, whereas the second policy offered broader coverage as to the type of clinical trials to which it would apply but offered more limited coverage regarding the scope of reimbursable items and services provided in the context of those trials. The older of these policies arose from an interagency agreement between the Food and Drug Administration (FDA) and the Health Care Financing Administra-

From: *Clinical Evaluation of Medical Devices: Principles and Case Studies, Second Edition*
Edited by: K. M. Becker and J. J. Whyte © Humana Press Inc., Totowa, NJ

tion (HCFA; the precursor agency to the Centers for Medicare and Medicaid Services [CMS]) in 1995. That policy authorized Medicare coverage for use of Category B devices in trials conducted pursuant to an FDA-approved investigational device exemption (IDE). The more recent policy evolved from an executive order issued in 2000 and allows for coverage of routine costs incurred incident to certain federally sponsored trials and certain drug trials.

It is important to reiterate that neither of these coverage policies serves to guarantee Medicare coverage for a particular clinical trial or a particular investigational item or service. Instead, the policies simply eliminate the experimental nature of the items or services as sufficient reason for denying coverage. Under these two policies, the investigational nature neither precludes nor guarantees coverage. Investigational items or services may only receive Medicare reimbursement under either of these clinical trial policies if the items and services meet all other criteria for Medicare coverage.

2. Medicare Coverage for Routine Costs

On June 7, 2000, President William J. Clinton issued an executive order authorizing Medicare coverage for routine costs associated with Medicare beneficiaries' participation in clinical trials. CMS promulgated a national coverage decision and issued a program memorandum to implement this policy, which became effective for services furnished on or after September 19, 2000.[1]

Medicare coverage of routine costs extends only to certain types of clinical trials that show real promise of giving rise to improved health outcomes for Medicare beneficiaries. Covered trials must be scientifically sound in terms of hypothesis, trial design, and implementation. Potentially covered trials must evaluate a diagnostic test or therapeutic intervention that—if proven safe and effective—is potentially reimbursable under the Medicare program. Studies involving services that are statutorily excluded from coverage (e.g., cosmetic surgery) or that do not fit in to an established Medicare benefit category are also not eligible for coverage under the routine costs policy.

CMS policy outlines three fundamental requirements that trials must satisfy to qualify for coverage:

1. The trial must evaluate an item or service that falls within a Medicare benefit category and is not statutorily excluded from coverage.
2. The trial must have a therapeutic intent and must not be designed solely to test toxicity or disease pathophysiology.
3. Trials of therapeutic interventions must enroll patients with diagnosed disease rather than healthy volunteers. Trials of diagnostic interventions may enroll healthy patients in order to establish a proper control group.

In addition to the three basic requirements listed above, eligible trials must also satisfy various indicia of scientific validity. There must be a legitimate scientific purpose for the study, as it must seek to answer an open scientific question that is relevant to the welfare of Medicare beneficiaries. The trial must also be designed to gather reliable data that may advance evidence gathering and analysis necessary to yield an answer to the underlying scientific inquiry. CMS has enumerated the following factors as characteristics that potentially covered trials must incorporate:

1. The principal purpose of the trial must be to test whether the intervention potentially improves the participants' health outcomes.
2. The trial must be well supported by available scientific and medical information or it must intend to clarify or establish the health outcomes of interventions already in common clinical use.
3. The trial must not unjustifiably duplicate existing studies.
4. The trial design must be appropriate to answer the research question it asks.
5. The trial must be sponsored by a credible organization or individual capable of executing the proposed trial successfully.
6. The trial must be in compliance with federal regulations relating to the protection of human subjects.
7. All aspects of the trial must be conducted according to the appropriate standards of scientific integrity.

CMS intends to implement a process for prospective trial sponsors to certify that the planned trial satisfies all of the above basic requirements and desirable characteristics, although a process has not been formalized. CMS plans to establish a Medicare clinical trials registry and implement a procedure for enrolling certified trials in it. Until CMS establishes this self-certification procedure, the only trials that are eligible for routine cost coverage are those that are deemed to be automatically qualified. Trials are deemed to qualify based on sponsorship of various respected federal agencies, FDA approval, or certain drug trials that are exempt from the requirement of FDA approval. CMS has established four categories of trials that are presumed to meet the desirable characteristics and are eligible for coverage under the routine costs policy:

• Trials funded by the National Institutes of Health, Centers for Disease Control and Prevention, Agency for Healthcare Research and Quality, CMS, Department of Defense, and Veterans Administration.
• Trials supported by centers or cooperative groups that are funded by the National Institutes of Health, Centers for Disease Control and Prevention, Agency for Healthcare Research and Quality, CMS, Department of Defense, and Veterans Administration.

- Trials conducted under an investigational new drug (IND) application reviewed by the FDA.
- IND-exempt drug trials.

Only sponsors of drug trials conducted pursuant to an IND application and IND-exempt drug trials are required to identify themselves to CMS before submitting claims for Medicare coverage of routine costs related to the trial. This notification may be accomplished simply by sending an e-mail to a specially designated address at CMS.

At present, only trials that are in one of four categories of deemed trials are eligible for coverage under the routine costs policy, although this will change in the near future. Besides CMS' plans to establish an algorithm for nondeemed trials to qualify for coverage of routine costs, Congress has added a fifth category of trials that will be deemed eligible under the policy. As of January 1, 2005, Medicare coverage will extend to routine costs incurred in clinical trials for Category A devices. (The IDE process and Category A designation are discussed in Section 3.)

2.1. Services That Do Not Qualify As Routine Costs

To fully appreciate the scope of services potentially covered under the routine costs coverage policy, it is instructive to first consider what the policy does and does not cover. Coverage under the routine costs policy specifically excludes the investigational item or service itself. This exclusion is fairly straightforward and is a primary distinguishing feature of this coverage policy. For example, a drug or biologic under investigation of a pharmaceutical trial or an investigation device in a device trial would not be eligible for coverage. The scope of coverage under the routine costs policy differs substantially from the coverage afforded under the Category B IDE policy, which allows Medicare coverage for the investigational device itself.

Besides the investigational item or service, the policy also explicitly precludes coverage for two other types of costs. The Medicare program will not cover items and services that do not serve a clinical management function but are provided solely to satisfy data collection and analysis demands of the trial. This means that heightened patient monitoring or additional laboratory studies conducted following an investigational intervention might qualify for Medicare coverage if they were necessary to ensure patient safety but would not be eligible for coverage if their sole purpose were to collect data to serve the investigational purpose of the trial.

Finally, the policy provides that the Medicare program will not cover items and services that are customarily provided free of charge by the research sponsors. This means that if the trial sponsor generally offers a diagnostic test or a medical supply gratuitously to all trial enrollees, the sponsor may not seek payment from Medicare for providing such items and services to trial subjects who are Medicare beneficiaries.

2.2. Services That Qualify As Routine Costs

Despite the above exclusions, the policy liberally defines the nature of routine costs that may be eligible for Medicare coverage. Routine costs include conventional medical care such as items and services that would be provided to the beneficiary regardless of participation in a clinical trial. Treatment of complications arising from participation in the trial is also covered under the routine costs policy. Although the policy does not allow direct coverage of the investigational item or service itself, it rather liberally allows coverage for certain costs that are directly related to providing the investigational item or service, even costs that are solely incurred to provide the investigational item or service. For example, although an intravenously administered investigational drug in a trial cannot be covered, the cost of administration may be eligible for Medicare reimbursement.

CMS imposes two simple procedural requirements on providers seeking Medicare coverage for routine costs incurred incident to qualifying clinical trials. Claims must be submitted with the QV procedure code modifier to designate that the item or service was provided as routine care in a Medicare qualifying clinical trial. Claims must also include a secondary International Classification of Diseases, 9th Revision, Clinical Modification [ICD-9-CM] diagnosis code V70.5, which indicates *health examination of defined subpopulations*. By including the QV procedure code modifier and ICD-9-CM diagnosis code V70.5 on Medicare claims forms, the provider attests that the services or items were provided as routine costs to Medicare beneficiaries participating in qualifying clinical trials and that the services and items meet the requirements for coverage under CMS' routine costs coverage policy.

Use of the QV modifier and V70.5 diagnosis code informs the Medicare contractor tasked with processing claims that the services or items rendered represent routine costs of a qualifying clinical trial. Contractors may then determine whether the items and services are covered in accordance with national Medicare policy. When coverage is left to contractor discretion, the contractor may determine whether or not coverage is appropriate regardless of the investigational nature of the delivery of care. If the items and services are covered, reimbursement is based on the same fee schedules and payment methodologies generally applicable outside of the clinical trial context. Items and services are either covered or not covered, and there is no reimbursement differential based on the investigational context. Thus, although in some circumstances a contractor may use discretion to deny coverage entirely, if the contractor decides to cover the item or service, it must abide by any fee schedules and payment criteria. It is not within the contractor's discretion to decide to cover an investigational item or service but reduce the payment level because of the investigational nature of the care provided.

3. Medicare Coverage of Category B Investigational Devices

In 1995, the FDA and HCFA promulgated an interagency agreement regarding reimbursement categorization of investigational devices.[2] Although somewhat less well recognized than its eponymous authority over food and drugs, the FDA also regulates medical devices. Despite the FDA's jurisdiction over medical devices commanding less attention from the public than the agency's oversight authority for food and drugs, regulation of medical devices remains an important part of the FDA's mission and a compelling public health concern. Similar to its more publicized oversight capacity for drugs, the FDA must ensure that medical devices marketed for use in the United States are both safe and effective.

The federal Food, Drug, and Cosmetic Act gives the FDA authority to determine which medical devices can be lawfully marketed in the United States. The FDA requires some types of medical devices to undergo clinical testing to establish safety and effectiveness before granting approval to market the device commercially. Because these unapproved devices have not yet been cleared for marketing, investigators hoping to use the unapproved device in the context of a clinical trial must first obtain permission from the FDA for an IDE. An IDE allows research sponsors to transport the unapproved devices without violating the federal Food, Drug, and Cosmetic Act and use the device in human subjects participating in the approved clinical trial. Prospective device trial sponsors must submit an application to conduct a trial under an IDE.

Before 1995, the FDA simply granted or denied these IDE applications. The interagency agreement added an additional categorization step to the FDA's IDE-approval process. For all approved IDE applications, the FDA now determines whether the device, for its proposed use in the particular trial, falls within Category A or Category B based on the level of risk to patients. Designation as Category A means that the device is both experimental and investigational. The initial questions of safety and efficacy of the device type have not yet been resolved and assignment of the IDE into Category A means that the FDA is uncertain whether the device is safe and effective for the intended use. Designation as Category B means the device is investigational but not experimental for its intended use. This means that the underlying questions of safety and effectiveness of the device type have been resolved and the FDA has determined that the device type can be both safe and effective. An IDE may be assigned to Category B if it involves an inherently low risk device. IDEs involving high-risk devices may also be assigned to Category B if the trial involves a device that is only slightly modified from a device that has been proven safe and effective or an application of the device that is similar to an established FDA-approved use.

Devices used pursuant to an FDA-approved IDE that are classified as Category B are eligible for full Medicare coverage. Unlike the limited Medicare coverage afforded by the routine costs policy, the Category B IDE policy authorizes Medicare coverage for the investigational item itself as well as for routine costs associated with the clinical trial. The investigational nature of Category B devices is neither sufficient reason to grant nor deny Medicare coverage. Such devices are covered only if they meet all other Medicare coverage requirements and coverage is not precluded by a national noncoverage policy. Although a Medicare contractor may deny coverage for several reasons, it cannot refuse to cover routine costs because of the investigational nature of the use of the device.

4. Impact of Clinical Trials Coverage Policies

Medicare's dual policies on coverage for clinical trials serve an important role in overcoming historical barriers to beneficiaries' access to clinical trials. In the short term, expanded coverage helps Medicare beneficiaries obtain potentially useful investigational therapies that may be available only in the clinical trial context. In the long term, coverage helps researchers gather sufficient reliable data on therapeutic and diagnostic interventions to yield generalizable scientific knowledge and determine whether their use improves health outcomes in the Medicare population. Facilitating Medicare beneficiaries' participation in clinical investigations offers substantial value not only to individual Medicare beneficiaries but also to the program as a whole.

References

1. *See* "Medicare Coverage—Clinical Trials: Final National Coverage Decision," available at http://www.cms.hhs.gov/coverage/8d2.asp and "Medicare Coverage— Clinical Trials: Program Memorandum," available at http://www. cms.hhs.gov/coverage/8d3.asp.
2. *See* "Implementation of the FDA/HCFA Interagency Agreement Regarding Reimbursement of Investigational Devices," September 15, 1995 (D95-2). Available at http://www.fda.gov/cdrh/d952.html.

9

Developing Drug–Device Combination Products With Unapproved Components

Guy Chamberland

1. Introduction

The development of combination products (CPs) is a challenging field that faces complex product jurisdiction and development issues. Despite these issues, companies are driven by the potential therapeutic success of combining two components into a single product. The driving force behind this new enthusiasm is the local delivery of drugs or biologics to a specific tissue, thereby minimizing the exposure of the whole body to potential toxic compounds as well as maintaining drug/biologic levels at a specific site within a tissue to obtain a desired therapeutic response. As a recent example, Cordis, a subsidiary of Johnson & Johnson, successfully commercialized the CYPHER™ Sirolimus-Eluting Coronary Stent, bringing cardiologists a promising new therapy for the prevention of restenosis. The CYPHER Sirolimus-Eluting Coronary Stent consists of a device that brings a drug component to the coronary tissue to maintain vessel patency by minimizing the occurrence of restenosis following stent implantation. The device component provides mechanical prevention of restenosis, whereas the drug component delivers a pharmacologic/biologic locally to assist in the vessel's recovery from any injury caused by the stent implantation. Cordis developed this product with both speed and regulatory finesse. The dramatic clinical trial results indicate the CYPHER Sirolimus-Eluting Coronary Stent is a very promising new therapy.[1]

This success should not mislead one to conclude that developing CPs is without challenges. In the case of the CYPHER Sirolimus-Eluting Coronary Stent, its development was relatively simple because both components, the bare stent and drug, were previously approved by the Food and Drug Administration (FDA). Before undertaking the development of a drug–device CP, the com-

From: *Clinical Evaluation of Medical Devices: Principles and Case Studies, Second Edition*
Edited by: K. M. Becker and J. J. Whyte © Humana Press Inc., Totowa, NJ

pany and/or investors should assess the risks involved in bringing such a product to market. These risks include delays resulting from regulatory requirements or ambiguities, toxicological risks of systemic exposure to a product intended for local delivery, and the increased scientific complexity associated with optimizing the performance of a product comprising both a medical device and a drug. A recent article by Nancy Patterson points to the increasing business risk associated with the development of CPs that consist of an increasing number of unapproved components.[2] The development of a drug or device as a single entity involves multiple risks, each of which can lead to the termination of product development. Therefore, a question that must be addressed is whether developing a drug–device CP that is composed of two unapproved components is a viable project. From a medical point of view, it certainly is; however, economically, the development of such a product is risky. This chapter discusses the complexity of developing a drug–device CP with unapproved single components, including the differing requirements in the United States, European Union, and Canada. Table 1 provides a summary of these requirements. Additionally, this chapter uses a drug-eluting stent (DES) with an unapproved drug component as a case study, because the regulatory pathways for these products are well established.

2. Determining Classification or Designation

Several informative articles have recently been published regarding the classification or designation of CPs.[3–5] The classification or designation of a CP should be undertaken as early as possible, because it is a requisite step in the development plan and regulatory strategy. The company must consult all applicable laws, regulations, guidance documents, agreements, and precedents. Classification or designation issues should also be discussed internally and/or with consultants before contacting the regulatory agency or submitting a request for classification or designation. In addition, the regulatory pathway of similar products should be considered when available. As more products are approved and the regulatory pathways become clearer, a classification or designation request to the regulatory agency will no longer be needed, although a submission should be considered if the product has unique characteristics. Discussions concerning the designation of a product in accordance with 21 CFR Part 32(e) need to account for the implications of that decision on the regulatory pathway. The designation of the product should not be decided prematurely, because the consequences of all pontential regulatory pathways need to be weighed and considered, such as Orphan Drug designation, premarket notification (501(k)), Quality System Regulation (QRS)[6] vs current Good Manufacturing Practice[7] (cGMP), and reimbursement issues. For example, a drug–device CP with a single premarket approval application (PMA) that is

Table 1

Combination Products: Regulatory Requirements and Directives

Country	Regulatory requirements	Definition
United States	Federal Food, Drug and Cosmetic (FD&C) Act, Section 201 [321]	Drugs, devices, and biological products are defined individually within the (FD&C) Act. Prior to 1991, the relatively small number of combination products (CPs) were regulated on an ad hoc basis.
	21 CFR Part 3 2(e)	A CP is comprised of two or more regulated components (i.e., drug/device, biologic/device, drug/biologic, drug/device/biologic) that are physically, chemically, or otherwise combined or mixed and produced as a single entity or when packaged separately, are CPs through labeling for use with the other component.
	Intercenter Agreements	The Intercenter Agreements are working agreements developed between the Food and Drug Administration (FDA) Centers (i.e., Biologics/ Drugs/Devices, Drugs/Biologics), which outline certain categories of products and how these products have been regulated by the FDA.
European Union	The Council of European Communities. Council Directives 93/42/EEC (Medical Device Directive [MDD]), 90/385/ EEC (Active Implantable Medical Device Directive [AIMDD]) and 2001/83/EEC (Medicinal Products Directive [MPD]) as amended.	Drugs and biological products are defined individually within Council Directive 2001/83/EEC. Devices are defined individually within Council Directives 93/42/EEC and 90/385/EEC.
	Medical Devices Guidance Document MEDDEV 2. 1/3, rev 2	A CP is regulated either by the MDD/AIMDD or MPD. This guidance document provides the criteria for demarcation between MPD and MDD/AIMDD.
Canada	Food and Drugs Act (R.S. 1985, c.F-27)	Drugs, devices, and biological products are defined individually within the Foods and Drugs Act. Before 1999, drug–device CP had to fulfill the requirements of both the Food and Drug regulations and the Medical Device regulations.
	Drug/Medical Device Combination Products Policy	A CP is a product that combines a drug and a device component, such that the distinctive nature of the drug component and device component is integrated in a singular product.

regulated as a medical device would not be entitled to an orphan drug designation,[8] even though the Orphan Drug Act applies to both drugs and devices. Orphan drug designation[8] applies to products regulated as drugs or biologics via a new drug application (NDA) or product license application, whereas products regulated as medical devices via a PMA or 510(k) submission may apply for humanitarian use device[9] status. Orphan drug designation does not apply to the device component but could be a major advantage for the drug component and ultimately the CP. Incorrect classification can lead to wasted time and resources, refusal of an orphan drug designation, and ultimately delayed approvals resulting in increased time to market, making subsequent rounds of financing more difficult.

2.1. US Designation Process and Critical Issues for the Combination Product

In the United States, the Office of Combination Products assigns the review responsibility for a CP based on its intended use, its primary mode of action,[10] and the FDA Intercenter Agreements.[11] These Intercenter Agreements provide guidance and identify products subject to regulation as devices and those subject to regulation as drugs, using a process that has been presented in many publications.[3,12,13] Assignment of the lead center for premarket review and regulation is determined based on the primary mode of action of the CP.[10] For some products, the designation process is simple, whereas for others, the decision is not as clear. A comparison of the impact of the designation is critical to understanding some of the issues resulting from the designation decision and its outcome. For many companies, obtaining a designation of primary review by the Center for Devices and Radiological Health (CDRH) as a medical device is favorable because fewer clinical trials are required for a PMA than an NDA. Although this is generally true, CDRH would require additional trials if for certain reasons establishing an effective dose was an issue with the CP. Thus, well-designed clinical trials are essential for a successful PMA, which may include only one or two clinical trials. The designation of two CPs that each has the same drug component are compared based on their primary mode of action. The first is a DES with a radioactive pharmaceutical as the drug component that is used for the prevention of restenosis, and the second product is a drug-coated occlusion coil (DCOC) used to heal aneurysms following embolization. For each of these products, it is important to determine the role of the device. For the DES, the drug and device components are "married" owing to the elution profile of the drug from the stent and the therapeutic properties of the stent itself (e.g., prevention of restenosis). For the DCOC comparator, the drug and device components are not "married," because the drug's physicochemical properties allow it to freely bind to the surface of platinum metal, thus any platinum alloy coil could be used as the device component.

From the drug point of view, the role of the device in both cases is simply to facilitate the delivery of the drug to the therapeutic site. The DES' primary role is attributed to the stent, because it is a permanent scaffold that maintains the opening of the coronary artery. The secondary role of the DES comes from the drug, as it acts to prevent restenosis induced by the angioplasty and stent. Changes (e.g., polymer coating, drug) to the stent will directly influence the degree of prevention of restenosis, because the outcome is influenced by inflammatory reactions. The term *new chemical entity* (NCE) will be used to refer to an investigational new drug. If an NCE is added to the stent to create the DES, the question that needs to be addressed is whether the experience with the stent outweigh the new safety issues raised by the NCE. These are the toxicity issues that arise from systemic exposure of the NCE and not the local toxicity issues associated with any DES. For the DCOC, the primary role is also attributed to the device, because the occlusion coil immediately induces embolization after implantation and acts as a permanent implant to maintain the plug. The secondary role is that of a drug, which facilitates the healing process of the aneurysm once the clot (embolization) has been formed. There are no critical device issues to review if a coating is not used to bind the drug to the platinum metal and if the physical properties of the coil (e.g., thrombogenicity) and delivery system are maintained. Another important question to consider for NCEs with a new therapeutic action or claim is whether the drug component has become more critical to the premarket approval process. Section VII.A.2 of the Intercenter Agreement[11] between the Center for Drug Evaluation and Research (CDER) and the CDRH states that a "Device incorporating a drug component with the CP having the primary intended purpose of fulfilling a device function" is a CP and market approval authority will be granted to CDRH using device authorities. In addition, it states that intercenter consultation will be required if "a drug has not been legally marketed in the USA as a human drug for the intended effect." Section VIII also needs to be consulted as it outlines the general criteria that CDRH and CDER will apply to reach a verdict regarding the designation. Device Criteria VIII.A.5 states that "A device containing a drug substance as a component with the primary purpose of the combination being to fulfill a device function is a CP and will be regulated as a device by CDRH." Both criteria VII.A.2 and VIII.A.5 support the decision to have CDRH regulate the DES and DCOC as devices. This holds true as long as the primary mode of action is associated with the device component of the CP. For the DES, this determination is positive, because CDRH is experienced with the premarket review of the bare device and has engineering expertise in this area, the stent directly influences the clinical outcome, and any changes to the stent or its delivery may affect restenosis. Although as the device component is a Class III medical device regulated under the PMA regulations, the advantages of a premarket notification (510(k)) application are not available. A major issue that may be encountered during the review of the CP is CDRH's limited experience with drug review; thus, the efficiency of the CP review team will be critical. Although CDER is not experienced with review of the bare device and does not have clinical experience with these types of products, their expertise lies in drug review, which will be an asset to the review team. For a product with an NCE, it will be important to obtain active participation from CDER reviewers for the chemistry, manufacturing and controls, and nonclinical safety issues. The DCOC's case is different, because the product's primary mode of action can arguably be a drug or device. Thus, it is important to look at the advantages and disadvantages of the designation verdict. CDRH is experienced with review of bare occlusion coil devices and has engineering expertise in this area. The properties of the occlusion coil are important for inducing embolization through thrombogenicity. On the other hand, CDRH has minor clinical expertise with occlusion coils, because many of these products are approved without clinical trial data through a premarket notification application. Occlusion coils are Class III medical devices that are regulated through premarket notification (510(k)).* The addition of a drug to this Class III medical device makes the DCOC a product that is regulated through the PMA regulatory pathway. A major issue that will be encountered during the review process is CDRH's inexperience with drug review; thus, the efficiency of the CP review team will be critical for a successful and timely filing, particularly with an NCE. In turn, CDER is not experienced with the review of this medical device and has no clinical expertise in the embolization of cerebral aneurysms. The DCOC's drug component has a major role, because it enhances the clinical performance properties of the medical device. As previously stated, the addition of a drug to this Class III medical device makes the DCOC a product that is regulated through the PMA regulatory pathway, removing the benefits of a 510(k) submission. The regulation of the product as a drug under the NDA pathway could be advantageous to the company, because ruptured aneurysms are considered a rare disease. The NDA pathway would allow designation of the drug component under the Orphan Drug Act. Thus, in cases in which the DCOC makes substantial drug claims, arguments should be made to designate the product with primary responsibility given to CDER.

3. Development of a Drug–Device Combination Product

From a scientific point of view and irrespective of the premarket review mechanism, a drug–device CP application should contain information regarding the safety and efficacy of the CP and each of its components. A company's

---

*Since the preparation of this chapter, these devices have been reclassified as Class II medical devices.

regulatory specialist needs to understand all of the preclinical issues that can influence the risk/benefit ratio before contacting a regulatory agency regarding the requirements for initiating a clinical trial and submitting a request for designation or classification. It is important to note that reviewers of any regulatory agency will use the information provided in a Pre-Investigational Device Exemption (IDE) or Pre-Investigational New Drug (IND) meeting package to assess the risk/benefit ratio of the CP. This information should include both in vitro and in vivo data for the drug component, results from the biocompatibility testing (e.g., ISO 10993)[14,15] for the CP and/or device component, physical testing (i.e., bench testing) demonstrating that the drug does not alter the properties of the device, chemistry, and manufacturing and controls information regarding the quality, purity, and identity of the drug component, sterility information on the device and its delivery system, and clinical information, when available. The ultimate issue for the regulatory agency is whether the potential efficacy of the CP outweighs the risks to the patient population. Companies tend to address this issue based solely on efficacy results from animal models that are also used to demonstrate an absence of local toxicity. Regulatory review teams desire additional information for CPs, because risk is not only associated with toxicity but can also be the result of manufacturing issues, such as sterility and impurities, or potential drug–device interactions. The following questions need to be addressed to help design a successful development program and prepare for discussions with a regulatory agency:

- What is the added medical benefit of the drug to the device?
- Does the CP achieve its goal or introduce new risks to the patient population? Some of these questions can be answered by the elution profile (kinetics) of the drug from the device over time.
- What is the systemic exposure that results from the elution?
- What is the justification for the dose selection?
- Is the toxicity of the drug known, and if not, how much nonclinical safety data does the company have to support the clinical trial in the intended patient population?
- If the drug component has an acceptable safety profile, should a phase 1 study in human volunteers be considered? If the drug is approved, one must assess the local effects owing to the greater local drug concentration and determine if the amount that elutes from the device can lead to toxicity to nontarget organs.
- Is the compound that binds the drug to the device toxic, and/or does it leach into the circulation over time?
- Does the polymer/binding agent change the efficacy of the approved device?
- Does the drug affect the properties of the device and/or its delivery system? Drug–device interaction studies are critical to address this issue.
- What are the impurities in the drug formulation, and is the formulation stable under the new use?

These issues are straightforward and should be addressed in the development program, because they will provide the agency review team with much needed information to assess the risk–benefit ratio of the CP. If a company does not have an experienced regulatory toxicologist or device specialist, it should employ an experienced person to consider these issues. Early discussions with the regulatory agencies regarding the preclinical development program should be limited, given the initial lack of information concerning the product's safety. Although early consultation with the drug division on specific drug development issues may sound appealing, the reviewers ultimately assigned to a product may differ from those initially consulted and any discussions regarding the nonclinical safety program to support an NDA are not binding.

The majority of the preclinical safety and efficacy issues should be addressed before submission of the pre-IDE or pre-IND information package. Moreover, discussions with a regulatory agency such as the FDA should focus on the clinical program and not on questions related to product development. The company should work with consultants who are experienced and qualified to assist them in the process and should not underestimate or delay addressing any safety issues voiced by the consultants. A lack of focus on rigorous project planning will only result in delays in commencing the clinical trial and ultimately increase time to market. Similarly, from a global perspective, when developing a CP, one should not underestimate the significance of safety issues identified by a reviewer, even if other countries do not share this concern. As more data is provided to the other agencies, they too may come to the same conclusion and thus delay the approval of the clinical trial.

3.1. Case Study: Drug-Eluting Stents

This section examines in detail the development of a DES. One of the first steps following the selection of the drug and device (i.e., stent) should be the characterization of the release kinetics of the drug in vivo. This preclinical study should be designed to assess potential systemic exposure to the drug. Unfortunately, the measurement of drug concentration is not always possible; therefore radio- or fluorescent-labeled compounds should be used and stents should be recovered at various time points after implantation to determine the amount of drug remaining on the stent and delivered to the artery. Drug concentrations in the whole blood, plasma, urine, coronary artery tissue (including proximal and distal to stent), and in the myocardium (i.e., where the stent is implanted and proximal and distal to stent) should be measured at various points until all the drug has eluted from the stent and been eliminated from the coronary artery and myocardium. The systemic exposure to the drug should be determined and compared to any available toxicology data. Because the route of administration in humans may be different than that used in the toxicology studies, the company should determine drug levels in major organs (i.e., liver, kidney, lung) to assure that the distribution is similar. In addition, the systemic exposure data should be converted into the human equivalent dose[16] and compared to doses obtained in a phase 1 clinical trial.The company needs preclinical data demonstrating the potential efficacy and characterizing the potential toxicity at the local (vascular wall) and regional (myocardium) sites of the stent implantation. The choice of animal model is not always simple, but one rule of thumb is to select a model that is recognized as having a predictable outcome in humans. A classic approach has been to conduct a 28-day overstretch-injury model in pigs to assess the effectiveness of the stent, because maximal neointimal hyperplasia is induced by 4 weeks after injury.[17] According to Teirstein,[18] these models have been used to evaluate different stent coatings and drugs and have revealed adverse effects, such as intimal hemorrhage, incomplete healing, intimal fibrin deposition, adventitial inflammation, and medial necrosis. Long-term animal studies are now required by some regulatory agencies to detect the presence of potential adverse effects, such as aneurysms, thrombosis, and fibrosis; however, the clinical relevance of this data is not known. Virmani et al.[19] address the issue of the comparability of human and animal studies as well as the predictability of the animal models for DES. The presence of a polymer, drug, or both modify the healing events that follow the deployment of a stent in the coronary artery. Virmani et al.[19] provide data demonstrating that the time to healing is longer in humans than in animals. According to their research, the CYPHER and paclitaxel DES showed efficacy at 28 d, with lack of benefit by 3 and 6 months; however, the significance of the long-term results for humans is not yet known.[19] At a time when toxicologists aim to reducing the number of animals used in research,[20] does the dramatic effectiveness at 9 months of the CYPHER DES[1] in humans cast a doubt on the validity of conducting 3- or 6-months animal studies? Only the long-term results of the CYPHER DES in humans will provide this answer along with information concerning the number of DES that were terminated because of unwanted effects in these animal studies. Regardless of this outcome, a 6-month animal study should still be part of the development program because of its potential to detect adverse effects.

Once an adequate animal model has been selected, the local effects (e.g., healing, endothelialization) of the DES must be assessed along with an evaluation of the myocardium for effects, such as infarction/thrombosis and cytotoxicity, associated with regional toxicity. This data will be required to support one or several doses for the first clinical trial. A pilot efficacy study in a recognized animal model, such as the porcine coronary artery overstretch, should be conducted using multiple dose groups. This animal model is widely used to evaluate in-stent restenosis, because porcine coronary arteries have a structure and physiology similar to human coronary arteries.[17] Ideally, the study should include high enough doses to induce toxicity. Using a dose associated with toxicity and an efficacious dose allows the company to estimate a safety margin for the clinical trial. A second dose–response study should be performed with more animals, using at least three dose groups to obtain safety and efficacy data. The recommendation of the preclinical concensus group[21] should be followed when designing studies. According to their recommendations, the dose–response study should include the following:

1. Bare stent.
2. Stent with polymer or binding material.
3. CP at clinical dose.
4. CP at 3 to 10 times the clinical dose.
5. Overlapping stents.

Although a third, intermediate dose group does not appear to be required by the FDA, it should be included when feasible to obtain additional dose–response information. The minimal requirements to initiate a feasibility/pilot study are 30- and 90-days animal data obtained both at the clinical dose and at 3 to 10 times the clinical dose; from a preclinical point of view, 30- and 180-days data at these dose levels would be required to initiate a pivotal trial.[22] The 90-day data are optional and only required if the company intends to initiate a feasibility study in the United States. Although conducting an animal study at three- to 10 times the clinical dose sounds reasonable, companies should be aware that achieving these levels may require the manufacture of a stent with an increased polymer thickness or increased surface area. In these cases, additional control groups would be required for evaluation of the drug-related effects, because the increased polymer thickness may induce unexpected or undesirable local toxicity. The efficacy/safety studies should evaluate all proposed clinical uses (e.g., overlapping stents, pre-exisiting; in-stent restenosis). Evaluating overlapping stents is critical, because this occurs in clinical practice and may lead to elevated drug concentrations delivered locally to the artery tissue. According to Schwartz and Edelman,[21] no animal model of human vascular disease exists and therefore a company should select an animal model based on its experience with recognized models. The coronary arteries of domestic crossbred swine and the iliac arteries of rabbits are adequate because their size and injury response are similar to that of human vessels. Regardless of the chosen model, the study should include the chronic administration of medications (e.g., aspirin, ticlopidine) that are usually administered to the patients. The concensus paper provides details on endpoints, time points, and methods to be used for evaluating the safety and efficacy of the DES.[21] All unexpected deaths should be fully investigated by a qualified veterinary pathologist, and all tissues should be examined histologically (including the myocardium and coronary artery).

The pathological examination should include histological sections from the proximal, middle, and distal regions of the stent as well as beyond the edges of the stent. Finally, when the DES reaches the clinical development phase, the company will be required to demonstrate an acceptable risk–benefit ratio in the intended patient population. In addition, it will be required to demonstrate that the drug provides an additional benefit over the clinical benefit provided by the stent alone. The clinical development plan may initially consist of first-in-human studies of an NCE administered parenterally to assess pharmacokinetic and safety issues. Alternatively, the plan could involve first-in-human studies with the CP. In either case, these pilot studies would be followed by dose-finding studies, the results of which will eventually be used to conduct a pivotal trial.

3.2. Development Issues for a Combined Product Containing a New Chemical Entity

The development of a drug–device CP involves understanding the properties of each component as well as the properties of the combination of the two components. The device component of the drug–device CP possesses its own therapeutic properties and/or acts as a delivery system for the drug component. The drug–device CP and each of its components induce local reactions in the body that must each be understood for the successful development of the CP. In the example of a DES, the bare stents possess a permanent therapeutic activity; however, the deployment of the stent causes injury to the coronary artery, resulting in restenosis. Moreover, the materials and design of the stent directly influence the degree of restenosis.[37] Therefore, at the discovery phase, the company must carefully choose the stent that it will use as part of its DES. Changing the stent component later in the development would involve recharacterizing the local effects of the new device. The development of the DES may involve the addition of a polymer to bind the drug to the metal scaffold of the stent. Because the simple addition of a polymer can induce local toxicity, the careful selection of this material is important to ensure biocompatibility. The selection of the drug component is a complex decision, as the ideal candidate must possess sufficient efficacy to combat the restenosis induced by the angioplasty and the stent deployment. The drug must be released from the stent surface over the desired period (i.e., therapeutic window) and reach the target tissue. The local delivery of a drug to the coronary artery results in significantly higher levels of the drug in the tissue relative to the administration of the same drug parenterally. This higher level may result in local toxicity issues that were not characterized during the development of the parenteral formulation. Both short- and long-term toxicity must be examined. There are many advantages to developing a DES using two approved components. In the case of the stent, the company will have both short- and long-term safety and efficacy data in humans. As for the drug component, the route of elimination, target organs, and toxicity profile (e.g., single- and repeat-dose toxicity, mutagenicity, reproductive toxicity) will be known. Ideally, the drug would be delivered from the stent at levels inferior to the no observed effect level (NOEL) for the parenteral route and what would be distributed systemically would not raise any safety concerns. The advantage of using an approved drug is that safety data in humans has already been established. Once the stent and drug component have been selected based on the animal efficacy models (with the desired release kinetics and efficacy), the design of the DES is frozen and the company begins investigating the preclinical efficacy and safety of the CP in animal models. Regarding the safety portion, the company first needs to elucidate the release kinetics of the drug and determine systemic exposure. Using appropriate scaling factors,[16] the human equivalent dose (HED) is determined and compared to the data collected from the parenteral dose in humans. Ideally, the HED is inferior to the systemic levels obtained with the parenteral formulation. If the HED were superior to the systemic level, the company would have to be certain that the preclinical safety program supports the use of this dose and should consider a phase 1 study in volunteers to collect safety data in humans. The development program becomes a lot more complicated when one or both components are not approved. It is not the lack of approval itself that increases the risk of bringing a CP to market but the numerous preclinical and clinical safety and efficacy data that are required to initiate clinical studies in humans. Although it is likely that the Office of Combination Products or the European Medical Device Competent Authority designated the CP in an identical classification to that of a DES with both components approved, the risk of taking this new product into a clinical setting is much higher.

Let us first look at developing a DES with an unapproved drug. Without considering the regulations, the review staff of the regulatory body will require the same type of preclinical information that is readily available to the company for the approved products. For example, a reviewer would ask the following questions:

- What are the target organs of toxicity?
- What is the route of elimination?
- What is the elimination half-life and kinetics?
- Is the toxicity reversible?
- What is the NOEL?
- Is the drug mutagenic?
- Will it affect the reproductive system?
- Does it alter hepatic metabolism?

Because a reviewer would not be able to determine the potential risks to a patient population without this information, a company developing a DES with a nonapproved drug must invest time and money to determine the toxicity profile of the drug component administered as a single component. The company needs to design a nonclinical safety program that will characterize the toxicity profile of the drug in order to support the clinical trials of the CP or a phase 1 trial in human volunteers to collect safety data in human subjects.

The International Conference on Harmonisation of Technical Requirements for Registration of Pharmaceuticals for Human Use (ICH) guidance documents should be used to design this program, which must involve a rodent and a nonrodent animal species. In addition to understanding the elution kinetics, systemic exposure, and elimination of the drug in the animal efficacy model, the ICH M3[23] guidance document should also be used as a starting point when designing the preclinical program. The company should consider conducting studies to characterize the affinity of the drug to receptors as well as screening studies such as a receptor-affinity profile. This latter data would provide information on the general pharmacological properties of the NCE. The ICH M3 guidance document outlines the safety pharmacology and toxicology requirements needed to support clinical trials and marketing approval. A cardiovascular and hemodynamic profile in an adequate nonrodent species will be required. Other safety pharmacology studies may also be required based on the toxicity profile of the drug. Careful consideration should be given to ICH S7A[24] and S7B.[25] The genetic battery of tests described in ICA S2A[26] and S2B[27] should be performed early on to avoid the development of a DES with a mutagenic drug. Similarly, assessing whether the NCE inhibits or induces the cytochrome P450 enzymes is important if the intended patient population is taking other medications. The potential effects of the NCE on reproductive organs (i.e.,the NCE's effect fertility or damage genital organs) is also important to determine. The timing of these studies must be carefully considered.

Finally, the company must perform both single- and repeat-dose toxicity studies in both a rodent and nonrodent species. The duration of the repeat-dose studies must support the duration of the clinical exposure that results from the DES, in accordance with ICH M3. This information will be required by regulatory agencies, because the amount of data required varies according to the NCE's class and/or toxicity profile. In terms of local toxicity, there is no difference between developing a DES with an NCE or an approved drug. The company will be required to perform animal studies as previously described in section 3.1.

3.3. Developing a Drug–Device Combination Product With a Radiopharmaceutical New Chemical Entity

Despite the association of intracoronary radiation with low rates of restenosis,[28] this form of therapy has had complications such as edge restenosis and late thrombosis. Thus, Angiogene Inc. has developed a DES using a radiopharmaceutical drug in an attempt to limit such complications.[29,30] Developing a DES with a radiopharmaceutical NCE presents unique challenges in many fields of drug development, including regulatory approval, nonclinical safety, and manufacturing. The development of radiopharmaceutical drugs is well established, with numerous products approved worldwide. Regulatory agencies have created a number of guidance documents for industry that describe the requirements for initiating clinical trials and obtaining marketing approval.[31–33] These documents must be consulted when developing a DES with a radiopharmaceutical NCE, because the radionuclide of the molecule affects how studies are conducted. In addition to the standard preclinical efficacy studies, regulatory agencies require an assessment of the radiation-absorbed dose that the coronary artery and surrounding tissue receive from the DES. An acceptable model must be used to determine the radiation-absorbed dose that is based on the geometry of the device and takes into account exposure of surrounding tissue.[54] This data is critical early in the development process to avoid complications, such as edge restenosis.[28]Administration of the NCE to animals, especially in acute or sub-acute studies, will be limited by occupational exposure of workers to the radiation. Therefore, an acceptable approach is to develop a nonclinical safety program in which some studies are performed with the radioactive NCE and others (e.g., repeat-dose studies) are performed with the nonradioactive form of the NCE. These latter studies allow the characterization of the toxicity profile of the molecule itself. This is acceptable for many radionuclides, because the effects of the radiation are well known and the effects in animals and humans are dependent on the dose the tissue receives. In addition, multiple-dose studies can be performed by administering the radioactive NCE on a weekly or biweekly basis, depending on the isotope's half-life and the NCE's pharmacokinetics.

Another unique aspect of this class of pharmaceuticals is that some molecules undergo radiolysis when the radionuclide decays, leading to degradation of the molecule. Because this phenomena also occurs in the body, the toxicity of the decayed NCE must be characterized. This can be done by performing nonclinical safety studies using NCEs that have decayed over several half-lives; four to five half-lives is usually sufficient. The identification of the fragments is required and important for the safety assessment. When feasible, the pharmacokinetics of the fragments should also be determined. Another aspect of the nonclinical safety program requires estimation of the radiation-absorbed doses that each organ will receive from the radiation and extrapolating this to humans.[53] When the radioisotope emits γ-radiation, the amount of preclinical data does not need to be as extensive, because sequential whole-body images of human subjects can be used to estimate the radiation-absorbed dose. In the case of β-radiation, such as phosphorus-32, whole-body imaging cannot be used, because the radiation does not penetrate as deeply. Therefore, regulatory agencies require more extensive preclinical data to support an application to initiate clinical trials for products emitting β-radiation. Tissue distribution studies in both a rodent and nonrodent species are required and the period of observation must be sufficiently long enough to adequately estimate the radiation-absorbed dose.

The decay of the NCE also impacts the manufacturing of the product, because the shelf-life must be based on the half-life of the radionuclide. Potential interaction of the radiation with the container-closure systems and any materials that the NCE comes in contact with should be determined. Compliance with federal and local radiation-protection regulations is also compulsory for the manufacturer. Each country has nuclear regulatory commission requirements, and the sponsor of a clinical trial must obtain appropriate clearance (i.e., license) or an exemption to import a product. In addition, the investigator who will receive and use the radioactive product must also comply with applicable country and institution regulations.[34]

3.4. Development Issues for a Combination Product With a Nonapproved Device

The issues encountered when developing a DES with a nonapproved device component are different from those encountered with approved devices. The nonclinical safety evaluation of the device is determined by performing biocompatibility studies in conformance with the International Standards Organization's (ISO's) standard ISO 10993.[4] Part 1 of ISO 10993 describes the principles for evaluating medical devices and the tests required as a function of the type of product (e.g., long-term implant). In addition, the FDA has created a guidance document that describes the biocompatibility, physical testing, and animal studies required for the development of a stent.[35] This guidance document should also be consulted for DES.

As a general rule, biocompatibility testing is required for all new materials and includes analysis of the following: sensitization assay, irritation, cytotoxicity, acute systemic toxicity, hemocompatibility, pyrogenicity, implantation, mutagenicity/genotoxicity, subchronic toxicity, and/or chronic toxicity/carcinogenicity. These tests may not be required if the material has a long history of use in marketed interventional devices, although the company will have to provide the FDA with sufficient information to establish that biocompatibility testing is not necessary. Because biocompatibility testing is designed to assess the toxicity of the product that will be used in humans, testing should be performed on the sterilized final product and any leachable material from the sterilized final product. Although identification and characterization of leachables is not required because the final product is tested, information regarding the composition of the leachables may help the company assess the toxicity associated with the product.A critical aspect of the new device evaluation is its physical properties, which are determined by performing physical testing. An FDA guidance document[35] describes the type of physical testing required for an intravascular stent; as with biocompatibility testing, these tests should be conducted on the final sterilized product. These tests include specification-conformance testing, stent integrity, and stent/catheter system testing. The specification-conformance testing is performed on material samples such as the metal wire and includes tests such as the determination of the tensile strength of the wire and its resistance to corrosion. Stent integrity testing is done on the final sterilized product after deployment with the proposed delivery system. Fatigue testing, stent recoil, stent expansion, and stent uniformity testing are examples of this physical testing. A final category of testing is required to demonstrate that the delivery system can safely and reliably deliver the stent to the intended location. This testing should be done on sterilized products with the stents mounted and includes establishing the balloon inflation and deflation time, stent crimping, and maximum pressure. When it comes to preclinical efficacy, the use of an unapproved device has a minor impact on the design of the studies required by the FDA. The evaluation should include evaluation of the bare stent itself and stent plus polymer(s) in order to separate the effects caused by the addition of the polymer(s). This is important because stent coatings have stimulated intense inflammatory responses in animal models[36] and surface materials and geometric configuration of stents can affect thrombosis or neointimal hyperplasia.[37]At this point, it is clear that the development of a DES with an unapproved device and NCE significantly increases the risk of bringing the product to the market. Each of the components brings regulatory challenges, and once these components are selected during the discovery phase, the characterization of the toxicity of the NCE will significantly impact the timing of bringing the DES into a pilot clinical trial.

4. Developing a Combination Product With a New Chemical Entity in Europe

Medical device companies tend to bring new devices into the clinic in Europe before the United States. Likewise, products are typically approved first in Europe and then in the United States, although it is not necessarily easier to bring a CP containing an NCE to clinics in Europe than in the United States.The European Community's (EC's) regulatory system differs greatly from that in the United States. The European system evolved because of the harmonization of many countries into one economic community, with the creation of directives[11] to address this complex relationship. When a company complies with a directive, it complies with the national laws of each country.[38]

However, companies should remember that the transposition of directives into national law by each country can be late and/or incomplete, interpretations of the directives can differ, and national parliaments and legislators may not understand the spirit of the directive.[39,40] There are several major directives for the development of pharmaceutical products in the EC (i.e., Medicinal Products Directive [MPD],[41] Medical Device Directive [MDD],[42] Active Implantable Medical Device Directive [AIMDD]).[43] As a general rule, a product is regulated by one of these directives.

In the EC system, each member state appoints a competent authority, which is usually a government agency (e.g., Medical Device Agency).[44] Some directives, such as the MDD and AIMDD, require independent assessment of products or systems by an organization that has been nominated by a government, which, in turn, notifies the EC of its choice.[46,47] Each competent authority can appoint one or more notified bodies (NB) to fulfill the role described in the directive.[45] NBs are accountable to the national competent authority, because they take the responsibility of overseeing the approval (CE Mark[2]) of medical devices.[38, 45, 47] One of the first steps in the designation or classification process is determining the role of each of the components of the CP and determining which component has the primary role. In Europe, this is analogous to establishing the primary mode of action (i.e., principal intended action) of the CP. In addition, two factors should be examined when deciding which regulation applies: the product's intended purpose and its method of commercialization (i.e., device and drug form a single integral unit).

Each of the three directives along with several other key guidance documents has to be examined to determine the appropriate regulatory path for product approval. The European Commission has created a guidance document that provides definitions of medical devices and accessories,[48] and the CP must fall within the scope of this definition to be regulated as a medical device. Rule 13, Classification Criteria, Annex IX MDD,[42] applies to devices that incorporate a medicinal substance. There is also a guideline that can be consulted for the classification of medical devices.[49] Similar to the guideline for medical de-

---

[1] Directives are: instructions to national governments to add, repeal or amend their existing laws and regulations.

[2] "CE Marking symbolises the conformity to all the obligations incumbent on manufacturers for the product by virtue of the Community Directives providing for its affixing. When affixed to products it is a declaration by the person responsible that: the product conforms to all applicable Community provisions, and the appropriate conformity assessment procedures have been completed."[38] CE refers to Conformité Europeene.

vices, the Medicines and Healthcare Products Regulatory Agency (MHRA) has created a guidance note that defines a medicinal product[50] and outlines the scope of products having similar principal intended actions. In addition, medical device guidance document MEDDEV 2.1/3 rev 2[51] should also be consulted as it provides the demarcation between each directive. This guidance document's goal is similar to that of the US FDA's Intercenter Agreements in that it provides the criteria for determining if a product should be regulated under the MPD, MDD, or AIMDD. The criterion based on the "principal intended action" is crucial in the definition of a medical device. Typically, the medical device function is fulfilled by physical means, whereas the action of a medicinal product is generally achieved pharmacologically. According to these documents, medical devices can be assisted in their function by pharmacological, immunological, or metabolic means, but once these means are no longer ancillary to the principal purpose of a product, the product becomes a medicinal product. This is specifically addressed in the MDD, which makes it clear that such products are devices, "provided that the action of the medicinal substance is ancillary to that of the device."[42] For some CPs, precedent makes classification relatively straightforward. However, it is important to submit a classification proposal to the appropriate Competent Authority if the product has any unique characteristics, such as an unapproved radiopharmaceutical drug. For example, the classification request should be sent to the medical device Competent Authority if a manufacturer believes it should be classified as a device. The NB may contact the Competent Authority to confirm the classification and/or discuss issues regarding the technical file. Selecting an NB is a key step; a company should consult an expert if it is unfamiliar with these organizations and their implications. Brooks and Johnston[44] describe the interaction between medical device companies and NBs in an article that serves as a good resource for those unfamiliar with the process. In the case of a DES with an unapproved radiopharmaceutical drug, the Medical Device Agency and the Medicines Control Agency agreed that this type of product would fall within the scope of the MDD (93/42/EEC) as a Class III medical device under Rule 13 of Annex IX.[29] This decision was mainly based on the radioactive NCE not being a sealed source. The agencies considered it a medicinal substance, because it would enter systemic circulation. For medical devices, the approval process and requirements are well defined and the NB is experienced with these issues. The normal route for bringing a CP to market involves appointing an NB and gaining quality system approval. The NB consults with the medicinal Competent Authority on the requirements for the drug–pharmaceutical section of the Technical File/Design Dossier. This is not an issue for a CP with an approved drug. In the United Kingdom, for example, the NB begins the process by submitting the form NBA 201 to the MHRA[52] without the company's participation. Thus, the selection of an NB for an NCE is an important decision, as the organization must have sufficient knowledge of NCEs and CPs to present and discuss requirements with the MHRA. The NB is not a competent authority and cannot independently dictate the requirements for the NCE. Hence, the majority of the research should be completed and ready to submit for approval (i.e., CE Mark) before selecting an NB. Some NBs may be willing to consult with the competent authority before selection. This is another aspect that differs between the EC and the FDA. In contrast to the United States, EC companies do not generally deal directly with the Competent Authority to discuss the requirements of the technical file. It may be worthwhile to identify consultants in the EC that can arrange for an informal discussion of the NCE with the MHRA before selecting the NB. Although the MHRA is not bound by the opinions it expresses in this setting, an informal meeting for a CP with an NCE is important, because the company will have to assess whether its preclinical and manufacturing information meets the Competent Authority's requirements.

5. Developing a Combination Product With a New Chemical Entity in Canada

Developing a drug–device CP in Canada is similar to the process followed in the United States in that companies can deal directly with the regulatory agencies that approve their product. On the other hand, the development of a CP covered by the medical device regulations is more similar to the EC system than the US system in that the medical device regulations require conformance to essential principles (called essential requirements in Europe) and ISO 13485 certification. Health Canada accepts only quality system certificates issued by third-party auditing organizations called the Canadian Medical Devices Conformity Assessment System (CMDCAS). Essential principles/requirements define the results that must be achieved and the safety issues that must be addressed but do not specify scientific or technical methods to achieve the goal.[38] In 1999, the Therapeutic Products Programme issued two policies that deal with CPs: Drug and Medical Device CP Decisions[55] and Drug/Medical Device Combination Products.[56] Before these policies, the drug–device CP had to fulfill the requirements of both the Food and Drug regulations and the Medical Device regulations. The creation of these policies simplified the development of CPs in Canada. The classification of a CP is an important step in its development and should be addressed early on. The first step in determining the classification of a CP is to consult the Drug and Medical Device CP Decisions policy.[55] This policy lists CPs classified by the Therapeutic Products Classification Committee (TPCC). If a similar product is not listed, obtaining classification becomes an essential step. The classification process in Canada differs from the designation process in the United States and usually involves a meeting with the regulatory agency. As with the United States, the classification procedure should be completed before submitting an Investigational Testing Application (ITA) or Clinical Trial Application (CTA). Do not assume that the classification will be identical to the designation in the United States or similar to the classification given in Europe as there are several products that do not fall into the same designation. The Therapeutics Programme defines a CP as follows: A *"CP is a therapeutic product that combines a drug component and a device component (which by themselves would be classified as a drug or a device), such that the distinctive nature of the drug component and device component is integrated in a singular product."*[56] This definition is key because the regulation does not apply to combinations of drugs and medical devices where the drug component and the device component can be used separately (e.g., products sold together in procedure packages and trays) and is only a CP through labeling. The policy entitled *"Drug/Medical Device Combination Products"*[56] provides criteria for determining the classification. According to this document, the CP will be subject to either the medical device regulations or the food and drug regulations according to the principle mechanism of action. The term used by the Canadian authorities is different than that used by the US FDA (primary mode of action) but basically results in the same verdict. To classify the product in Canada, the company should determine the principal mechanism of action of the CP and subsequently request, in writing, a face-to-face meeting with the Bureau responsible for the regulation of products that fall within this classification. The meeting package should contain a synopsis of the CP, and each component, along with information that describes the principle mechanism of action and all relevant safety and efficacy data obtained to date. When making the request, the company should request the presence of the other relevant Bureau at the CP classification meeting, although it is the responsibility of the Bureau receiving the meeting request to invite the other relevant Bureau to the meeting. For complex products, one should not expect a verdict to be reached at the meeting. If a decision cannot be reached because of lack of consensus regarding the classification of the product, the decision will be referred to the Therapeutic Products Classification Committee (TPCC). The TPCC issues a final verdict within 30 days of receipt from the bureau that received the classification request. The company has 30 days to send a letter of intent to appeal the decision to the director general if it disagrees with the verdict. Determining the classification of the product is important, because the regulatory agency will issue a Screening Deficiency Notice if an ITA or CTA for a product that has not been classified is submitted. Once the classification is established, the company proceeds as with any other product, although it must address the issues raised by both bureaus, because the review team will consist of reviewers from each.

6. Issues a Start-Up Company Faces When Considering the Development of a Drug–Device Combination Product

Most start-up companies have either device or drug expertise in-house but not both. Acquiring this expertise is not simple and brings many complications from a financial, human resources, and design philosophy viewpoint. A start-up drug company is usually composed of scientists, whereas a device company predominantly consists of engineers. Because scientists and engineers are not trained to think the same way, their approach to developing a product is vastly different. This difference can be seen by comparing the drug cGMP regulations[6] to the QSR[5] for medical devices (ISO 13485 in Europe and Canada). Although cGMPs dictate documentation of methods and procedures, training of employees, validation of equipment, and other requirements, they do not impose any design control. In fact, scientists do not encounter design control during their formal educational training. In contrast, user needs and design input, output, process, and review are all part of QSR, with design control being a part of the basic educational training of an engineer. Cross training is not a productive alternative, as it usually results in significant delays. Thus, a start-up drug company should either outsource the device aspects of the development or partner early on with a device firm. Trying to bring the device expertise in house is not a viable economic option and attempting to do so will have significant consequences on the time to market.

Device companies face fewer problems when developing a drug–device CP that has a primary mode of action of a device, because the development staff is accustomed to design control and will perform all of the required activities,[57] including defining and documenting user needs, performing risk analyses, design input, design output, and formal design reviews. These latter activities are absent in start-up companies that have no employees with device experience or that do not use device consultants. A start-up drug company typically discovers the requirements of QSR and ISO 13485 well into the development of the drug–device CP. At that point, the company must hire qualified consultants to guide it through the implementation of QSR and ISO and seek ISO 13485 certification. This brings new hurdles, because the company must train its staff, perform retrospective documentation of certain activities, and even consider outsourcing much of the device development. The management of the start-up company also faces new challenges, as they must now formally become involved in the development activities as part of the design review process—an activity not required by cGMPs. Device companies are in a more favorable position than drug companies are because the cGMP drug aspects of development do not impact the design of the CP. cGMPs assure the quality, purity, and identity of the compound and only apply to the production of clinical-grade material. Animal safety studies require compliance only with Good Laboratory Practices,[58] which does not regulate design-related activities. As for animal studies with a device, biocompatibility studies are performed with the final sterilized product and safety and efficacy studies are performed with a product that has at least been through the design process. Even so, the results of the animal studies can lead to a new design loop and modification of the product.

7. Conclusion

CPs will bring therapeutic solutions to many unmet medical needs. However, undertaking the development of a CP requires a thorough understanding of the development and regulatory issues, as well as skilled project planning. Determining the primary mode of action and subsequently obtaining designation and classification of the drug–device CP is one of the first steps in the development program, because this dictates the requirements for premarket approval. CPs that consist of unapproved entities present additional scientific and regulatory hurdles. These issues must be defined early on and addressed in the development plan. Project teamwork is even more critical with this type of product, because the team will have to address issues related to both drugs and devices. Despite these challenges, CPs that consist of unapproved components can be brought successfully to the clinic and subsequently commercialized.

References

1. Food and Drug Administration, Health and Human Services. 2002. Cordis presentation: CYPHER™ Sirolimus-eluting Coronary Stent at the FDA Circulatory System Devices Panel on October 22.
2. Patterson, N. 2003. Faster approvals seen for drug/device combination products. *BBI Newsletter* 26:247–252.
3. Inose, C. and Brown, M. May 2002. CPs in the US: navigating the regulatory jungle. *Reg AFF Focus*.
4. Garrison C. May 2002. CPs in Europe: a case study. *Reg AFF Focus*.
5. Mohammed F. 2003. International filing requirements and strategies for combination medical device/drug products. *N.O.C.* (newsletter of the Canadian Association of Professional Regulatory Affairs). 58:15–17.
6. Food and Drug Administration, Health and Human Services. 2002. Quality System Regulation. 21 CFR §820.
7. Food and Drug Administration, Health and Human Services. 2002. Current Good Manufacturing Practice for Finished Pharmaceuticals. 21 CFR §211.
8. Food and Drug Administration, Health and Human Services. 2002. Orphan Drugs. 21 CFR §316.
9. Food and Drug Administration, Health and Human Services. 2002. Humanitarian Use Devices. 21 CFR §814(H).
10. Food and Drug Administration, Health and Human Services. 2002. Product jurisdiction. 21 CFR §3.

11. Food and Drug Administration, Health and Human Services. 1991. Intercenter Agreement Between The Center for Drug Evaluation and Research and The Center for Devices and Radiological Health.

12. Byrd, G. N. CPs and regulatory strategy. November 1997. *Reg AFF Focus.*

13. Leichter, L. February 2001. Medical device combination products. *Reg AFF Focus.*

14. Food and Drug Administration, Health and Human Services. 1995. Required Biocompatibility Training and Toxicology Profiles for Evaluation of Medical Devices (G95–1).

15. International Organization for Standardization. 1992–2003. ISO 10993 Biological Evaluation of Medical Devices, parts 1 to 17.

16. Food and Drug Administration, Health and Human Services. 2002. Guidance for Industry and Reviewers (Draft), Estimating the Safe Starting Dose in Clinical Trials for Therapeutics in Adult Healthy Volunteers.

17. Babapulle, M. N. and Eisenberg, M. J. 2002. Coated stents for the prevention of restenosis: part 1. *Circulation* 106:2734–2740.

18. Teirstein, P. S. Living the dream of no restenosis. 2001. *Circulation* 104:1996–1998.

19. Virmani, R., Kolodgie, F. D., Farb, A., and Lafont, A. 2003. Drug-eluting stents: are human and animal studies comparable? *Heart* 89:133–138.

20. Weekley, L. B., Guittin, P., and Chamberland, G. 2002. The International Symposium on Regulatory Testing and Animal Welfare: recommendations on best scientific practices for safety evaluation using nonrodent species. *ILAR J.* 43:S118–S122.

21. Schwartz, R. S. and Edelman, E. R. 2002. Drug-eluting stents in preclinical studies: recommended evaluation from a consensus group. *Circulation.* 106:1867–1873.

22. Food and Drug Administration, Health and Human Services. 2002. *Best Practices to Get an IDE Approved for a Drug-Eluting Coronary Stent.* Presented at the FDA Questions and Answers session, Cardiovascular Radiation Therapy 6th Annual Meeting.

23. International Conference on Harmonisation of Technical Requirements for Registration of Pharmaceuticals for Human Use. 2000. ICH M3 Nonclinical Safety Studies for the Conduct of Human Clinical Trials for Pharmaceuticals.

24. International Conference on Harmonisation of Technical Requirements for Registration of Pharmaceuticals for Human Use. 2000. ICH S7A, Safety Pharmacology Studies for Human Pharmaceuticals.

25. International Conference on Harmonisation of Technical Requirements for Registration of Pharmaceuticals for Human Use. 2002. ICH S7B, Safety Pharmacology Studies for Assessing the Potential for Delayed Ventricular Repolarization (QT Interval Prolongation) By Human Pharmaceuticals.

26. International Conference on Harmonisation of Technical Requirements for Registration of Pharmaceuticals for Human Use. 1995. ICH S2A, Guidance on Specific Aspects of Regulatory Genotoxicity Tests for Pharmaceuticals.

27. International Conference on Harmonisation of Technical Requirements for Registration of Pharmaceuticals for Human Use. 2002. ICH S2B, Genotoxicity: A Standard Battery for Genotoxicity Testing of Pharmaceuticals.

28. Teirstein, P. S., Massullo V., Jani S., et al. 1997. Catheter-based radiotherapy to inhibit restenosis after coronary stenting. *N. Engl. J. Med.* 336:1697–1703.

29. Chamberland G. July 30, 2003. Navigating US and international pathways for products with no approved component. Presentation given at the CPs conference, hosted by Barnett International.

30. Gobeil, J. F., Leclerc, G., Martel, R. et al. October 20–24, 2001. Pharmacokinetics of a stent-based local drug delivery of a short DNA oligonucleotide coupled with P32 (Oliglow) in a swine model. Canadian Cardiovascular Congress, Halifax, Nova Scotia (Abstract).

31. Food and Drug Administration, Health and Human Services. 1981. Guidelines for the Clinical Evaluation of Radiopharmaceutical Drugs.

32. The European Agency for the Evaluation of Medicinal Products. 1991. Guideline on Radiopharmaceuticals.

33. The Council of European Communities. 1989. Council Directive 89/343/EEC.

34. Holder L. September 1999. Clinical trials of medical devices containing radioactive isotopes. *Reg AFF Focus.*

35. Food and Drug Administration, Health and Human Services. 1995. Guidance for the Submission of Research and Marketing Applications for Interventional Cardiology Devices: Intravascular Stents.

36. van der Giessen, W. J., Lincoff, A. M., Schwartz R.S., et al. 1996. Marked inflammatory sequelae to implantation of biodegradable and nonbiodegradable polymers in porcine coronary arteries. *Circulation.* 94:1690–1697.

37. Campbell, R. and Edelman, E. R. 1995. Endovascular stent design dictates experimental restenosis and thrombosis. *Circulation.* 91:2995–3001.

38. European Commission. 2000. Guide to the Implementation of Directives Based on the New Approach and the Global Approach. European Communities.

39. Dieners, P. July 2000. Recent developments in European regulatory affairs. *Reg AFF Focus.*

40. Williams, M. H. January 1999. An overview of the European Union and how it functions. *Reg AFF Focus.*

41. The Council of European Communities. 2001. Council Directive 2001/83/EEC as amended.

42. The Council of European Communities. 1993. Council Directive 93/42/EEC as amended.

43. The Council of European Communities. 1990. Council Directive 90/385/EEC as amended.

44. Brooks P. and Johnston M. November 2001. Medical device manufacturers and their notified body: how to make the partnership a success. *Reg AFF Focus.*

45. European Commission. 2001. Designation and Monitoring of Notified Bodies within the Framework of EC Directives on Medical Devices (MEDDEV 2.10-2 Rev 1).

46. Ruston, R. March 1999. The European legislative scene for devices. *Reg AFF Focus.*

47. Medical Devices Experts Group. 2002. Report on the Functioning of The Medical Devices Directive.

48. European Commission. 1994. Guidelines Relating to the Application of the Council Directive 90/385/EEC on Active Implantable Medical Devices and the Council Directive 93/42/EEC on Medical Devices (MEDDEV 2.1/1).
49. European Commission. 2001. Guidelines for the Classification of Medical Devices (MEDDEV 2.4/1 Rev. 8).
50. Medicines and Healthcare Products Regulatory Agency. 2003. MHRA Guidance Note No. 8. A Guide to What is a Medicinal Product.
51. European Commission. 2001. Guidelines Relating to the Application of the Council Directive 90/385/EEC on Active Implantable Medical Devices and the Council Directive 93/42/EEC on Medical Devices (MEDDEV 2.1/3 rev. 2).
52. Medicines and Healthcare Products Regulatory Agency. 2003. MHRA Guidance Note No. 18. Guidance for Notified Bodies: Devices Which Incorporate a Medicinal Substance.
53. Stabin, M. G. 1996. MIRDOSE: A personal computer software for internal dose assessment in nuclear medicine. *J. Nucl. Med.* 37:538–546.
54. Janicki, C., Duggan, D. M., Coffey, C. W., et al. 1997. Radiation dose from a phosphorus-32 impregnated wire mesh vascular stent. *Med. Phys.* 24:437–445.
55. Health Canada. 1999. Drug and Medical Device CP Decisions.
56. Therapeutic Products Programme, Health Canada. 1999. Drug/Medical Device Combination Products.
57. Food and Drug Administration, Health and Human Services. 1997. Design Control Guidance for Medical Device Manufacturers.
58. Food and Drug Administration, Health and Human Services. 2002. Good Laboratory Practice for Nonclinical Laboratory Studies. 21 CFR §58.
59. Schievink, W. I. 1997. Intracranial Aneurysms. *N. Engl. J. Med.* 336:28–40.
60. Raymond, J, Leblanc, P., Desfaits, A. C. 2002. In situ beta radiation to prevent recanalization after coil embolization of cerebral aneurysms. *Stroke.* 33:421–427.

10

Wall Street's Perspective on Medical Device Evaluation
Innovation Investing

Adam K. Galeon

> "The key to investing is not assessing how much an industry is going to affect society, or how much it will grow, but rather determining the competitive advantage of any given company and, above all, the durability of that advantage. The products or services that have wide, sustainable moats around them are the ones that deliver rewards to investors."—Warren Buffet

1. Sustainable Competitive Advantage: The "Moat" That Surrounds the Castle

Corporations create value for their shareholders by earning a return on invested capital that exceeds its cost of capital. A company's ability to do this depends largely on its competitive advantage. In the medical device industry, success is categorically linked to new product flow, transitively marrying innovation and shareholder value creation. The medical device industry follows the punctuated equilibrium interpretation of Darwinism rather than the traditional theory of gradualism. The puncuated equilibrium theory states that evolution tends to happen in fits and starts, sometimes moving fast, sometimes moving slowly or not at all. A particular device market tends to chug along with little change to speak of as companies introduce iterative, "me-too" products; however, every so often, a game-changing new technology emerges that revolutionizes the standard of care. As a result, innovation invariably governs the competitive dynamics in medical devices, with the industry constantly trying to address unmet clinical needs or improve existing treatments to create a sustainable competitive advantage.

From: *Clinical Evaluation of Medical Devices: Principles and Case Studies, Second Edition*
Edited by: K. M. Becker and J. J. Whyte © Humana Press Inc., Totowa, NJ

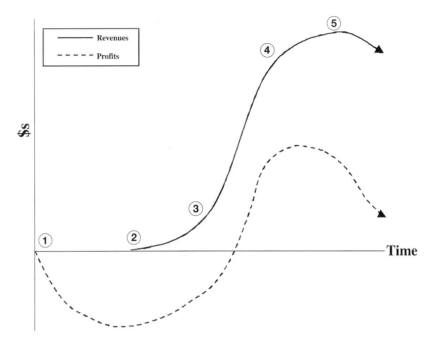

Fig. 1. New technology life cycle.

2. New Technology Life Cycle

Sizing up the market opportunity for a new technology is much more than just *units x price*. There other moving parts to consider. First and foremost, investors must be able to identify where a product or technology is within its life cycle (*see* Fig. 1). We segment the new technology lifecycle in devices into five stages:

1. *Proof of Concept*. Stage 1 begins when a company identifies an unmet clinical need and initiates a program to develop a method of capitalizing on the opportunity. Research and discovery (R&D) dollars are dedicated to basic science and bench tests to determine the feasibility of the project and whittle down the possibilities to a particular angle of pursuit. The company moves on to animal tests for its technology and, after becoming comfortable with its approach, to human clinicals. Toward the end of stage 1, companies build out their sales infrastructures to prepare for commercialization.

2. *Introduction*. Approvals are granted in Europe (CE Mark) and the United States (Food and Drug Administration [FDA] approval), and the market begins to take shape as products are launched and companies generate revenue. The technology is quick to penetrate thought leaders and tech-savvy early adopters, although the majority of the clinical community is much slower to catch on. Thus, companies

spend a lot of money on raising awareness, training physicians, and providing high-level service.

3. *Growth*. Use passes some rate-limiting step (e.g., physicians find a comfort zone with the technique, awareness spikes following a major clinical symposium, reimbursement is put in place). Feedback from the clinical community in stage 2 helps manufacturers refine the original designs and improve their marketing message. Competition often emerges toward the middle of stage 3, but the pie is still big enough for each participant to grow. Profitability of the device improves as companies get better at manufacturing and benefit from increased volumes. Toward the middle of stage 3, the late adopters—the less tech-savvy physicians—start to capitulate as the clinical data become robust and irrefutable, and the technology becomes the standard of care. Finally, helping to offset decelerations in the US and European markets, the devices garner approval in Japan, where units pale in comparison but the opportunity is attractive owing to incredibly high average selling prices.

4. *Survival*. The penetration curve begins to slow, and the market starts to feel crowded. The technology curve flattens as new product introductions become more of a tweak of previous versions than anything radically new. Growth becomes a market-share game.

5. *Maturity*. Penetration peaks, competing technologies become nearly indistinguishable. Market value and profit margins erode as discounting becomes an effective tool for maintaining share. What was once a growth driver now weighs on the company's overall growth rate, presenting a difficult decision of what to do with the franchise. Withdrawing the product or divesting the business is accretive to the company's growth rate; however, management teams must take into consideration the importance of the brand to the success of other business segments and the impact of eliminating the associated stream of cash flow.

Importantly, this model is just the starting point in delineating the lifecycle of a new technology. In some cases, the curve may look considerably different. One shortcoming of this approach is that time is not the only independent variable; rather, the evolution of the market can be influenced by factors such as management decisions and public policy. There are two key considerations that influence the shape of the curve: *amplitude* and *slope*.

2.1. Amplitude = Market Opportunity

The first part of predicting the curve's peak is sizing the addressable target population. From an investment perspective, one of the nice things about devices is that demand is relatively predictable. Based on demographics and historical trends, we can, for example, estimate fairly well how many heart failure diagnoses and hip replacements there will be in a given year. This annual occurrence of a condition is known as *incidence*. The highest reaching curves are often for new technologies, which have an addressable market that is still driven by *prevalence*, or the total number of cases existing within the population. For example, when continuous positive airway pressure (CPAP) was first introduced, it had the full population of 18 million patients with obstructive sleep apnea to treat. Similarly, new trials have exponentially increased the target population for implantable defibrillators, pushing the curve's peak higher and sliding the technology further back down the curve from stage 3 to 2.The last component that bears mention is *pricing*. Robust pricing is one of the biggest reasons why devices are such an economically attractive industry. There are a few reasons for this:

- *Low threat of new entrants.* The barriers to entry in devices are as large as in any sector. The top competitors in each market have a headlock on the industry's top talent in a number of important areas. As a result, success begets success, creating the ability to fortify walls in existing markets and build new ones in others.

- *Low buyer bargaining power.* New, large market opportunities in devices are typically associated with higher-tech, higher-profit procedures, which enables physicians to use their product(s) of choice and help sustain pricing.

- *Low degree of rivalry.* Device companies tend to cooperate more than most industries, focusing on profits rather than chasing customers indiscriminately through discounts and channel incentives. One important contributor is the industry's low-industry concentration. A study published by economist Joe Bain found that manufacturing industries in which the eight largest competitors comprised more than 70% of sales were nearly twice as profitable as those in which the eight largest comprised less than 70% of sales. Of note, the biggest three or four companies in most device markets represent over 70% of sales, and in some cases (e.g., stents, pacemakers, defibrillators), over 90%.

2.2. Slope = Length of Product Cycles

"Better" medical devices are generally characterized as those that improve patient outcomes, reduce invasiveness, and are more convenient for the operator. Conveniently for the industry, innovation and economics usually go hand-in-hand, as "better" devices often reduce complication rates, hospital stays, recovery times, and, in turn, total costs. Moreover, patients are increasingly demanding specific therapies as information becomes more readily available in today's Internet world, and fear of malpractice keeps physicians aware of the latest and greatest technology. This confluence of factors is a windfall for the industry, driving faster penetration (i.e., a steeper stage 3 to 4 slope) than most industries. On the whole, we can say that owing to the aforementioned phenomena, product cycles are generally much shorter in devices than in most industries, amplitude clearly has a role. In medical devices, two more critical considerations are *technique* and *reimbursement*. Specifically, how much of a departure is the new technique from the standard of care. If it represents a

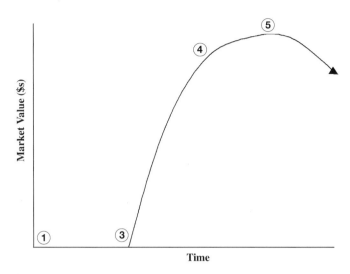

Fig. 2. Drug-eluting stent life cycle.

considerable change, such as the transition from coronary artery bypass graft (CABG) to percutaneous transluminal coronory angioplasty (PTCA), the curve in stage 2 may be slow to climb. In the case of drug-eluting stents, which tripled the market for stenting with absolutely no change in technique, penetration soared right out of the gates and eliminated stage 2 altogether, as depicted in Fig. 2.

Investors are focused on growth, not absolute returns, and thus the slope of the curve in stage 4 becomes critical to determine how big of a hole a company needs to fill to offset deteriorating fundamentals in one of its target markets. An important factor to consider is the speed of price erosion, which is largely a function of the number of competitors and their track records for competing on price. One important distinction between the medical device industry and the pharmaceuticals industry is the low threat of substitutes in medical devices. Generic competition generally is not an issue in devices, principally because technologies are constantly tweaked to come up with new and slightly improved iterations. Consequently, by the time the patents expire on a particular device, it is likely to have already been replaced by several generations of products. Thus, we rarely see an abrupt collapse in stage D similar to when a drug loses market exclusivity.

3. The Reward for Innovation is Bigger and Faster Than in Most Industries

The byproduct of the high peaks and steep slopes of new technology cycles in medical devices is that the economic returns are large and quick, as illustrated in Fig. 3.

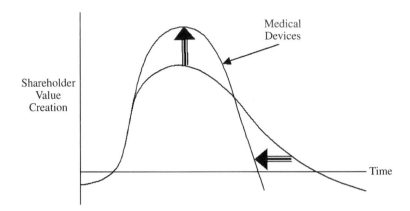

Fig. 3. The economic reward for innovation.

The challenge for investors is that the forward-looking capital markets capture value faster than companies generate the actual returns. Investors stay well attuned to the steps to commercialization, each of which provides an important catalyst to understanding the amount of shareholder value created by the new technology. Whether caused by a new perception of the curve or just discounting a different probability of success, the expectations embedded in companies' valuations fluctuate with each key data presentation and regulatory milestone. The challenge for companies is managing expectations and maintaining a sustainable growth trajectory. As the curve for a successful new technology begins to plateau, investors' insatiable appetite for growth begs the question of *"what's next?"* Accordingly, management teams must be mindful of having new opportunities in the wings to offset decelerations in their more mature franchises. Naturally, the intensity of this challenge increases with the amplitude and brevity of the product cycle. To maintain a continuum of value creation, companies must (1) defend their competitive advantages, (2) poach others', and (3) have enough shots on goal to maximize its probability of success in both. The following principles underscore these concepts.

4. The Babe Ruth Effect

Columbia Business School Professor Michael Mauboussin describes a phenomenon in which a portfolio manager's performance is not necessarily tied to the success rate of his investments. The disconnect is an exercise in probability: "the frequency of correctness does not matter; it is the magnitude of correctness that matters." If the values of three of a fund's four investments decline slightly but the fourth goes through the roof, the fund's overall performance is likely to be quite good. Mauboussin calls this *The Babe Ruth Effect* (the Babe had 1330 strikeouts for his 714 home runs, yet he is considered one of the most prolific hitters of all time). There is a high degree of risk inherent to new product development in all industries, particularly in health care owing to the extraordinarily high clinical, regulatory, and competitive barriers to entry. A "phase transition probability study" by the Tufts Center for the Study of Drug Development shows that 71% of new chemical entities examined in phase I make it to phase II; 44% in phase II make it to phase III; and 69% in phase III reach FDA approval. That equates to a 22% hit rate for all programs the industry feels are promising enough to fund development. No such data are available for devices, but another reason for the economic appeal of the industry is that the success rate is quite a bit higher. This is caused in large part by a concept prevalent in devices that we call *R&D Economies of Scope*. Often, concepts applicable to one area give birth to ideas in others, and so researchers have a better idea of what they are going to get in developing a device than a drug. For example, the advent of the coronary stent led to stents for the peripheral vasculature, and pacemaker technology formed the foundation of neurostimulation devices.

5. If You Can't Make It, Buy It

An investment decision in medical devices is as much an evaluation of a company's pipeline as it is a measure of the prospects of its existing product portfolio. Often, the pipeline is not sufficiently robust to instill investor confidence in a company's longer-term prospects. Alternatively, a start-up company becomes visible with a technology that threatens an established company's competitive advantage. In either case, augmenting one's pipeline via acquisitions becomes an important strategy, one that many companies consider an integral part of their R&D efforts. Importantly, these deals are usually a "win–win" situation for both parties. Smaller companies gain access to clinical, regulatory, and sales and marketing firepower that help leverage their technology beyond what they could ever do on their own. Furthermore, their shareholders gain a quick exit strategy for their investment. One of the benefits for the acquirer is that the development risk is mitigated, as the company usually only becomes involved once there is adequate evidence of the product's commercial viability. For safe measure, the transactions are often constructed with downside protection, structured initially as a collaboration that includes milestone payments, earn-outs, and ultimately, an option to buy if prespecified milestones are met.

6. Innovation Fitness: "Sustaining" vs "Disruptive"

This chapter opened with the notion that a key to investing is identifying the sustainability of a company's competitive advantage. For product-driven industries, a fundamental yardstick in this exercise is the "fitness" of a particular innovation. Innovations across all sectors have historically been categorized as incremental, radical, evolutionary, or revolutionary, depending on the magnitude of departure from previous offerings. In delineating the attractiveness of new business opportunities and the sustainability of growth, Clayton Christensen, in his 1997 book, *The Innovator's Dilemma*, separates innovation into two new distinct categories: sustaining and disruptive. Sustaining innovations are those that improve on current solutions and can be either incremental or radical. These innovations improve upon existing products, enabling the manufacturer to sell them to its best customers at higher prices. Although a sustaining innovation may put a company ahead by a nose in a particular market, competitors, which have a lot to lose, will surely take a fight over a flight approach. As such, the incumbents, which typically have superior competitive resources (e.g., capital, patent estates, engineering, manufacturing expertise), will almost always prevail in the long run. One need not be a veteran of medical devices to have seen this phenomenon play out time and time again.

Products based on disruptive technologies, in stark contrast to sustaining ones, are developed targeting features such as smaller size, ease of use, and cost reduction. Often, disruptive technologies often get off to a slow start, as the success rates and/or feature sets typically lag behind those of established technologies. In effect, sustaining innovations improve on existing products, and disruptive ones—at least at the outset—are worse. However, Christensen finds disruptive innovations as the best way to depose established franchises and create the most sustainable competitive advantage. For example, the PC disrupted the mainframes and minicomputers in the early 1980s, but a minicomputer user could not switch to a PC because it could not support the necessary applications when it was first launched. As it evolved and its range of capabilities became increasingly robust, the PC completely supplanted its predecessors, and companies like Digital Equipment Corporation were displaced by the likes of Apple Computer and IBM. A similar scenario played out with traditional health insurers and health maintenance organizations.

Our mission is not to dismiss the significance of sustaining technologies. Enhancing one's product offering is critical to a company's strategy and in building near-term shareholder value; however, Christensen's innovator's dilemma lies in that fact that to managers, sustaining innovations are enormously attractive, and investing in disruptive ones often seems counterintuitive. On the other hand, investing in a solution that is often simpler, cheaper, and carries lower margins does not seem to make much sense at first blush. Moreover, many companies base their investments on customer feedback and intense market research, both of which offer transparency into the reception and payoff on a sustaining innovation. Conversely, disruptive innovations typically represent a paradigm shift, making them impossible for customers and market research to fully appreciate. As a result, sound management skills that may be a hallmark of a company's success—such as listening to customers, calculated investing for improving near-term returns, and managing profit and loss (P&L) metrics to impress shareholders—can also lead to that company's demise.

7. Case Study: CABG to PTCA

In the 1970s, the concept of coronary angioplasty went unnoticed as companies such as Bentley Laboratories (eventually bought by American Hospital Supply, which was then bought by Baxter and spun out as part of Edwards Lifesciences) and Cobe (bought by Sorin Biomedica) fought it out in the market for CABG. CABG is a highly invasive procedure that involves physical access into the thoracic cavity, and at the time, was the standard of care for treating occlusions in coronary vessels. A catheter-based percutaneous solution was first introduced in the 1960s and was conceived to reduce scarring, hospital stays, and, in turn, costs. In the decade to follow, smaller players such as Scimed, Schneider (now both part of Boston Scientific), and ACS (Guidant's vascular franchise) developed the technique and launched their first products in the United States in 1980. PTCA flew under the radar of the established players, as the technology needed to be refined for several years before penetration climbed to a formidable level. Exemplifying the innovator's dilemma, the established players found sustaining innovations so attractive that they did not develop an appreciation for the disruptive technology until it became clear that they were about to lose the game. Today, more than 1.5 million stents are implanted annually in the United States alone, compared to approximately 400,000 CABG procedures.

8. Conclusion: Innovation Investing

At the end of the day, an investor's goal—like the innovating companies themselves—is to make a return on an investment. One of the principles of investing in publicly traded companies is that one has to listen to what the stock is saying; that is, every share price has certain embedded growth expectations, which are a reflection of consensus views, and appreciation or depreciation is contingent on those expectations being revised up or down. In other words, an investment decision comes down to determining the embedded expectations, comparing them to one's own, considering reasons for discrepancy and prospects for revision, and deciding to bet "the over" or "the under." To anticipate that, one has to understand the pressure points of the company's growth profile. Lower-tech, less-innovative medical devices businesses (e.g., hospital supplies) generally participate in more mature markets. These companies are not exposed to the steep slopes of medical device product cycles; rather, the key value drivers are typically market share, pricing, and capital efficiency. The growth profiles of higher-tech medical device companies are invariably married to innovation and new product flow. To that end, device stocks can be relatively volatile, as the emergence of each new technology is associated with a news-intensive clinical and regulatory pathway, with sharp revisions in expectations. As pointed out earlier, shareholder value is created well before the companies generate the actual returns. As a result, an innovation investor must recognize a new product's potential before the market fully appreciates it and anticipate a revision to embedded expectations. The magnitude of shareholder value creation—or destruction—is a reflection of the extent to which consensus views change. To that end, contrary to what intuition may suggest, the greatest value appreciation is typically not captured anywhere on the steep slope of the product cycle, because investors already have a good sense of the fate of the market and its participants. Rather, the point on the curve where investors may need to make the most drastic revisions to their models is toward the tail end of what has been defined as stage 1. Early clinical trials (early stage 1) are harbingers of things to come, but they can be misleading. Most early clinical work is conducted in a small number of institutions, may not be randomized, and encompasses "easy-to-treat" patients—that is, these trials tend to exclude patient populations with risk factors associated with the disease and/or therapy. As a result, the results may not be emblematic of a real-world setting, which is why the FDA requires a pivotal trial be designed to embody to support approval. The data from the pivotal trial is a relatively transparent premonition of the product's eventual labeling and positioning and can lead to drastic revisions in what investors expect as it relates to regulatory timing, market penetration, and market share.

Suggested Readings

1. DiMasi, J. A. 2002. The economics of pharmaceutical innovation: trends in costs, risks, and returns. *CPSA Digest*. Available from: www.milestonedevelopment. com/CPSA/2002/tuoa1.html. Accessed August 4, 2005.
2. Bain, J. S. 1951. Relation of profit rate to industry concentration: American manufacturing, 1936–1940. *Quart. J. Econ.* 65:293–324.
3. Mauboussin, M. 2002. The Babe Ruth effect: frequency vs magnitude. *Credit Suisse First Boston: The Consilient Observer* 1:1–4.
4. *Fortune*. November 1999.
5. Christensen C. M. 1997. *The Innovator's Dilemma*. Boston: Harvard Academic Press.

II

CASE STUDIES

11

Challenges in Conducting Implantable Device Trials
Left Ventricular Assist Devices in Destination Therapy

Ursula Maria Schmidt-Ott, Alan J. Moskowitz, Annetine C. Gelijns, Julie C. Choe, Michael Parides, and Deborah V. Davis Ascheim

1. Introduction

The medical device industry is characterized by a high level of innovation and the development of a continuous stream of devices that promise to enhance the life expectancy and quality of life of patients. With changes in US Food and Drug Administration (FDA) regulatory policies and increased demand for more rigorous clinical evidence by payors, physicians, and patients alike, there has been more emphasis on randomized trials to evaluate new devices as they move from bench to bedside.

The randomized controlled trial (RCT) is widely regarded as the most powerful and sensitive tool for comparing therapeutic interventions and, hence, the most persuasive evidence for adopting new technology.[1] Many of the differences between drug and device innovation, however, complicate the translation of RCTs' methodology from pharmaceuticals to devices. Device trialists must contend with unique technical (e.g., surgical), ethical, and methodological issues,[2] which are readily evident in the case of left ventricular assist devices (LVADs). LVADs fall into the highest of three risk categories defined by the Medical Device Amendments of the federal Food, Drug, and Cosmetic Act[1] and are one example of an important innovative device. Advances in mechanical circulatory support technology are particularly important as nonsurgical treatment methods for end-stage heart failure are palliative at best, with little effect on the profoundly high mortality rates seen in this disease.

This chapter discusses the challenges of LVAD trials, highlighting the methodological, ethical, and economic issues that arise in designing and conducting clinical trials that compare surgical deployment of high-risk medical devices with medical therapy.

From: *Clinical Evaluation of Medical Devices: Principles and Case Studies, Second Edition*
Edited by: K. M. Becker and J. J. Whyte © Humana Press Inc., Totowa, NJ

2. LVADs and the Treatment of End-Stage Heart Failure

Chronic heart failure is a leading cause of death in the United States and much of the developed world.[3] It affects an estimated 5 million Americans, with 550,000 new cases diagnosed each year at an annual treatment cost of nearly $26 billion.[4] The annual mortality of heart failure of all classes in aggregate is 18.7%[4] and that of advanced heart failure is 75%. Medical therapy has proven to be beneficial for mild to moderately severe heart failure, but for patients with advanced heart failure, cardiac transplantation and, more recently, LVADs constitute the only effective treatment option. Cardiac transplantation has 5-year survival rates of approximately 65%.[5] Yet, the success of cardiac transplantation remains limited by the complication of long-term immunosupression, the development of allograft coronary artery disease, and most importantly, the serious shortage of donor organs.[6] An estimated 20 to 40% of transplant candidates die awaiting a donor heart.[7] Biological solutions to this shortage will not be available for many years, and efforts aimed at increasing the supply of donor organs[8]—currently about 2700 hearts annually worldwide[5]—have failed to ameliorate the shortage, underscoring the crucial need for alternatives to cardiac allotransplantation. Mechanical support by means of ventricular assist devices is at present the most promising alternative. Ventricular assist devices are mechanical pumps that take over most or all of the function of the damaged pumping chamber (ventricle) and restore normal hemodynamics and blood flow to vital organs. An LVAD was first used clinically in 1963,[9] and an implanted total artificial heart was first used as a bridge to transplantation in 1969.[10] Since the National Institutes of Health (NIH) established the Artificial Heart Program in 1964 to promote the development of the total artificial heart and LVADs,[11] a variety of circulatory support devices have been developed for the temporary support of patients with end-stage heart failure. LVADs are intended to augment the function of the left ventricle in several clinical circumstances. They may be used to provide short-term circulatory support in patients who have transient cardiac failure after cardiac operations or myocardial infarction or during viral myocarditis.[1] They are also used to as a "bridge" to cardiac transplantation, providing circulatory support to transplantation candidates who deteriorate awaiting a donor organ.[1] Another use is as a permanent alternative to transplantation (i.e., destination therapy) for patients who are not transplant candidates.[1] The first device approved for destination therapy was evaluated in the Randomized Evaluation of Mechanical Assistance for the Treatment of Congestive Heart Failure (REMATCH) trial, which we discuss in the following section.

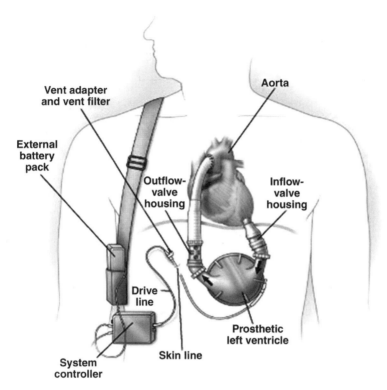

Fig. 1. Components of the HeartMate vented electric left ventricular assist device.[12] The inflow cannula is inserted into the apex of the left ventricle, and the outflow cannula is anastomosed to the ascending arota. Blood returns from the lungs to the left side of the heart and exits through the left ventricular apex and across an inflow valve into the prosthetic pumping chamber. Blood is then actively pumped through an outflow valve into the ascending aorta. The pumping chamber is placed within the abdominal wall or peritoneal cavity. A percutaneous drive line carries the electrical cable and air vent to the battery packs (only the pack on the right side is shown) and electrical controls, which are worn on a should holster and belt, respectively.

3. Features of the REMATCH Trial

The REMATCH trial compared long-term implantation with the Thoratec HeartMate® implantable LVAD (*see* Fig. 1) to optimal medical therapy in patients who were not candidates for transplantation.[12] The research was divided into two phases: a multicenter randomized pilot study that had limited statistical power and a definitive multicenter RCT.

3.1. Phase I: The Pilot Study—PREMATCH

The pilot study had several major objectives.[6] The first was to obtain preliminary data on patient outcomes and device reliability, which was needed to design the large-scale clinical trial. Although there was considerable experience using this device to support transplantation candidates, the LVAD had never been implanted in the older and sicker patients, who were ineligible for transplantation. Second, the HeartMate vented electric (VE) device was still undergoing major modifications and the pilot study served to further the experience of the surgeons with the device. Third, a randomized pilot study was needed to determine whether randomization was feasible in a surgical trial of a life-threatening condition in which treatment options were so vastly different. Finally, because there were no explicit guidelines for optimal medical management of patients with end-stage heart failure, part of the pilot mission was to develop standards for both the medical and surgical treatment of these patients.

An important design issue in the pilot study was whether the device was to be compared with standard community heart failure care or optimal medical management as delivered by heart failure specialists. The former would give results that were more easily generalized, whereas the latter would provide a more stringent test of the device. It was the consensus of the investigators that the extraordinary nature of the risks and lifestyle complications of living with an LVAD merited a more rigorous test; therefore, optimal medical management (OMM) would be the more appropriate comparison. The pilot study was conducted from April 1996 through April 1998. Twenty-one patients from five centers were randomized: 10 OMM and 11 LVAD patients who had an average of 276 days and a maximum of 607 days of LVAD support. Despite the apparent differences among the treatment arms in this small cohort, patients and physicians agreed to the randomization scheme.[6]Equipoise, the principle of equivalent uncertainty regarding the merits of two or more treatments, is required of investigators conducting ethical research; it would not be ethical to conduct an RCT for a device determined from initial testing to be in the breakthrough realm for the population being considered.[1] Although initially challenged, the position of equipoise for REMATCH was strengthened by the analysis of pilot data, which demonstrated a 3-month mortality of almost 30% in both treatment groups.[1] In addition to supporting the notion of equipoise for REMATCH, the pilot study demonstrated that operative deaths were not excessive in this older and sicker population. This piece of information was important for designing the analytical plan for the full-scale clinical trial, because a high operative mortality rate in a trial comparing surgical and medical therapies might lead to crossing survival curves should the survivors of surgery experience a lower mortality rate than their counterparts. This would mean that the mortality rates were not proportional and would preclude use of the Cox proportional hazards model, a mainstay of survival analysis. Although the survival curves in PREMATCH did not violate the proportional hazards assumption, mortality rates were uniformly high in both treatment arms, which was owing to heart failure in the medical therapy arm and infection in the LVAD arm. This latter piece of information led to modification of the device driveline, which was believed to be the portal for infection; clinical management modifications related to antimicrobial use; aseptic techniques in the operating room; and management of driveline infections. During the course of the pilot study, REMATCH protocol design was finalized and the pivotal trial was initiated in May 1998.

3.2. Phase II: The Randomized Controlled Trial—REMATCH

3.2.1. Organization and Funding of the Trial

The study was conducted in 20 experienced cardiac transplantation centers with the expertise to provide sophisticated treatment of advanced heart failure. The trial was supported through a cooperative agreement with Columbia University, NIH's National Heart Lung and Blood Institute, and Thoratec Inc., who provided supplemental financial support for data collection and donated the LVADs for the trial. Patients' care costs, except for the initial LVAD implantation admission, were supported by the Centers for Medicare and Medicaid Services (CMS), which also provided access to its data files for purposes of cost and cost-effectiveness analysis. The trial was supervised by a steering committee and conducted by an independent Coordinating Center (International Center for Health Outcomes and Innovation Research, Columbia University). An independent morbidity and mortality committee adjudicated causes of death and adverse events. The NIH appointed a Data and Safety Monitoring Board to review trial progress. The FDA granted an investigational device exemption for the pivotal phase III trial, and participating institutional review boards approved the protocol and informed consent documents at each participating institution.

3.2.2. Study Patients

The target population was adults with end-stage heart failure who were ineligible for cardiac transplantation because of their advanced age or comorbid clinical conditions. Patients needed to fulfill the following entry criteria: (1) New York Heart Association (NYHA) Class IV symptoms of heart failure for at least 90 d, despite attempted therapy with angiotensin-converting enzyme inhibitors, diuretics, and digoxin; (2) left ventricular ejection fraction of no more than 25%; and (3) peak oxygen volume of no more than 12 mL/kg per minute or documented failure to wean intravenous inotropic therapy owing to symptomatic hypotension, decreasing renal function, or worsening pulmonary congestion. β-Blocker therapy was acceptable if administered for at least 60 out of 90 days before randomization. At the encouragement of the investigators, enrollment criteria were relaxed slightly after 18 months, reducing the requirement of symptoms of NYHA Class IV to at least 60 days and a peak oxygen consumption to 14 mL/kg per minute. In addition, patients who had been in NYHA Class III or IV for at least 28 days and who had received at least 14 days of aggressive support with an intra-aortic balloon pump or who had demonstrated a dependence on intravenous inotropic agents, with two failed attempts to wean pharmacological support, were now also eligible. Only five patients (three in the group that received an LVAD and two in the medical therapy group) were enrolled who met the broadened criteria. Transplantation was contraindicated in the study population for at least one of the following reasons: more than 65 years of age, insulin-dependent diabetes mellitus with end organ damage, chronic renal failure with a sustained serum creatinine, of more than 2.5 mg/dL for at least 90 days before randomization, or other significant co-morbidities. Detailed exclusion criteria were previously reported.[6]

3.2.3. Study Design

The trial used a parallel design in which patients were randomly assigned to the VE Thoratec LVAD or OMM in a 1:1 ratio. The randomization was stratified by center and blocked to ensure ongoing equivalence of comparison group size. Eligibility was determined by site investigators and confirmed by the Coordinating Center "gatekeeper," who reviewed all source documentation supporting eligibility before randomization. Surgical management followed guidelines developed and updated by a surgical management committee, which addressed pre-, intra-, and postoperative measures. OMM followed guidelines developed by the medical management committee, with the goal of optimizing organ perfusion and minimizing symptoms of congestive heat failure. The unethical nature of sham surgery and inability to mask device function precluded double-blind design, although investigators were masked to overall outcome data throughout the trial. Only the statisticians of the Coordinating Center and the trial Data and Safety Monitoring Board were privy to the full data set throughout the course of the study. To comply with FDA reporting requirements, Thoratec received ongoing data for LVAD patients but was masked to all OMM data. Outpatients were followed monthly throughout the duration of the study.

3.2.4. Statistical Analysis

The primary endpoint was all-cause mortality analyzed using the log-rank test. Cox proportional hazards regression was used to estimate hazard ratios and 95% confidence intervals (CIs) and to adjust for differences in baseline outcome predictors. Analyses were conducted according to the intention-to-treat principle, in which patients' outcomes were analyzed according to their treatment group, regardless of whether they actually received that therapy or crossed over to the parallel therapy. The trial was designed to enroll 140 patients and to continue until 92 deaths had occurred. These figures were estimated on the basis of the following assumptions: a review of the literature suggested that the 2-year mortality rate for similar patients receiving medical management was approximately 75%, and it was hypothesized that use of the LVAD would reduce this rate by one-third (i.e., the 2-year mortality rate among the patients in the medical therapy group would be 75%; treatment with a LVAD would reduce the risk of death by 33%). The analyses were designed to have a 90% power (two-sided test with $\alpha = 0.05$) to detect a significant difference between the treatment groups. Interim analyses were scheduled at three points (after 23, 46, and 69 deaths had occurred), using a two-sided significance test with the O'Brien-Fleming spending function and a type I error rate of 5%.Secondary endpoints included the incidence of serious adverse events, the number of days in and out of the hospital, quality of life, symptoms of depression, and functional status. Frequency of adverse event occurrence was analyzed by Poisson regression. Quality of life and functional status were assessed using the Minnesota Living with Heart Failure questionnaire, two prespecified subscales (i.e., physical function and emotional role) of the 36-item Medical Outcomes Study Short-Form General Health Survey (SF-36), and the NYHA heart failure classification. Symptoms of depression were assessed with the Beck Depression Inventory. Analysis of covariance was used to test for significant differences in the mean quality of life among surviving patients, after adjustment for baseline values. Economic evaluation was based on costing data collected directly during the trial. To determine the hospitalization costs, the REMATCH data were combined with data from CMS, standard billing forms, and line-item bills provided directly by clinical centers. All costs before patient randomization were eliminated to provide a uniform starting point for trial-related treatment costs. Institution-specific cost reports were used to calculate ratio of cost to charges for each major research category. The LVAD costs were estimated to be $60,000. Fees for professional services were not included in the analysis, in part because many of the physicians in the trial did not bill for their services. Implantation hospitalization costs began at randomization and ended at discharge from the acute care facility. Rehospitalization costs included readmission to acute care facilities or intermediate care centers. Costing data were compared by student *t*-test after log transformation and predictors determined by regression.

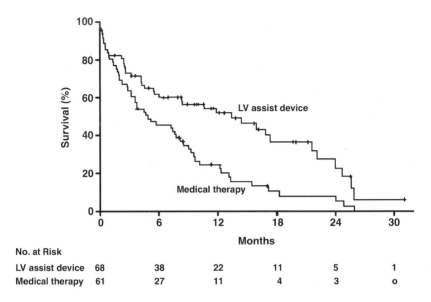

Fig. 2. Kaplan-Meier Analysis of survival in the group that received the left ventricular (LV) assist devices and the group that received optimal medical therapy.[12] Crosses depict censored patients. Enrollment in the trial was terminated after 92 patients died; 95 deaths occurred by the time of final analysis.

3.2.5. Results

3.2.5.1. BASELINE

One hundred twenty-nine patients were enrolled in the study from May 15, 1998 to July 27, 2001. Sixty-eight patients were randomized to LVAD implantation, and 61 patients were randomized to medical management. The two groups were similar with regard to baseline characteristics. Advanced age (>65 year) was the most common reason for ineligibility for cardiac transplantation. All 129 were included in the primary endpoint analysis. Two patients from the medical therapy group withdrew from the trial 1 and 6 months after randomization. All patients who were assigned to receive a LVAD had the device implanted. No patients in either group crossed over during the trial or underwent cardiac transplantation.

3.2.5.2. SURVIVAL

The clinical data set used for the analysis that follows was closed when the REMATCH trial met its predetermined mortality endpoint in July 2001. Figure 2 shows the Kaplan-Meier survival curves. A reduction of 48% in the risk of death from any cause—the primary endpoint—in the group that received a

LVAD was demonstrated, as compared with the medical therapy group (relative risk, 0.52; 95% CI, 0.34–0.78; $p = 0.001$). The Kaplan-Meier estimates of survival at 1 year were 52% in the device group and 25% in the medical therapy group ($p = 0.002$) and at 2 years, they were 23% and 8% ($p = 0.09$), respectively. After all the patients had completed the 24-month follow-up period, the 2-year survival rate in the LVAD group reached statistical significance. As of July 2001, the median survival was 408 days in the device group and 150 days in the medical therapy group. At the time of final analysis in July 2001, 41 deaths had occurred in the device group and 54 in the medical therapy group. Terminal heart failure caused the majority of deaths in the medical therapy group, whereas the most common cause of death in the device group was sepsis (41% of deaths) and device failure (17% of deaths). A prespecified analysis with stratification according to age (18–59 years, 60–69 years, and >70 year) showed that there was a significant reduction in the risk of death among patients in the device group who were ages 60 to 69 year (relative risk, 0.49; 95% CI, 0.25–0.95), and a trend toward a benefit in the younger age-group (relative risk, 0.47; 95% CI, 0.17–1.28) and the older age-group (relative risk, 0.59; 95% CI, 0.31–1.15). In the overall group of 22 patients who were younger than 60 year, the 1-year survival rate was 74% in the device group and 33% in the medical therapy group.

3.2.5.3. QUALITY OF LIFE

All patients completed the baseline assessments of quality of life, and there was no significant difference between groups. Five of the 11 patients in the medical therapy group who were alive at 1 year did not complete the questionnaires. All but 1 of 24 patients in the device group completed the questionnaire. At 12 months, the LVAD group had significantly better mean scores on the SF-36 physical function scale ($p = 0.01$) and the SF-36 emotional role scale ($p = 0.03$). The Minnesota Living with Heart Failure score was also better in the device group than in the medical therapy group at 1 year, but the difference was not significant ($p = 0.11$). The Beck Depression Inventory scores favored the LVAD group at every interval and were significant at 12 months ($p = 0.04$).

3.2.5.4. ADVERSE EVENTS AND EVENTS RELATED TO THE DEVICE

Adverse event analyses were done when the trial met its predetermined mortality endpoint in July 2001. Because of differences in survival between groups, adverse events were reported as rates per patient-year. Overall, the LVAD group was more than twice as likely to experience an adverse event than the group receiving medical management (rate ratio, 2.35; 95% CI: 1.86–2.95). Specific serious adverse events that were significantly more likely in the

LVAD group included nonneurological bleeding (rate ratio, 9.47; 95% CI: 2.3–38.90) and neurological dysfunction (rate ratio, 4.35; 95% CI: 1.31–14.50). A higher rate of sepsis in the LVAD group was nearly significant, with a rate ratio of 2.03 and 95% CI from 0.99 to 4.13. Rate ratios exceeded 2.0 for supraventricular arrhythmia and peripheral embolic events, but 95% confidence intervals overlapped 1.0. Events specifically related to the device were also reported as rates per patient-year. The rates for specific types of events were suspected malfunction of the LVAD (0.75), perioperative bleeding (0.46), infection of the driveline tract or pocket (0.41), other pump component infections (0.23), right heart failure (0.17), failure of the LVAD system (0.08), and thrombus in the LVAD (0.06). The probability of device failure was reported at 35% at 2 year. The device required surgical replacement in 10 patients.

3.2.5.5. HOSPITALIZATION

The median number of days spent in the hospital was 88 for the LVAD group and 24 for the medical management group. Median time spent out of the hospital was 340 days for the LVAD patients and 106 days for patients managed medically.

3.2.5.6. COSTS

Detailed cost records were obtained for 52 of 68 LVAD patients (77%).[13] The remaining 16 patients were not in the CMS database, and the participating hospitals were unable to provide cost data for them. The mean costs for the initial implant plus related hospitalization were $210,187 plus $193,295. When implantation hospitalization costs were compared between hospital survivors and nonsurvivors, the mean costs increased from $159,271 plus $106,423 to $315,015 plus $278,713. Sepsis, pump-housing infection, and perioperative bleeding were the major risk factors driving LVAD costs. In the patients who survived the procedure ($n = 35$), bypass time, perioperative bleeding, and late bleeding were significant risk factors. The average annual readmission cost per patient for the overall cohort was $105,326.[13]

3.2.6. Discussion

This trial demonstrated that long-term support with an LVAD resulted in substantial improvement in survival in patients with severe heart failure who were not candidates for cardiac transplantation. The implantation of an LVAD was associated with a relative reduction in the risk of death of 48% during the entire follow-up period and an absolute reduction in the mortality rate of 27% at 1 year. These findings imply that the implantation of an LVAD could avert at least 270 deaths annually for every 1000 patients with end-stage heart failure. Early experience with artificial hearts suggested that any potential survival benefit would be achieved at an unacceptable cost to the quality of life.[14] In the REMATCH trial, measurements of the quality of life at baseline and during the study in the medical therapy group reflected the severe physical, emotional, and functional impairment of these terminally ill patients. Although the scores on the physical-function and emotional-role subscales of the SF-36 were significantly better at 1 year in the device group than in the medical therapy group, they were not the scores of healthy people in the general population.[15] However, the physical-function scores were comparable to those reported for patients receiving long-term hemodialysis and ambulatory patients with heart failure[15] and the emotional-role scores were better than those reported for patients with clinical depression but similar to those for ambulatory patients with heart failure. Despite the substantial survival benefit, the morbidity and mortality associated with the use of the LVAD were considerable. In particular, infection and mechanical failure of the device were major factors in the 2-year survival rate of only 23%. The device used in the trial has a large-bore percutaneous line, which can become a conduit for bacterial and fungal infection. Investigators also found that malnutrition was a problem in these chronically ill patients, predisposing them to infection and other complications. Failure of the LVAD was the second most frequent cause of death in the device group, leading to modifications of the device. Still, the decreased mortality rates, despite the increased rates of adverse events in the device group, represented an acceptable tradeoff, given the natural history of end-stage heart failure, although the frequency of complications indicates the need for further improvements in patient care and device design. In all, these findings established the LVAD as a new long-term myocardial replacement therapy, joining cardiac transplantation as a treatment option for end-stage heart failure. Although the 1-year survival rate of more than 80% for transplantation far exceeds the survival rate for LVADs in the REMATCH trial,[5] the outcomes for transplantation do not include the substantial mortality rates among patients who are awaiting transplantation. Thus, the combination of the availability of LVADs and the encouraging 1-year survival rate of 74% in patients younger than 60 suggests that a comparison of the long-term use of improved versions of these devices and transplantation may eventually be appropriate.

4. Post-REMATCH Era and Development of New Devices

Although the HeartMate LVAD has been approved by the FDA as a bridge to transplantation since 1994 and used in more than 4000 patients worldwide, destination therapy now offers a new treatment option for the much larger end-stage heart failure population.[16] Following the REMATCH trial, Thoratec submitted a premarket approval application for HeartMate's use as destination therapy in patients who are ineligible for cardiac transplantation. The FDA approved this application on November 6, 2002, and the HeartMate VE assist device became the first LVAD approved for use as a long-term solution for the growing worldwide incidence of end-stage heart failure.[16] Four months after the FDA approval, the Medicare Coverage Advisory Committee met to consider payment options for the device.[17] In March 2003, the committee voted 6 to 1 that the device offered "substantial benefit to patients with severe congestive heart failure."[18] The group recommended restricting the use of the device to patients who met the study criteria and advised that only heart transplantation centers be authorized to perform the surgery.[17] On the basis of the advisory committee's recommendations, the CMS changed their reimbursement policies in October 2003 to cover not only patients awaiting heart transplantation as they had previously but also to provide reimbursement for patients who undergo the procedure for destination therapy at designated LVAD facilities.[17] There is great expectation that the coming years will see substantial reduction in morbidity and mortality in heart failure, driven in part by innovative new devices. At present, four classes of new generation devices are at advanced stages of clinical development. One is a new generation of pneumatic, direct mechanical ventricular assist devices that are wrapped around the heart and do not directly contact the blood. Others consist of a totally implantable LVAD and compact axial-flow pumps that may decrease the risk of infection. The fourth class is represented by the pneumatic total artificial heart, which may have a role as a bridge to transplantation for a subgroup of patients with severe biventricular dysfunction and a fixed increase in pulmonary vascular resistance in whom left ventricular support may not be sufficient.[19] In the long run, continued innovations and technical advances will safely extend the use of mechanical circulatory-support systems to a broader patient population, rather than the current restrictions of implantation in the terminally ill.

5. Insights From REMATCH and Future Directions

Comparing LVAD implantation with medical therapy in the REMATCH trial prompted consideration of a number of methodological, economic, and ethical issues.[6] Conducting the pilot study proved essential, in part because it provided the patient and device-reliability data necessary to plan the larger-scale pivotal trial. The position of equipoise was strengthened by the analysis of the pilot data, which indicated no major difference in early outcomes between the device and OMM.[1] A Coordinating Center gatekeeper was implemented into the study design to guarantee eligibility of each patient enrolled. This was particularly important when evaluating such an expensive intervention. Randomization in the pilot study was shown to be feasible.[6] After randomization in a trial comparing surgery with medical therapy, both the investigator and the patient will immediately know the treatment assignment and have expectations regarding outcome. Patients with life-threatening diseases consenting to trials are likely biased toward the procedure and thus may perceive randomization to the control arm as a loss of hope, with possible deleterious impact on enrollment and potentially on outcome. To ameliorate these potential effects of randomization to a control arm in a nonblinded trial, an option to receive active device therapy after the achievement of a predefined time or intermediate endpoint can be implemented.[1] This feature may encourage recruitment and retention, while realigning incentives for the patients and physician to continue full efforts after randomization to a control arm. In the REMATCH trial, OMM patients were given the option of crossing over to LVAD therapy in June 2001, when the primary endpoint criteria was reached and LVAD therapy was recognized to offer a survival advantage. Three of five patients crossed over, one patient withdrew from the trial, and one patient opted to continue OMM alone. Regardless, trial enrollment remained challenging owing to the reluctance of patients and families, as well as physicians, to accept randomization in the setting of a life-threatening illness when a new therapy might be lifesaving. The economic challenges and their impact on conducting innovative device trials were well illustrated in the REMATCH trial.[12] The NIH and Thoratec supported the costs of data collection and analysis, the NIH covered costs of the operation, and the company provided devices free of charge. CMS agreed to support all treatment costs, except the implant hospitalization, which remained the responsibility of the participating hospitals. It was considered unethical to charge patients, especially because they had a life-threatening condition and were therefore particularly vulnerable. The economic burden of the implant hospitalization caused centers to decline participation in the trial, and participating centers imposed restrictions on enrollment. In the last quarter of 1999, the company provided extra financial support for patient care, which had a small positive effect on enrollment, but more importantly, it kept participating centers in the trial. Unlike pharmaceutical innovation, medical device research is often developed in small companies without previous product revenue to support clinical research. Thus, financial impediments have complicated the design of a sufficient research infrastructure and have profoundly impaired the conduct of clinical trials of innovative devices, for which there have been substantial treatment costs.[1] The financial disincentives to enrollment increase the duration of the trial and the overall costs, delaying the time to potential recovery of development costs. This underlines the need for options of adequate clinical trial funding. A promising strategy may be conditional coverage, in which insurers pay the treatment costs for patients in an approved research protocol and sponsors cover the costs of conducting and analyzing the research. A new CMS policy, introduced September 2000, approved Medicare coverage of the routine health care costs for beneficiaries involved in clinical trials, particularly those sponsored or approved by federal agencies.[20] In addition, an advisory panel to the NIH supported public–private partnerships between industry and having NIH fund clinical trials,[21] as was the case in the REMATCH trial. Innovative, implantable devices are increasingly subject to rigorous coverage and reimbursement decisions by the payor community. The criteria that payers (i.e., public and private insurers) use to make these decisions vary. Although CMS has no specific statutory mandate to account for cost and cost-effectiveness when making coverage decisions, increasing fiscal pressure on CMS has made it a topic of much interest among CMS representatives. This underlines the crucial need for designing clinical trials to incorporate measures of cost and cost-effectiveness.

As LVADs for destination therapy are an emerging procedure, opportunities exist for reducing costs and improving outcome with experience. First, the REMATCH analysis revealed that sepsis was the most important predictor of cost and the most common cause of death in the device group.[13] Improvements in the design of devices offer a sizeable opportunity to reduce infections and consequent costs of future patients. Newer devices have smaller, more flexible drivelines or use a totally implantable device, which eliminates this portal of infection. Malnutrition, another major factor that predisposes this population to infection, may be reduced with newer axial flow devices, which do not have an intra-abdominal component, and with newer approaches to nutritional management and chronic inflammation.[13] Second, improved surgical proficiency and innovative approaches to management of bleeding address important factors that could also improve outcomes and reduce overall costs. Device-related bleeding has already been reduced by two significant modifications to the approved HeartMate device. Third, device reliability is essential to improve the outcome and minimize the cost of readmisssion. The REMATCH experience, as of July 2001, revealed that 17 devices in 52 patients needed replacement. In addition to sepsis, the leading causes of replacement were mechanical pump failure and inflow valve incompetence.[13] There have been more than 40 modifications made to the device that was initially employed in the REMATCH trial. These changes were approved by the FDA[13] and may continue to improve device reliability and outcomes. Modifications to devices that may alter the cost and outcome of RCTs take place during the trial and continue thereafter. Even over the 2-year course of the REMATCH trial, device modifications and implementation of an infection management guideline were associated with a significant improvement in survival and occurrence of adverse events in patients enrolled during the second half of the trial, as compared with those enrolled in the first half of the trial. This suggests that we will continue to see further improvements as we mature in our experience with LVAD as destination therapy. Finally, patient selection offers another opportunity to improve outcome and reduce costs. Survivors of the index hospitalization after LVAD implantation cost noticeably less to manage than nonsurvivors do. Moreover, those surviving more than 1 year had substantially less hospital resource utilization than the rest of the cohort, although in the predictive model of the REMATCH trial, there were no significant predictors of costs among the baseline patient characteristics.[13] This might be in part a result of the small sample size. Similar to the bridge to transplant trials, in which preimplant patient characteristics have been identified that are independent risk factors for survival, larger LVAD destination therapy trials or continued postmarketing data collection in the future could help identify patient profiles of high-risk and high-cost patients that would assist in the patient selection process. With the growing number of different destination therapy trials, the recognized success of LVAD therapy for end-stage heart failure may be followed by a cycle of improving results and expanding the patient population definition.

Surgical therapies for heart failure carry front-loaded risk that is easier to absorb for patients expecting high early mortality.[1] As survival and improved function are realized by these desperate patients, the procedure is then sought by patients at earlier stages of the disease. These patients are more likely than the initial subjects to obtain good results from the procedure at lower costs.[1] Regarding the LVAD population, it can be assumed that with increasing reliability of devices, in the long term, the bridge to transplant and the destination-therapy population will merge and the distinction of short- and long-term indications for therapy will become less relevant. Thus, in the future, those bridge to transplant patients not receiving a donor organ may stay on long-term LVAD support as an alternative to transplantation. Of the estimated 60,000 patients who could benefit from cardiac transplantation each year but are not necessary candidates, it is conjectured that approximately 20% would be candidates for long-term LVAD therapy at present.[13] The number of devices and patients who form the basis of approval is of necessity relatively small, and extensive further experience is required to optimize the clinical use of new devices in long-term circulatory support. Because rare events or complications that require years to develop may not yet be appreciated and the duration in observation is more limited when severity of illness is higher (as in current populations with end-stage heart failure), efforts have focused on trying to shorten the premarketing clinical trial and FDA review processes, while shifting more emphasis to rigorous postmarketing studies and implementing registries both for advanced heart failure and LVADs. The former creates the potential for a larger number of patients to be rapidly exposed to a newly approved product, whereas registries would track changes in patient management and device modifications, as well as patient outcomes, as we continue to learn to care for LVAD patients and as device modifications continue to be introduced into practice.

References

1. Stevenson, L. W., Kormos, R. L., et al. 2001. Mechanical cardiac support 2000: current applications and future trial design. *J. Am. Coll. Cardiol.* 37:340–370.
2. Moskowitz, A., Reemtsma, K., Rose, E., et al. Clinical outcomes in surgery. In: Sabiston, D. C., ed. 1997. *Textbook of Surgery: The Biological Bias of Modern Surgical Practice.* WB Saunders, Philadelphia, 36–53.
3. Frazier, O. H. 2000. Mechanical cardiac assistance: historical perspectives. *Semin. Thorac. Cardiovasc. Surg.* 12:207–219.
4. *Heart Disease and Stroke Statistics—2004 Update.* 2003. American Heart Association, Dallas.
5. Hosenpud, J. D., Bennett, L. E., Keck, B. M., et al. 2001. The Registry of the International Society for Heart and Lung Transplantation: eighteenth official report— 2001. *J. Heart Lung Transplant.* 20:805–815.
6. Rose, E. A., Moskowitz, A. J., Packer, M., et al. 1999. The REMATCH trial: rationale, design, and end points. Randomized Evaluation of Mechanical Assistance for the Treatment of Congestive Heart Failure. *Ann. Thorac. Surg.* 67:723–730.
7. Abouawdi, N. L., Frazier, O. H. 1992. The HeartMate: a left ventricular assist device as a bridge to cardiac transplantation. *Transplant. Proc.* 24:2002–2003.
8. McLoughlin, M. P., Chapman, J. R., Gordon, S. V., et al. 1991. "Go on—say yes": a publicity campaign to increase commitment to organ donation on the driver's license in New South Wales. *Transplant. Proc.* 23:2693.
9. Hall, C. W., Liotta, D., Henly, W. S., et al. 1964. Development of artificial intrathoracic circulatory pumps. *Am. J. Surg.* 108:685–692.
10. Frazier, O. H. Long-term mechanical circulatory support. 1991. In: Edmunds, H. L., ed. *Cardiac Surgery in the Adult.* McGraw-Hill, New York, 1477.
11. The artificial heart program: current status and history. 1991. In: Hogness, J. R., VanAntwerp, M., eds. The artificial heart: prototypes, policies, and patients. National Academy Press, Washington, DC, 14–25.
12. Rose, E. A., Gelijns, A. C., Moskowitz, A. J., et al. 2001. Long-term mechanical left ventricular assistance for end-stage heart failure. *N. Engl. J. Med.* 345:1435–1443.
13. Oz, M. C., Gelijns, A. C., Miller, L., Wang, C., et al. 2003. Left ventricular assist devices as permanent heart failure therapy: the price of progress. Ann. Surg. 238:577–583; discussion 583–585.
14. The Schroeder Family, Barnette, M. 1987. The Bill Schroeder story: an artificial heart patient's historic ordeal and the amazing family effort that supported him. William Morrow, New York, 383.
15. Ware, J. E., Jr., Snow, K. K., Kosinski, M., et al. 1993. *SF-36 Health Survey: Manual and Interpretation Guide.* New England Medical Center Health Institute, Boston.
16. Thoratec Inc. What is Destination Therapy? Available from: www.thoratec.com/ /heartmate-destination-therapy/heartmate_faqs.htm. Assessed September 9, 2004.
17. Gillick, M. R. 2004. Medicare coverage for technology innovations—time for new criteria? *N. Engl. J. Med.* 350:2199–2203.

18. Medicare Coverage Advisory Committee meetings for ventricular assist devices as destination therapy. Minutes. March 12, 2003. Available from www.cms.hhs.gov/mcac/id79-2.pdf?origin=globalsearch&page=/med/viewmcac.asp&id=79. Accessed July 16, 2004.

19. Goldstein, D. J., Oz, M. C., Rose, E. 1998. Implantable left ventricular assist devices. *N. Engl. J. Med.* 339:1522–1533.

20. Centers for Medicare and Medicaid Services, Health and Human Services. Medicare Coverage for Clinical Trials. Available from: www.cms.hhs.gov/coverage/8d2.asp. Accessed September 7, 2004.

21. National Institutes of Health Director's Panel on Clinical Research, Executive Summary. Available from: www.nih.gov/news/crp/97report/execsum.htm. Accessed September 7, 2004.

12

BELIEF

A Randomized, Sham-Procedure-Controlled Trial of Percutaneous Myocardial Laser Therapy

Jan Erik Nordrehaug and Janet M. Fauls

1. Introduction

A majority of patients with angina related to coronary artery disease respond to medical management, percutaneous coronary intervention (PCI), or coronary artery bypass grafting (CABG). There is, however, a growing number of patients who have medically refractory angina caused by diffuse coronary artery disease who are not eligible for PCI or CABG owing to diffuse, distal disease.[1] Treatment options are limited for this subset of patients.

In 1995, before the completion of well-controlled trials, the Norwegian Ministry of Health (NMOH) prohibited surgical transmyocardial laser revascularization (TMR) as a routine clinical option for patients with severe angina owing to concerns regarding procedural morbidity and mortality. After the availability of randomized medically controlled trial results, this decision was revisited in 1999 through an extensive review of the nonclinical and clinical literature by an independent expert consensus panel of cardiothoracic surgeons, cardiologists, and physiologists commissioned by the NMOH.[2] The panel concluded that TMR consistently provides significant symptomatic relief; however, the mechanism for this clinical benefit is not understood. As a result, the ban was withdrawn, although the lack of evidence discounting the placebo effect as the primary mechanism for the observed clinical improvement limited the routine clinical use of the technology. To address the panel's findings, and because a sham TMR trial is not ethically possible, we designed a randomized, two-arm, double-blind clinical trial known as the Blinded Evaluation of Laser Intervention Electively For Angina Pectoris (BELIEF) trial to control for patient bias (the placebo effect) and investigator bias in determining the clinical benefit of percutaneous myocardial laser (PML) treatment.

From: *Clinical Evaluation of Medical Devices: Principles and Case Studies, Second Edition*
Edited by: K. M. Becker and J. J. Whyte © Humana Press Inc., Totowa, NJ

Table 1
Percutaneous Myocardial Laser Systems

Characteristic	Biosense DMR system	Axcis PML system	Slimflex PML system
Laser type	Ho:YAG	Ho:YAG	Ho:YAG
Catheter system	Single	Dual	Dual
Mechanical tissue penetration	None	3 mm	3 mm
Laser spot diameter	0.3 mm	1.6 mm	1.0 mm
Laser spot area	0.07 mm^2	2.0 mm^2	0.8 mm^2
Laser pulses delivered	1	4	3
Bench testing available	Yes	Yes	Yes
Open-label RCT available	No	Yes	Yes

Ho:YAG, holmium:yttrium-aluminum-garnet; RCT, randomized controlled trial.

2. Trial Design and Methods

2.1. The Device System

During the planning phase for this trial, three PML device systems were clinically available from the manufacturers (i.e., Biosense Webster, NOGA/ DMR system; CardioGenesis Corporation, Axcis PML system; and Eclipse Surgical Technologies, Slimflex PML system). The technological characteristics of these systems differ substantially (Table 1). To objectively choose a device for this trial, we identified several specific criteria. The device had to be well characterized and shown to be safe in bench and in vivo testing, receive or be under evaluation for European authorization, and have been demonstrated to provide significant clinical benefits in a randomized, controlled, open-label trial. The system chosen based on these criteria was the CardioGenesis Axcis PML system.[3]

The device system is composed of the holmium:yttrium-aluminum-garnet laser, an electrocardiography (ECG) monitor, and the sterile delivery system (Fig. 1). The laser is a portable, self-contained unit that has a maximum therapeutic output of 6 W and consists of a laser console and a footswitch. During clinical usage, this laser emits infrared radiation (2.1 μm wavelength) in four 2 J pulses, each 350-μsec in width and 35-msec in frequency. The ECG monitor is a microprocessor-controlled unit that generates a trigger signal synchronized to each R-wave of the patient's ECG, which is then transmitted to the laser. When the laser is in the "ready" mode, the combination of a depressed footswitch on the laser and the trigger pulse from the ECG monitor causes the laser to produce the required energy, typically 75 to 150 msec, from the

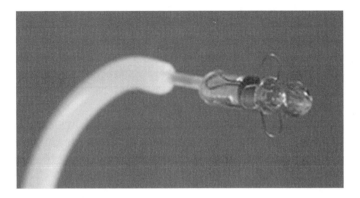

Fig. 1. Distal tip of the percutaneous myocardial laser catheter.

R-wave trigger. The laser energy is delivered to the target tissue via a single-use, coaxial assembly composed of two catheters: a laser catheter and an outer aligning catheter. The laser catheter is a nominal 6 F braid-reinforced catheter with a shaped distal tip that contains an optical fiber. This catheter terminates distally with a lens that delivers energy at a 1.6-mm spot size to create channels that are approximately 1 mm in diameter. Four nitinol petals located 3 mm from distal tip of the lens serve as depth stops to retard physical penetration of the lens, although the laser energy continues to penetrate the tissue to a depth of approximately 5 to 6 mm. For this reason, clinical application is limited to eligible areas of the left ventricular wall that are at least 8 mm thick to reduce the potential for pericardial perforation. The optical fiber is connected to the laser source via a standard military adapter connector at its proximal end. The aligning catheter is a nominal 9 F braid-reinforced catheter with a soft distal tip that is available in a series of tip configurations according to patient anatomy. This component provides passage for the laser catheter to access the left ventricle.

2.2. Financial Support

Owing to the nature of this trial and its potential impact on the field, funding for the trial was provided through a grant from the Bergen Norway Heart Foundation. All aspects of the trial conduct and follow up were conducted independently from the manufacturer and its representatives. None of the investigators had any financial interest in the manufacturer of the investigational device. Patients received no compensation for participating in the trial.

2.3. Participating Institutions

Two of the largest teaching and training institutions in Northern Europe and the largest in Norway participated in this trial: Haukeland University Hospital (Bergen, Norway) and Ullevääl University Hospital (Oslo, Norway). These are large institutions, with more than 1100 beds each, and are recognized as regional Centres of Excellence with funding directly provided by the government through the annual national budgets. All aspects of this trial were conducted under the governance of the institutional Ethics Review Committees and according to the recommendations for biomedical research as set forth in the Declaration of Helsinki.

2.4. Trial Design and Primary Endpoint

The conduct of any sham-controlled study in human patients demands that the benefits derived from the study outweigh the risks from exposure to the sham procedure. Accordingly, the ethics committees of both institutions required that the trial be rigorously designed to address the most relevant clinical question with the minimum number of patients. Considering the NMOH expert consensus panel findings that (1) this therapy is targeted to patients who have exhausted their options for relief of debilitating angina; (2) a clear mechanism of action is not known; (3) angina improvement is a well-established, clinically relevant endpoint in these patients; and (4) angina improvement has been shown to be subject to placebo effect in the literature,[4] this trial was designed to specifically control for the two primary sources of bias in clinical trials involving innovative technology: patient bias (placebo effect) and assessment bias. As such, the trial was designed as a randomized, two-arm (1:1), double-blind, sham-controlled trial intended to detect a clinically relevant difference in the proportion of patients with at least a one class improvement in Canadian Cardiovascular Society (CCS) angina class, assuming 20% improvement in the sham group. Using a two-sided significance level (α) of 0.05 and 80% power, 39 patients in each group were calculated to be necessary, based on the uncorrected chi-square statistic for analysis. To account for possible dropouts, the total trial enrollment was established at 82 patients. The trial was not powered for secondary assessments, which included ejection fraction, medication usage, exercise time, and quality of life. This sample size reflects a balance between limiting patient exposure to potential hazards from a sham procedure while producing the opportunity for a statistically and clinically meaningful result. All assessments were conducted by blinded, independent assessors. In addition, the trial duration was set at 1 year with interim follow up at 3 and 6 months.

2.5. Statistical Analyses

The CCS angina classification scale provides ordinal (not interval-ratio) data. Per the prospectively defined protocol, the comparison in improvement in angina class between the PML and control groups is made using the Mann-Whitney U test. Results are also presented as proportions improving from baseline and are analyzed using Fisher's exact test. Baseline characteristics are compared using a two-sided chi-square test with correction for continuity or Fisher's exact test for categorical data. The two-sided, two-sample student's *t*-test is used for analysis of continuous data compatible with a normal distribution. For cardiac event-free survival data, differences between groups were compared with the log-rank test. Two-sided *p* values of 0.05 or lower were considered to indicate statistical significance.

2.6. Patient Criteria

Inclusion criteria were stable CCS Class III or IV angina refractory to maximally tolerated doses of at least two antianginal medications, evidence of reversible myocardial ischemia in target areas on exercise testing or technetium sestamibi stress myocardial perfusion scanning, ejection fraction of at least 0.25, and myocardial-wall thickness at least 8 mm in the targeted region(s) by echocardiography. In addition, an expert consensus committee consisting of board-certified cardiologists and cardiothoracic surgeons reviewed each patient angiographically to verify that each patient was not a candidate for PCI or CABG. Exclusion criteria included myocardial infarction (MI) within 6 weeks, symptomatic heart failure with exercise limited by dyspnea, significant ventricular arrhythmias requiring chronic therapy or an implantable automatic defibrillator, suspected or observed ventricular thrombus; unstable angina requiring hospitalization within 14 days before consent or necessitating a significant change in medication, inability to perform exercise testing without significant risk, no anginal pain during exercise-tolerance testing, previous cardiac transplant, aortic valve stenosis (valve area <1.5 cm^2), prosthetic valve in the aortic position, significant peripheral vascular disease precluding bilateral femoral access, being an otherwise poor candidate for interventional cardiac procedures, and calcification that interferes with fluoroscopic visualization of the heart. Minors, pregnant women, and prisoners, as well as patients who were institutionalized, unable to give informed consent, and with certain other medical or psychiatric problems were also excluded from the study.

2.7. Baseline Tests and Assessments

After providing written informed consent, patients referred for possible participation in the trial underwent eligibility assessments to verify complete agreement with all inclusion and exclusion criteria. Board-certified cardiologists conducted all baseline and follow-up exercise testing and angina assessments. Patients meeting all criteria were then asked to complete a quality of life questionnaire and were scheduled for the procedure. Randomization occurred at the time of the procedure.

2.8. Procedural Blinding and Randomization

The randomization code was generated independently from the investigational sites and personnel by the Biostatistics Department of the University of Bergen, Institute of Medicine. Individual, sealed patient envelopes were prepared and provided to the laser operator, who maintained these envelopes in a locked safe. Throughout this trial, only one individual participating in the procedures, the laser operator, was aware of the randomized assignments. To assure blinding of the patients, treating investigators, and staff during the procedure, the laser console and the laser operator were placed behind an opaque curtain out of view in the cardiac catheterization laboratory. Two laser catheters were pre-calibrated to produce an identical response from the laser. At the initiation of every procedure, the patient was instrumented with a sterile guiding catheter and one of the laser catheters. The distal end of the other laser catheter was placed into a lead box behind the opaque curtain. The laser technician, positioned behind the curtain, then opened the treatment assignment envelope. Depending on the group assignment, the proximal end of the sterile catheter positioned in the patient's left ventricle (PML assignment) or of the catheter terminated to the lead box (sham assignment) was connected to the laser console by the laser operator. In the PML group, pulsed-laser energy was delivered to the endocardium through the laser catheter to create a nontransmural channel. For the sham group, the procedure was identical except there was no laser energy delivered to the patient when the interventionalist activated the footswitch. Throughout the procedure, music was played. There was no visual or audible feedback from the laser system that could reveal the randomized assignment.

2.9. Procedure Description

The PML procedure has been described in detail in the literature.[5] Briefly, a 9 or 10 F introducer sheath was inserted into one of the femoral arteries through which the guiding catheter was inserted. After establishing access, a heparin bolus was administered to achieve an activated clotting time of greater than 250 seconds and is measured at least every 30 minutes during the procedure.

Lidocaine (bolus plus infusion) was administered during the procedure to suppress potential dysrhythmias which may be induced by catheter manipulation.

Coronary angiographic images were obtained to provide a guide for positioning the fiberoptic tip. The guiding catheter was loaded onto a basally diagnostic pigtail catheter and introduced into the left ventricle over a J-tipped guidewire. After removing the pigtail catheter, a pressurized heparin drip was administered through the guiding catheter. The laser catheter was then inserted through the guiding catheter under fluoroscopic visualization. Once in the left ventricle, the investigator manipulated the coaxial catheters through individual or *en bloc* adjustments to appropriately target the various regions of the heart. Following the optimization of fluoroscopy projections, both coronary angiography and ventriculography were performed to visualize laser catheter distal lens placement at the targeted region of the endocardium. The position of the laser catheter is confirmed continuously during the procedure by fluoroscopy, using projections that provide orthogonal views. During the procedure, contact with the myocardial wall is verified as a change from a slight reciprocating movement to a noticeable beating motion when contact is established. Fluoroscopy was also used to appropriately space channels (approximately 1-cm spacing). The location of each channel, laser time, and total procedure time were documented.

When the channel placement procedure was completed, the laser and guiding catheters were removed and routine posttreatment procedures for the management of cardiac catheterization patients, including removal of sheaths, were followed. Immediately after the procedure and again before discharge, an echocardiogram was performed to exclude the occurrence of significant pericardial effusion and to assess wall motion, and blood samples were drawn for cardiac enzyme analyses. Blood samples were frozen and not analyzed until after completion of the trial to ensure blinding. Patients were also assessed after the procedure with serial electrocardiograms to rule out serious cardiac adverse events.

2.10. Patient Follow-Up

Following discharge, patients were to return at 3 months, 6 months, and 1 year for a physical exam, assessment of adverse events, and independent assessment of the primary endpoint variable, CCS angina class. Secondary assessments also to be made at these time intervals included medication usage, patient-reported quality of life per the Seattle angina questionnaire,[6] ejection fraction, and exercise time per a chronotropic protocol.[7]

2.11. Maintenance of the Blind

Throughout the trial, the only individual aware of the randomized assignment was the laser operator. This individual followed strict protocol require-

Table 2
Baseline Characteristics of Randomized Groups

	PML (*n* = 40)	Sham (*n* = 42)	*p* Value
Age (yr)	65 ± 9.3	67 ± 9.8	0.44
Men	38 (95%)	37 (88%)	0.43
CCS Class III/IV	36 (90%)/4 (10%)	35 (83%)/7 (17%)	0.52
Ejection fraction	0.64 ± 0.12	0.63 ± 0.12	0.65
History of congestive heart failure	1 (2.5%)	1 (2.4%)	1.00
History of diabetes mellitus	5 (13%)	8 (19%)	0.61
History of systemic hypertension	19 (48%)	20 (48%)	1.00
History of myocardial infarction	25 (66%)	29 (69%)	0.86
History of smoking	29 (73%)	32 (76%)	0.63
Prior CABG or PCI	36 (90%)	37 (88%)	0.86
Serum cholesterol (mmol/L)	6.7 ± 1.9	7.0 ± 2.1	0.47
Triple vessel disease	36 (90%)	32 (76%)	0.28
Medications ACE inhibitor	8 (20%)	11 (26%)	0.60
Aspirin	35 (88%)	33 (78%)	0.38
β-blocker	35 (88%)	38 (90%)	0.73
Calcium antagonist	20 (50%)	25 (59%)	0.26
Diuretic	4 (10%)	9 (19%)	0.35
Lipid-lowering agent	34 (85%)	40 (95%)	0.15
Nitrate	37 (93%)	37 (88%)	0.71

PML, percutaneous myocardial laser; CCS, Canadian Cardiovascular Society; CABG, coronary artery bypass grafting; PCI, percutaneous coronary intervention; ACE, angiotensin-converting enzyme.

ments for nondisclosure. The patient, treating physician, staff, and independent assessors were unaware of the assignments. After the 6-months follow-up, a prespecified analysis of results was conducted by an independent statistician at the University of Bergen Institute of Medicine. Summary results were made available without identification of randomized assignments. After the 1-year follow-up, a prespecified analysis of results again was conducted by the independent statistician. Summary results were made available with full disclosure of randomized assignments. At this time that patients were notified of their randomized assignments.

3. Trial Results

3.1. Patient Demographics

Eighty-two patients (40 PML and 42 sham) qualified for the trial and were randomized between March 1999 and June 2000. All key baseline characteristics were comparable between groups (Table 2). Typical of this advanced disease state, the majority of patients were diagnosed with triple vessel disease (83%), and patients presented with a significant history of prior coronary artery interventions (89%) and risk factors, including MI (68%) and hypertension (48%).

3.2. Procedural Outcomes

Procedural outcomes were similar among the randomized groups in terms of the total (skin-to-skin) procedure time (63 ± 27 minutes, PML; 63 ± 17 minutes, sham [$p = 0.69$]), the laser/sham procedure time (36 ± 16 minutes, PML; 37 ± 12 minutes, sham [$p = 0.66$]), and number of laser/sham channel placements made (19 ± 4.5, PML; 20 ± 3.6, sham [$p = 0.30$]). The primary target region within the left ventricle was the lateral wall ($n = 67$, 82%), followed by the inferior wall ($n = 51$, 62%), and anterior wall ($n = 31$, 38%), with the majority of patients (65%) treated in two regions. Procedural success, defined as the completion of the entire procedure through patient discharge in the absence of death, MI, or perforation, was 97.5% in the PML group and 97.6% in the sham group. In the PML group, one patient underwent pericardiocentesis. In the sham group, one patient who received a total of 12 sham "channels" in the anterolateral wall was observed on ECG and echocardiography to have ST-segment elevation and general hypokinesia. The patient died approximately 1-hour postprocedure. Autopsy showed no evidence of perforation, tamponade, or death related to the device. Other complications in the PML group included arrhythmia treated medically and in the sham group included pericardial effusion and transient ischemic attack (TIA).

Predischarge ECGs showed no evidence of MI. Likewise, none of the postprocedure peak cardiac enzyme values (creatinine phosphokinase [CPK], creatinine kinase-myocardial bradykinin fraction [CK-MB], troponin I) determined after trial completion were indicative of MI. However, the mean peak values for PML-treated patients were significantly higher than those for sham-treated patients ($p < 0.05$) owing to the laser treatment (Fig. 2A, B, and C): CPK (IU/L): PML, 211 ± 119; sham: 159 ± 82 L; CK-MB (μg/L): PML, 28 ± 60; sham, 12 ± 7; troponin I (μg/L): PML, 6.0 ± 8.3; sham, 1.9 ± 1.4. These results further validate the effectiveness of the randomization and demonstrate the significant effect of the laser treatment.

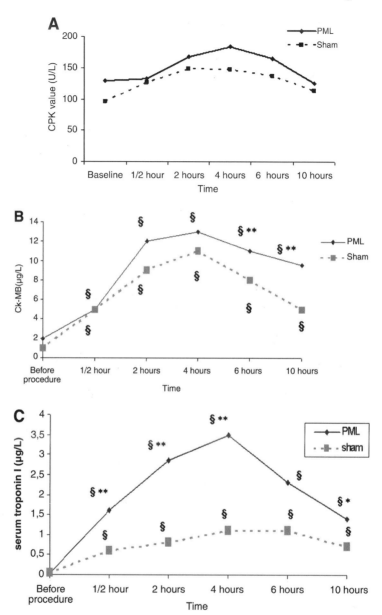

Fig. 2. Peak cardiac enzyme concentrations at baseline and postprocedure. (**A**) Creatinine phosphokinase; (**B**) creatinine kinase-myocardial bradykinin fraction, between group comparison: *$p < 0.05$, **$p < 0.01$, within group comparison: §$p < 0.05$ from baseline, repeated measures analysis between group comparison: $p = $ ns; (**C**) troponin 1, between group comparison: *$p < 0.05$, **$p < 0.01$, within group comparison: §$p < 0.05$ from baseline, repeated measures analysis between group comparison: $p < 0.01$.

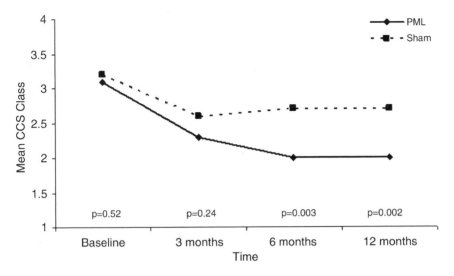

Fig. 3. Mean angina class improvement in randomized groups.

3.3. Follow-Up Compliance

Of the 82 patients randomized, 96% ($n = 79$) were available for follow-up through 1 year. There were no crossovers of patients from the sham to the treated group, no withdrawals, and none of the patients received additional revascularization procedures during the study. Two sham patients who died within 3 months of randomization and one PMR patient who suffered a stroke at 3 months were not available for the 6-month or 1-year CCS angina class assessment. Adverse event data are complete on 100% of randomized patients.

3.4. Primary Efficacy Endpoint Results

All patients were in CCS Class III or IV at baseline, with distributions (mean ± standard deviation) comparable between groups (PML, 3.1 ± 0.3; sham, 3.2 ± 0.4; $p = 0.52$). Results through 1 year, analyzed per protocol with worst-case imputation (class IV) for missing patients, are shown in Fig. 3. At 3 months, angina class improved to 2.3 ± 1.0 in the PML group and to 2.7 ± 0.9 in the sham group ($p = 0.24$), the latter of which signifies the effect of patient bias on this early outcome; however, this effect was shown to have reached a plateau at 6 months. Angina class was reduced significantly more in the PML group than in the sham group (PML, 2.0 ± 1.1; sham, 2.7 ± 0.9 [$p = 0.003$]), with significantly more PML than sham patients improving by at least one class (63% vs 36%, $p = 0.03$) or by at least two classes (40% vs 12%, $p <$

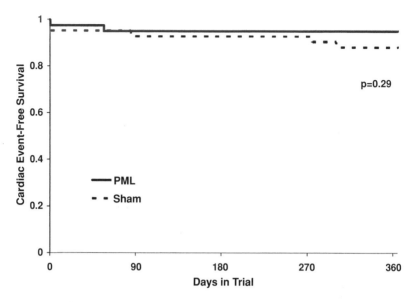

Fig. 4. Kapan-Meier cardiac event-free survival at 1 year.

0.01) from baseline. This significant improvement and between-group difference was sustained at 1 year in terms of distribution (PML, 2.0 ± 1.0; sham: 2.7 ± 0.9 [*p* = 0.002]), one or more class improvement (63% vs 38%, *p* = 0.04), and two or more class improvement (35% vs 14%, *p* = 0.04) from baseline. In a multivariate analysis of baseline predictors of two or more angina class improvement at 1 year, treatment assignment was the only significant independent predictor. The estimated odds ratio improvement at 1 year, after adjustment for baseline class, is 3.8 (95% confidence interval: 1.2–12).

3.5. Primary Safety Endpoint Results

At 1 year, cardiac event-free survival (defined as the absence of death, MI, perforation, stroke/TIA, and angina requiring rehospitalization) was similar between groups (*p* = 0.29 [log rank], Fig. 4). In the PML group, there were no deaths or MI. In addition to the single perforation, one patient suffered a stroke at 3 months and was unable to travel for further follow-ups. Three other patients were hospitalized for angina during follow-up. In the sham group, two patient deaths occurred: one procedurally and the other at 3 months owing to a suspected MI. One patient each was reported with MI, TIA, and atrial fibrillation. Three other patients were hospitalized for angina during follow-up. All-cause mortality at 1 year was nonsignificantly reduced in the PML group (PML, 0%; sham, 5%; *p* = 0.17 [log rank]).

3.6. Secondary Assessments

3.6.1. Medication Usage

There were no significant between-group differences in medical therapy usage at baseline. As per trial design, medical therapy remained stable in both groups with no significant dose changes in any cardiovascular medication through 1 year.

3.6.2. Quality of Life

The Seattle angina questionnaire has five components, each assessed on a 100-point scale, and was designed specifically to assess functional status and health-related quality of life in patients with angina. Scores were comparable among groups at baseline and improved in both groups during follow-up. Notably, scores for the angina-specific components were significantly higher ($p < 0.05$) in the PML group than in the sham group at 6 months, with these results sustained at 1 year: angina stability (PML, 60 ± 26; sham, 48 ± 25) and angina frequency (PML, 52 ± 33; sham, 35 ± 28).

3.6.3. Ejection Fraction

Baseline ejection fractions were comparable between groups and did not change during follow-up.

3.6.4. Exercise

The chronotropic assessment exercise protocol was designed for use in functionally limited patients who have implanted pacemakers, involving a warm-up, modest work increments, and slowly increasing speed and grade. This protocol was chosen primarily to satisfy the enrollment criteria (patient must experience angina pain on exercise) and to place the least burden on the patient in an effort to reduce the possibility of drop-outs. Patients in the study typically were accustomed to cycling rather than walking. Because of some patients' strong desire to perform cycle rather than treadmill tests, testing during follow-up was performed in approximately one-third of patients using bicycles and treadmills and in two-thirds of patients using treadmills alone. Test termination occurred at general exhaustion, ST segment depression of 3 mm or more, moderate or severe chest pain or dyspnea, drop of blood pressure, signs of circulatory insufficiency, arrhythmias with hemodynamic consequences, or patient wanting to stop. Exercise times were comparable between groups at baseline (PML, 610 ± 222 s; sham, 585 ± 235 s; $p > 0.5$). Results showed no significant between-group difference at 1 year in total time (PML, 620 ± 245 s; sham, 604 ± 229 s; $p > 0.1$).

4. Discussion

Symptomatic placebo effects in cardiovascular disease are common, not well understood in terms of mechanism, and are known to vary widely.[8] In short-term (average 10 weeks) cardiovascular drug trials, improvement in chronic, stable angina pectoris has been reported to occur in 30 to 80% of placebo-treated patients.[9] Notwithstanding this observation, long-term rates of improvement attributable to placebo (i.e., 6 months and beyond) have not been well characterized.

The focused objective of this randomized, double-blind, sham-controlled trial of PML in patients who suffer from debilitating, medically refractory angina was to control for the two predominant sources of bias in evaluation of long-term angina improvement: patient bias (placebo effect) and assessment bias. We showed that PML therapy using this particular device is superior to sham intervention for the stated primary endpoint. The improvement in the placebo group (12% [6 months] and 14% [1 year]) indicates that a placebo effect in the absence of assessment bias contributes to angina improvement but does not explain the complete clinical benefit following PML treatment using this device. Our unique design and excellent follow-up mitigated the influence of other potentially confounding factors, such as changes in medical therapy or incomplete ascertainment.

Our independently determined findings support the significant treatment effect observed in the unblinded trials that used the identical PML system.[3,10] As is well recognized in clinical trial design, groups can be reliably compared only within a trial and not across trials. As a result, the negative findings of the DMR in Regeneration of Endomyocardial Channels trial,[11] which used different interventional methods and test devices, single blinding, and a different patient population, should be considered on its own. A still different randomized study reported recently in a CABG-eligible chronic total occlusion population failed to observe a significant treatment effect beyond placebo at 6 months; however, this trial is significantly flawed owing to bias introduced by investigator unblinding and incomplete ascertainment (50%).[12]

Although sham-surgery trials of TMR cannot be ethically conducted, 1-year results from five independently conducted, randomized controlled trials of TMR as sole therapy[13–17] have shown significantly consistent improvement in angina symptoms, quality of life, and cardiac event-free survival. Longer-term follow-up of the cohorts among two of these trials to a maximum of 5 years have determined significantly persistent angina improvement, reduced rehospitalizations, and, in the longer of the trials, increased survival correlated to symptomatic improvement.[18,19] The overwhelmingly positive 1-year results, as well as the reported persistent significant angina relief beyond 3 and 5 years following TMR mitigates the concern of placebo effect as a primary mechanism for clinical improvement. Based on an assessment of the cumulative results from multiple randomized trials, the recently updated American College of Cardiology/American Heart Association guideline[20] and Society of Thoracic Surgeons guideline[21] have determined that the weight of the evidence favors the use of TMR in the treatment of stable, medically refractory, no-option angina patients.

5. Conclusion

The scientific validity of the evaluation of any new therapy mandates that the expectations and beliefs of the treating physician, the patient, and the assessor are excluded and/or controlled to the extent possible and as ethically permitted. Our randomized, double-blind, sham-controlled trial of PML followed this mandate, demonstrating that the clinical benefit at 1 year is attributable to the effect of treatment, above and beyond placebo.

References

1. Muhkerjee, D., Bhatt, D., Roe, T., et al. 1999. Direct myocardial revascularization and angiogenesis—how many patients might be eligible? *Am. J. Cardiol.* 84:598–600.
2. Wiseth, R., Korfang, K., Ilebekk, A., et al. 2001. Myocardial laser revascularization in the year 2000 as seen by a Norwegian specialist panel. The process of evaluating and implementing new methods in clinical practice. *Scand. Cardiovasc. J.* 35:14–18.
3. Oesterle, S. N., Sanborn, T. A., Ali, N., et al. 2000. Percutaneous transmyocardial laser revascularization for severe angina: the PACIFIC randomised trial. *Lancet.* 356:1705–1710.
4. Byerly, H. 1976. Explaining and exploiting placebo effects. *Perspect. Biol. Med.* 10:423–435.
5. Kim, C. B., Kesten, R., Javier, M., et al. 1997. Percutaneous method of laser transmyocardial revascularization. *Cathet. Cardiovasc. Diag.* 40:223–228.
6. Spertus, J. A., Winder, J. A., Dewhurst, T. A., et al. 1995. Development and evaluation of the Seattle Angina Questionnaire: a new functional status measure for coronary artery disease. *J. Am. Coll. Cardiol.* 25:333–341.
7. Wilkoff, B. L., Corey, J., Blackburn, G. 1989. A mathematical model of the cardiac chronotropic response to exercise. *J. Electrophysiol.* 3:176–180.
8. Bienenfeld, L., Frishman, W., Glasser, S. P. 1996. The placebo effect in cardiovascular disease. *Am. Heart. J.* 132:1207–1221.
9. Glasser, S. P., Clark, P., Lipicky, R., et al. 1991. Exposing patients with chronic, stable, exertional angina to placebo periods in drug trials. *JAMA.* 265:1550–1554.
10. Gray, T. J., Burns, S. M., Clarke, S. C., et al. 2003. Percutaneous myocardial laser

revascularization in patients with refractory angina pectoris. *Am. J. Cardiol.* 91:661–666.

11. Leon, M. B., Baim, D. S., Moses, J. W., et al. 2000. A randomized blinded clinical trial comparing percutaneous laser myocardial revascularization (using Biosense LV mapping) in patients with refractory coronary ischemia. *Circulation.* 102:II–565.

12. Stone, G. W., Teirstein, P. S., Rubenstein, R., et al. 2002. A prospective, multicenter, randomized trial of percutaneous transmyocardial laser revascularization in patients with nonrecanalizable chronic total occlusions. *J. Am. Coll. Cardiol.* 39:1581–1587.

13. Allen, K. B., Dowling, R. D., Fudge, T. L., et al. 1999. Comparison of transmyocardial revascularization with medical therapy in patients with refractory angina. *N. Engl. J. Med.* 341:1029–1036.

14. Frazier, O. H., March, R. J., Horvath, K. A. 1999. Transmyocardial revascularization with a carbon dioxide laser in patients with end-stage coronary disease. *N. Engl. J. Med.* 341:1021–1028.

15. Burkhoff, D., Schmidt, S., Schulman, S. P., et al. 1999. Transmyocardial revascularization compared with continued medical therapy for treatment of refractory angina pectoris: a prospective randomized trial. *Lancet.* 354:885–890.

16. Schofield, P. M., Sharples, L. D., Caine, N., et al. 1999. Transmyocardial laser revascularization in patients with refractory angina: a randomized controlled trial. *Lancet.* 353:519–524.

17. Aaberge, L., Nordstrand, K., Dragsund, M., et al. 2000. Transmyocardial revascularization with CO2 laser in patients with refractory angina pectoris. Clinical results from the Norwegian randomized trial. *J. Am. Coll. Cardiol.* 35:1170–1177.

18. Aaberge, L., Rootwelt, K., Blomhoff, S., et al. 2002. Continued symptomatic improvement three to five years after transmyocardial revascularization with CO2 laser: a late clinical follow-up of the Norwegian randomized trial with transmyocardial revascularization. *J. Am. Coll. Cardiol.* 39:1588–1593.

19. Allen, K. B., Dowling, R. D., Angell, W., et al. 2004. Transmyocardial revascularization: five-year follow–up of a prospective, randomized, multicenter trial. *Ann. Thorac. Surg.* 77:1228–1234.

20. Gibbons, R., Abrams, J., Chatterjee, K., et al. 2003. ACC/AHA 2002 guideline update for the management of patients with chronic stable angina (summary article): a report of the American College of Cardiology/American Heart Association Task Force on Practice Guidelines (Committee on the Management of Patients with Chronic Stable Angina). *Circulation.* 107:149–158.

21. Bridges, C. R., Horvath, K. A., Nugent, B., et al. 2004. Society of Thoracic Surgeons Practice Guideline: Transmyocardial Laser Revascularization. *Ann. Thorac. Surg.* 77:1484–1502.

13

Clinical Trials of an Anti-Adhesion Adjuvant

Michael P. Diamond

1. Introduction

Postoperative adhesions are scar tissues, which develop at the site of tissue injury after surgery.[1] Their potential significance is immense because of the frequency with which they form, as well as their consequences of potential morbidity and mortality.[2] Yet, many surgeons minimize their significance because of the general inability to identify adhesions without performing a repeat surgery, as there are not good serologic markers or accurate imaging modalities (e.g., ultrasound, computed tomography, or magnetic resonance imaging scans) and the perceived inability to do anything about them.

Despite the tendency of surgeons to minimize their importance, adhesions develop in the overwhelming majority of individuals who undergo surgery, regardless of the part of the body in which it is conducted. Data from intra-abdominal surgeries performed by experienced gynecologists and general surgeons suggest an incidence rate of 55 to 100%, with an average of more than 80% of patients.[3] Although data has been collected in these specialties because of the ability to perform second-look surgeries to directly examine the occurrence, extent, and severity of adhesions, their occurrence is also found following cardiothoracic, orthopedic, neurosurgical, urological, ophthalmological, and nasal surgical procedures.

Although consequences of adhesions are dependent on their location, they can contribute to difficult reoperative procedures, leading to an increase in the time required to perform a procedure (and the associated increased costs), as well as an increased potential for intra-operative complications, such as bleeding, nerve injury, and viscus perforation. Additionally, intra-abdominal adhesions are the leading noncancerous cause of small bowel obstruction, are associated with stretching of serosal surfaces causing abdominal and pelvic

From: *Clinical Evaluation of Medical Devices: Principles and Case Studies, Second Edition*
Edited by: K. M. Becker and J. J. Whyte © Humana Press Inc., Totowa, NJ

pain, alteration of the normal anatomical relationship of the fallopian tube and ovary leading to infertility, restriction of movement and peristalsis of the bowel resulting in bowel injury following radiation therapy, and limitation of flow of fluids throughout the abdominal cavity, which can impair the benefits of intraperitoneal chemotherapy and peritoneal dialysis.[1–4] Consequences elsewhere include adhesions following cardiothoracic surgery, which can result in attachment of the myocardium to the sternum, thereby increasing the risk of heart injury at a repeat surgery; stretching of nerve roots causing pain or sphincter disturbances; adherence of ligaments and tendons to their sheaths, thereby limiting the ability to flex or extend a joint; ocular muscle adhesions causing strabismus; and nasal sinus adhesions interfering with respirations.

2. Clinical Trial of an Anti-Adhesion Agent

In view of the high frequency with which adhesions develop and their potential life-altering consequences, the obvious question becomes why doesn't someone develop an agent to reduce their occurrence? The answer is multifaceted but represents the difficulties experienced in identifying a surgical model (which likely will vary greatly based on unique characteristics of the agent and other factors) in which the benefits of an efficacious agent could be demonstrated.

Critical questions that need to be considered when designing an anti-adhesion clinical trial include:

1. Primary and secondary endpoints (e.g., reduction of adhesions, patient report of improvement, or objective assessment).
2. Choice of investigator.
3. The patient population to be studied.
4. The method of entry into the surgical site (e.g., for gynecological procedures, use of laparotomy, laparoscopy, vaginal entry).
5. Method of assessment (e.g., surgeon report or video).
6. Scoring system to be used.
7. Significant inclusion and exclusion criteria.
8. Investigative center recruitment.
9. Standardization of procedure.
10. Choice of control group.

2.1. Primary and Secondary Endpoints

The overwhelming majority of clinical studies on anti-adhesion adjuvants that have been submitted to regulatory bodies (or for that matter for publication) have used identification and characterization of adhesions at second-look surgery, *per se*, as the primary endpoint. Although some have argued that clinical outcomes such as bowel obstruction, pain, or fertility would be better—as they represent the ultimate goal of the patient and the care provider—use of such endpoints is fraught with problems. For example, even though bowel obstruction could occur in the immediate postoperative period, it may not occur until 1, 5, 10, or 20 years or even longer after surgery; thus, such studies would be unacceptably long.[2] Pain could be used, but it is subjective because stoicism of individuals varies, the ability to work/play/function with pain can be greatly variable owing to other issues, and there is little ability to quantifiably validate reports of pain.[4] Fertility also has great limitation as an endpoint because of variation in age of the female partner, coexistence of other fertility factors, such as the male's semen characteristics (e.g., sperm count, motility, morphology, vitality), use of ovulation-induction medications, performance of artificial insemination, and time until couples attempt in vitro fertilization.[2] Additionally, use of fertility as an endpoint can be complicated by the couples' relationship, because divorce or relationship troubles can be secondary to the stress placed on their relationship by infertility.

2.2. Choice of Investigator

Choice of the investigator for a surgical trial must be made carefully and is likely to be much more difficult than when conducting a drug trial. For example, for patients with hypertension, it is unlikely that a given patient's outcome will differ as a function of the care provider who dispensed the antihypertensive agent. By contrast, surgical outcome is likely to be dependent in part on the skill and expertise of the surgeon and his or her surgical technique. (It should be noted that the reductions in adhesion scores following adhesiolysis were reportedly owing to experienced surgeons who were chosen in part because of their skills; such individuals have been able to significantly reduce adhesion scores by adhesiolysis, even though, as noted before, the vast majority of patients develop adhesions.)[5–7] However, surgeons with poor surgical technique have patients who develop worse adhesions (personal observations), and thus the outcomes are in part surgeon-dependent.

2.3. Patient Population to be Studied

Because redevelopment of adhesions after adhesiolysis occurs more frequently than adhesion development at sites of surgical incisions, the likelihood of postoperative adhesion development varies between patients who have different amounts of pre-existing adhesions. Additionally, adhesions are more likely to develop when more sites of pathology are treated. Obviously, this creates difficulty in assessing the effectiveness of an anti-adhesion adjuvant, as the final amount of adhesions at the time of a second-look procedure would be a function not only of use of the adjuvant but also of the amount of pre-existing adhesions and pathology present.

In an ideal world, with a sufficiently large patient population, subject's randomization would eliminate potential biases from variation in initial adhesions or pathology. Unfortunately, in the real world, randomization will not always be successful. Variation can be minimized by strict entry criteria limiting the initial amount of adhesions or pathology, allowing for more uniform patient population characteristics; however, this is at the expense of a more prolonged period of subject recruitment, as a smaller number of patients will meet the limited entry criteria. Thus, there is a need to try to achieve a happy medium between the needs of rapid subject enrollment and achieving a homogeneous group of patients.

An example demonstrating the significance of the patient population to be studied is a recent clinical trial conducted at centers in the United States and Europe in which the efficacy of Intergel™ was tested.[8] Although the sponsor reported a beneficial action of this agent at centers on both continents, the adhesion scores in Europe in treated patients declined. However, in the Unites States, the adhesion scores of the treated patients increased from the initial to second-look procedure, but the increment in increase in the treated patients was smaller than in the control subjects. This difference likely stemmed from differences in the patient population. In the United States, many patients underwent uterine myomectomy and had relatively little initial adhesions. In contrast, in Europe, fewer patients had myomectomies, but more patients had pre-existing adhesions, thus providing a better opportunity to allow demonstration of a reduction in adhesion score by Intergel.

2.4. Method of Entry Into the Surgical Site

Regulatory bodies may expect that use of anti-adhesion adjuvants during or following procedures will require separate studies for each method of entry into the abdominal cavity. For example, efficacy of an agent in reducing postoperative adhesions after treating endometriosis would have to be studied separately for procedures performed by laparotomy and laparoscopy.

Although some have suggested that minimally invasive surgery would reduce postoperative adhesion development, a recent meta-analysis did not demonstrate such a benefit.[5] Nonetheless, characteristics of some anti-adhesion adjuvants may make them difficult or impossible to deliver to the surgical sites by one mode of entry or may be altered by delivery, such that efficacy demonstrated by one mode of entry may not be manifested by an alternative mode of entry.

2.5. Method of Assessment

Scoring of adhesions at each surgical procedure is necessary to characterize the initial status of the surgical site, as well as the status at the time of the second-look procedure. Even though the simplest and most straightforward way to accomplish is to have operating surgeon do the scoring, this introduces the potential conscious or unconscious bias of the surgeon if he or she knows which patients are assigned to the control and treatment groups. Although some agent may be used in an unidentifiable placebo group, this is not usually the case.

When the treatment and control arms are readily identifiable, adhesion scoring can be conducted by an individual not present when the adjuvant is introduced. In such cases, the blind must be maintained in clinical records that may be reviewed by the individual who will be the examiner at second-look.

Alternatively, videotapes can be made of the surgical procedures and edited to exclude the portion of the procedures when the adjuvant is applied. In this way, a reviewer blinded to treatment group assignment can review the videotapes and score the adhesions.

2.6. Scoring System To Be Used

There is not an ideal scoring system for grading adhesions. The system must reflect the surgical model being examined and can be limited to specific locations (as was done for Interceed and Seprafilm) or examine broad areas throughout the abdominal and pelvic cavities (as was done for Sepracoat and Intergel).

Characteristics to be captured include the presence or absence of an adhesion at a site, the extent to which the surface of that site is involved in adhesions, and the severity of the adhesion. Severity can be depicted as both the thickness and opaqueness of the adhesion, as well as the degree of vascularity (if any).

Adhesion-scoring systems, which have attempted to allow some degree of quantitation of adhesions, are the American Fertility Society adhesion scoring system[9] (which primarily involves the ovaries and tubes) and the More Comprehensive Adhesion Surgery Method system,[10] which examines 23 sites throughout the abdomen and pelvis. The latter has been validated for reproducibility between observers. Other investigators have modified each system.

2.7. Significant Inclusion and Exclusion Criteria

Inclusion and exclusion criteria are chosen to allow enrollment of a selected patient population. As described earlier, having a homogeneous group of patients will minimize the introduction of variability, which will cause altered results. Patients could be limited based on the amount of pre-existing adhesions, the presence or absence of endometriosis and ovarian endometriotic cysts, and the occurrence, location, and size of uterine fibroids. Additionally, inclusion of all such pathologies (and others) raises the question of how outcomes can be assessed considering these variable pre-existing pathological processes.

However, limiting the patient population not only will increase the time needed to enroll subjects, as noted previously, but also raises the question of how effective the agent would be in other circumstances. Additionally, limiting the patient population to be studied may also result in future limitations in labeling approval from regulatory bodies.

Other inclusion and exclusion criteria may limit use of the experimental device in patient populations who may be at increased risk of complications either because of their special characteristics and/or unique aspects of the agent. Some variables include method of clearance, co-existing disorders, and allergies.

2.8. Investigative Center Recruitment

The timeliness of subject recruitment and the quality of the data generated are crucial elements of any clinical trial. For the sponsor, time is money. The number of allowable, investigative centers is often a limiting factor; thus, selection of an investigative center that fails to recruit, or recruits poorly, limits progress of the trial. Although it is impossible to completely predict which centers will recruit well, good prognostic signs are a prior track record of good recruitment in a similar study, an experienced investigator and study coordinator, and centers that are responsive to and enthusiastic about initial interactions.

Equally important is for the site to carefully adhere to the protocol and to accurately record the data generated. An experienced study coordinator, who is familiar with all aspects of study conduct as well as national and local regulatory requirements, will greatly enhance the ability to meet these goals.

2.9. Standardization of Procedure

To minimize variation between surgeons at different surgical centers, the study protocol can try to standardize aspects of the procedure. Ideally, this limits variation in factors known (or thought) to impact on postoperative adhesion development. Examples could include use of nonpowder gloves, choice of suture type and size, use of other anti-adhesion adjuvants, type of stitching, and avoidance of introduction of foreign bodies into the abdomen.

2.10. Choice of Control Group

An important consideration is choice of the control group. A strong case could be made for a placebo, particularly in situations in which the control group would be indistinguishable from the treatment arm. Because this is a rare occurrence, choice for the control group is usually either a crystalloid solution (e.g., saline) or a no-treatment control. The latter has the advantage of being the method of treatment if the patient was not participating in a trial and there was no alternative agent. On the other hand, use of saline (particularly in a study in which the agent is a liquid or gel) provides the opportunity for comparison with an equal volume of fluid (although available data do not demonstrate a benefit of crystalloids in reducing postoperative adhesions). Additionally, installation of such fluid provides additional time to identify (and treat) non-homeostatic sites, as occurs in the treatment arm.

When the model used has an approved anti-adhesion adjuvant, the other adjuvant can be selected as an active control. In such cases, a study design to show equivalence rather than superiority of the test agent may be considered.

3. Conclusion

As noted earlier, choice of study design needs to be individualized for each test agent. Stated differently, there is not a "best" model. In addition to the variables introduced by the sponsor's business plan, characteristics of the agent to be studied are also critical. Does the agent comprise a material barrier, a gel, or a liquid? Can it be incorporated into one of these types of compositions? Is it a device, a drug, or a biologic? Will it be given locally or systemically? What is its half life and the half-life of its effects, if different? How is it cleared or degraded? Are these processes altered in disease states? What is its permeability into underlying tissues? How is it retained at the site of delivery, and for how long?

Thus, the study design is a critical component of the clinical trial and will likely significantly impact the ultimate outcome. If not properly designed, developed, and executed for the specific agent to be tested, it may not be possible to demonstrate efficacy of the agent, even when it in fact exists. Consequently, great individualized consideration must be given to optimize the opportunity to demonstrate adjuvant efficaciousness, which includes taking into account unique attributes of that specific adjuvant and how it will be used.

References

1. Saed, G. M., Diamond, M. P. 2004. Molecular characterization of postoperative adhesions: the adhesion phenotype. *J. Am. Assoc. Gynecol. Laparoscop.* 11:307–314.
2. Diamond, M. P., Freeman, M. L. 2001. Clinical implications of post-surgical adhesions. *Hum. Reprod. Update.* 7:567–576.
3. Diamond MP. Incidence of postoperative adhesions. In diZerega, G. S., DeCherney, A., Diamond, M. P., Ellis, H., et al, eds. 2000. *Peritoneal Surgery.* Springer-Verlag, New York, 217–220, 2000.
4. Hammoud, A., Gago, L. A., Diamond, M. P. 2004. Adhesions in patients with chronic pelvic pain: role of adhesiolysis? *Fertil. Steril.* 82:1483–1491.
5. Diamond, M. P., Daniell, J. F., Feste, J., et al. 1987. Adhesion reformation and de novo adhesion formation following reproductive pelvic surgery. *Fertil. Steril.* 47:864–866.

6. Diamond, M. P., Daniell, J. F., Johns, D. A., et al. 1991. Postoperative adhesion development following operative laparoscopy: evaluation at early second-look procedures. *Fertil. Steril.* 55:700–704.

7. Wiseman, D. M., Trout, J. R., Diamond, M. P. 1998. The rates of adhesion development and the effects of crystalloid solutions on adhesion development in pelvic surgery. *Fertil. Steril.* 70:702–711.

8. Johns, D. B., Keyport, G. M., Hoehler, F., diZerega, G. S. 2001. Intergel Adhesion Prevention Study Group: reduction of postsurgical adhesions with Intergel adhesion prevention solution: a multicenter study of safety and efficacy after conservative gynecologic surgery. *Fertil. Steril.* 76:595–604.

9. The American Fertility Society. 1988. The American Fertility Society Classifications of Adnexal Adhesions, Distal Tubal Occlusion, Tubal Occlusion Secondary to Tubal Ligation, Tubal Pregnancies, Mullerian Anomalies, and Intrauterine Adhesions. *Fertil. Steril.* 49:944–955.

10. Diamond, M. P., Bachus, K., Bieber, E., et al. 1994. Improvement of interobserver reproducibility of adhesion scoring systems. *Fertil. Steril.* 62:984–988.

14

Use of Multiple Imputation Models in Medical Device Trials[1]

Donald B. Rubin and Samantha R. Cook

1. Abstract

Missing data are a common problem with data sets in most clinical trials, including those dealing with devices. Imputation, or filling in the missing values, is an intuitive and flexible way to handle the incomplete data sets that arise because of such missing data. Here we present several imputation strategies and their theoretical background, as well as some current examples and advice on computation. Our focus is on multiple imputation, which is a statistically valid strategy for handling missing data. The analysis of a multiply imputed data set is now relatively standard, for example in SAS and in Stata. The creation of multiply imputed data sets is more challenging but still straightforward relative to other valid methods of handling missing data. Singly imputed data sets almost always lead to invalid inferences and should be eschewed.

2. Introduction

Missing data are a common problem with large databases in general and with clinical and health care databases in particular. Subjects in clinical trials may fail to provide data at one or more time points or may drop out of a trial altogether, for reasons including lack of interest, untoward side effects, change of geographical location, and success of the procedure with no interest in follow-up assessments, etc. Data may also be "missing" due to death, although the methods described here are generally not appropriate for such situations because such values are not really missing (*see* Little and Rubin[1], example 1.7, and Zhang and Rubin[2]).

---

[1] A similar version of this chapter appears in cursory form as an entry in *The Encyclopedia of Clinical Trials*.

From: *Clinical Evaluation of Medical Devices: Principles and Case Studies, Second Edition*
Edited by: K. M. Becker and J. J. Whyte © Humana Press Inc., Totowa, NJ

An intuitive way to handle missing data is to fill in (i.e., impute) plausible values for the missing values, thereby creating completed data sets that can be analyzed using standard complete-data methods. The past 25 years have seen tremendous improvements in the statistical methodology for handling incomplete data sets using imputation. After briefly discussing missing data mechanisms, we present some common imputation methods, focusing on multiple imputation.[3] We then discuss computational issues and present some examples.

3. Missing Data Mechanisms

A missing data mechanism is a probabilistic rule that governs which data will be observed and which will be missing. Little and Rubin[1] and Rubin[4] distinguish three types of missing data mechanisms. Missing data are missing completely at random (MCAR) if missingness is independent of both observed and missing values of all variables, almost random dart throwing at the data matrix. MCAR is the only missing data mechanism for which "complete-case" analysis (i.e., restricting the analysis to only those subjects with no missing data) is generally acceptable. Missing data are missing at random (MAR) if missingness depends only on observed values of variables and not on any missing values. For example, if the value of blood pressure at the end of a trial is more likely to be missing when some previously observed values of blood pressure are high, and given these the missingness is independent of the value of blood pressure at the end of the trial, then the missingness mechanism is MAR.

If missingness depends on the values that are missing, even after conditioning on all observed quantities, the missing data mechanism is not missing at random (NMAR). Missingness must then be modeled jointly with the data— the missingness mechanism is "nonignorable." Nonignorable missing data present challenging problems because there is no direct evidence in the observed data about how to model the missing values.

The specific imputation procedures described here are most appropriate when the missing data are MAR and ignorable (*see* Little and Rubin[1] and Rubin[4] for details). Multiple imputation can still be validly used with nonignorable missing data, although it is more challenging to use it well. Multiple imputation is still more straightforward to use than other valid methods of handling the nonignorable situation.

4. Single Imputation

Single imputation refers to imputing one value for each missing datum. Singly imputed data sets are straightforward to analyze using complete-data methods, which makes single imputation an apparently attractive option with incomplete data. Little and Rubin[1] offer the following guidelines for creating imputations. They should be: (1) conditional on observed variables; (2) multivariate, to reflect associations among missing variables; and (3) randomly drawn from predictive distributions rather than set equal to means, to ensure that correct variability is reflected.

Unconditional mean imputation, which replaces each missing value with the mean of the observed values of that variable, meets none of the three guidelines listed above. Regression imputation can satisfy the first two guidelines by replacing the missing values for each variable with the values predicted from a regression (e.g., least squares) of that variable on other variables. Replacing missing values of each variable with the mean of that variable calculated within cells defined by categorical variables is a special case of regression imputation. Stochastic regression imputation adds random noise to the value predicted by the regression model, and when done properly, can meet all three guidelines.

Hot deck imputation replaces each missing value with a random draw from a donor pool of observed values of that variable; donor pools are selected, for example, by choosing individuals with complete data who have "similar" observed values to the subject with missing data, e.g., by exact matching or using a distance measure on observed variables to define "similar." Hot deck imputation, when done properly, can also satisfy all three of the guidelines listed above.

Even though analyzing a singly imputed data set with standard techniques can be straightforward, such an analysis will nearly always result in estimated standard errors that are too small, confidence intervals that are too narrow, and *p* values that are too significant, regardless of how the imputations were created. The reason is that imputed data are treated as if they were known with no uncertainty. Thus, single imputation is almost always statistically invalid, although the multiple version of a single imputation method will be valid if the imputation method is "proper." Proper imputations satisfy the three criteria of Little and Rubin.

4.1. Properly Drawn Single Imputations

Let Y represent the complete data, i.e., all the data we would observe in the absence of missing data, and let $Y = \{Y_{obs}, Y_{mis}\}$, where Y_{obs} is the observed data and Y_{mis} is the missing data. For notational simplicity, assume ignorability of the missing data mechanism. Also, let θ represent the (generally multicomponent) parameter associated with an appropriate imputation model, which consists of both a sampling distribution on Y governed by θ, $p(Y|\theta)$, and a prior distribution on θ, $p(\theta)$. A proper imputation is often most easily obtained as a random draw from the "posterior predictive distribution" of the missing data given the observed data, which formally can be written as:

$$p(Y_{mis}|Y_{obs}) = \int p(Y_{mis}, \theta|Y_{obs})d\theta = \int p(Y_{mis}|Y_{obs}, \theta)p(\theta|Y_{obs})d\theta. \qquad (1)$$

This expression effectively gives the distribution of the missing values, Y_{mis}, given the observed values, Y_{obs}, under a model governed by θ, $p(Y|\theta)p(\theta)$. This distribution is called "posterior" because it is conditional on the observed Y_{obs}, and it is called "predictive" because it predicts the missing Y_{mis}.

If the missing data follow a monotone pattern (*see* Section 4.1.1.), drawing random samples from this distribution is straightforward. When missing data are not monotone, iterative computational methods are generally necessary, as described shortly.

4.1.1. Theory With Monotone Missingness

A missing data pattern is monotone if the rows and columns of the data matrix can be sorted in such a way that an irregular staircase separates Y_{obs} and Y_{mis}. Figures 1 and 2 illustrate monotone missing data patterns. Missing data in clinical trials are often monotone or nearly monotone when data are missing due to patient dropout, and once a patient drops out, the patient never returns.

Let Y_0 represent fully observed variables, Y_1 the incompletely observed variable with the fewest missing values, Y_2 the variable with the second fewest missing values, and so on. Proper imputation with a monotone missing data pattern begins by fitting an appropriate model to predict Y_1 from Y_0 and then using this model to impute the missing values in Y_1. For example, fit a regression of Y_1 on Y_0 using the units with Y_1 observed, draw the regression parameters from their posterior distribution, and then draw the missing values of Y_1 given these parameters and Y_0. Next, impute the missing values for Y_2 using Y_0 and the observed and imputed values of Y_1. Continue until all missing values have been imputed. The collection of imputed values is a proper imputation of the missing data, Y_{mis}, under this model, and the collection of univariate prediction models is the implied full imputation model. When missing data are not monotone, this method of imputation as described cannot be used directly.

4.1.2. Theory With Nonmonotone Missingness

Creating imputations when the missing data pattern is nonmonotone generally involves iteration because the distribution $p(Y_{mis}|Y_{obs})$ is often difficult to draw from directly. However, the data augmentation algorithm (DA),[5] a stochastic version of the EM algorithm,[6] is often straightforward to implement.

Briefly, DA involves iterating between randomly sampling missing data given a current draw of the model parameters and randomly sampling model parameters given a current draw of the missing data. The draws of Y_{mis} form a Markov Chain whose stationary distribution is $p(Y_{mis}|Y_{obs})$. Thus, once the Markov Chain has reached approximate convergence, a draw of Y_{mis} obtained by DA is effectively a proper single imputation of the missing data from the correct target distribution $p(Y_{mis}|Y_{obs})$, the posterior predictive distribution of

Y_{mis}. Many of the programs discussed in Section 5.2. use DA to impute missing values. Other algorithms that use Markov Chain Monte Carlo methods for imputing missing values include variations such as Gibbs sampling[7] and Metropolis-Hastings[8,9]. See, e.g., Gelman, et al.[10] for more details.

As discussed previously, analyzing a singly imputed data set using complete-data methods usually leads to anticonservative results because imputed values are treated as if they were known, thereby underestimating uncertainty. Multiple imputation corrects this problem, while simultaneously retaining the advantages of single imputation.

5. Multiple Imputation

Described in detail in Rubin,[11] multiple imputation is a Monte Carlo technique that replaces the missing values Y_{mis} with $m > 1$ plausible values, $\{Y_{mis,1}, \ldots, Y_{mis,m}\}$ and therefore reveals and quantifies uncertainty in the imputed values. Each set of imputations creates a completed data set, thereby creating m "completed" data sets: $Y^{(1)}, \ldots, Y^{(l)}, \ldots, Y^{(m)}$, where $Y^{(l)} = \{Y_{obs}, Y_{mis,l}\}$. Typically, m is fairly small; $m = 5$ is a standard number of imputations to use. Each of the m completed data sets is then analyzed as if there were no missing data and the results combined using simple rules described shortly.

Obtaining proper multiple imputations is no more difficult than obtaining a single proper imputation—the process for obtaining a proper single imputation is simply repeated independently m times. When the missing data pattern is not monotone, this involves generating m sequences of $\{Y^{(t)}_{mis}\}$, each with different starting values. Approximately independent multiple imputations can also be obtained from a single sequence by using only every pth draw of Y_{mis}, provided p and the length of the sequence are sufficiently large.

5.1. Combining Rules for Proper Multiple Imputations

As in Rubin[11] and Schafer[12], let Q represent the estimand of interest (e.g., the mean of a variable, a relative risk, the intention-to-treat effect, etc.), let Q_{est} represent the complete-data estimator of Q (i.e., the quantity calculated treating all imputed values of Y_{mis} as known observed data), and let U represent the estimated variance of $Q_{est} - Q$. Let $Q_{est,l}$ be the estimate of Q based on the lth imputation of Y_{mis}, with associated variance U_l—that is, the estimate of Q and associated variance are based on the complete-data analysis of the lth completed data set, $Y_l = \{Y_{obs}, Y_{mis,l}\}$, $l = 1, \ldots, m$.

The multiple imputation estimate of Q is simply the average of the m estimates: $Q_{Mlest} = \Sigma^m_{l=1} Q_{est,l}/m$. The estimated variance of $Q_{Mlest} - Q$ is found by combining between and within imputation variance, as with the analysis of variance: $T = U_{ave} + (1 + m^{-1})B$, where $U_{ave} = \Sigma^m_{l=1} U_l/m$ is the within imputation variance, and $B = \Sigma^m_{l=1}(Q_{est,l} - Q_{Mlest})^2/(m - 1)$ is the between imputation variance. The quantity $T^{-1/2}(Q–Q_{MIest})$ follows an approximate t_v distribution with degrees of freedom $v = (m − 1)(1 + U_{ave}/((1 + m^{-1})B))^2$. Rubin and Schenker[13] provide additional methods for combining vector-valued estimates, significance levels, and likelihood ratio statistics, and Barnard and Rubin[14] provide an improved expression for v with small complete data sets (*see also* ref. *1*).

5.2. Computation for Multiple Imputation

Many standard statistical software packages now have built-in or add-on functions for multiple imputation. The S-plus libraries NORM, CAT, MIX, and PAN, for analyzing continuous, categorical, mixed, and panel data, respectively, are freely available[12] (http://www.stat.psu.edu/~jls/), as is MICE[15] (http://web.inter.nl.net/users/S.van.Buuren/mi/hmtl/mice.htm), which uses regression models to impute all types of data. SAS now has procedures PROC MI and PROC MIANALYZE; in addition IVEwear (Raghunathan et al.)[16] is freely available and can be called using SAS (http://support.sas.com/rnd/app/da/new/dami.html). WinBUGS,[17] a stand-alone software package for fitting Bayesian models, imputes missing values when data are incomplete, assuming missing values are MAR. New software packages have also been developed specifically for multiple imputation. Examples are the commercially available SOLAS (www.statsol.ie/solas/solas.htm), which has been available for years and is most appropriate for data sets with a monotone or nearly monotone pattern of missing data, and the freely available NORM, a stand-alone Windows version of the S-plus function NORM (www.stat.psu.edu/~jls/). Recently, Stata announced that it supports analyses of multiply imputed data sets. (For more information, *see* www.multiple-imputation.com or Horton and Lipsitz.[18])

6. Examples

6.1. Lifecore

Intergel® solution is a medical device developed by Lifecore Biomedical to prevent surgical gynecological adhesions. A double-blind, multicenter randomized trial was designed for the US Food and Drug Administration (FDA) to determine whether Intergel significantly reduces the formation of adhesions after gynecological surgery. The data collection procedure for this study was fairly intrusive: patients had to undergo a minor abdominal surgery (a laparoscopy) weeks after the first surgery in order for doctors to determine the primary endpoint, the number of gynecological adhesions. The trial suffered from missing data because not all women were willing to have another surgery, despite having initially agreed to do so. Medical device trials (and clinical trials in general) often suffer from missing data when data collection methods are invasive.

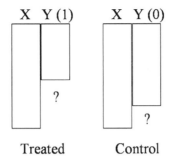

Fig. 1. Pattern of missing data for Intergel® trial.

The original proposal from the FDA for imputing the missing values (the counts of adhesions) was to fill in the worst possible value (defined to be 32 adhesions) for each missing datum, which should lead to "conservative" results because there were more missing data in the treatment arm than in the placebo arm. This method ignores observed information when creating imputations; for example, most patients with observed data had 10 or fewer adhesions. Furthermore, because the imputed values were so much larger than the observed values, the standard errors based on these worst-possible value imputations were inflated, making it unlikely to be able to get significant results, even when the two treatments were substantially different. Figure 1 displays the general pattern of monotone missing data in this case, with X representing covariates, $Y(0)$ outcomes under placebo, and $Y(1)$ outcomes under Intergel. The question marks represent missing values.

Colton, Piantadosi, and Rubin[19] instead used a multiple imputation hot deck procedure to impute missing values. Donor pools for each patient with missing data were defined by treatment group and covariates: treatment center and three measures of baseline seriousness of adhesions, which were observed for all patients. For each patient whose outcome was missing, the donor pool consisted of the two patients in the same treatment group and treatment center who had the closest baseline adhesion scores. "Closeness" was defined by the Mahalanobis metric, a corrected Euclidean squared distance, calculated for the baseline adhesion measures. For each patient with missing data, the first imputation consisted of a random draw from that patient's donor pool. The remaining value in the donor pool was used for the second imputation. The small number of imputations was deemed acceptable because less than 6% of the outcomes were missing.

Formally, this method is improper, but the limited donor pools should still make the method conservative because the matches are not as close as they would be with bigger sample sizes, or as they could be if a smooth model were used to create the imputations. The donor pool approach also has the advantage that imputations can be created without using any outcome data while remaining blind to treatment group labels, meaning that there is no opportunity to create imputations that influence results in a particular intended way.

6.2. Genzyme

Fabrazyme® is a drug developed by Genzyme Corporation to treat Fabry's disease, a rare and serious X-linked recessive genetic disease that occurs due to an inability to metabolize creatinine. Preliminary results from a phase III FDA trial of Fabrazyme vs placebo showed that the drug appeared to work well in patients in their 30s who were not yet severely ill, in the sense that it lowered their serum creatinine substantially. A similar phase IV trial involved older patients who were more seriously ill. Because there was no other fully competitive drug, it was desired to make Fabrazyme commercially available earlier than initially planned, a decision that would allow patients randomized to placebo to begin taking Fabrazyme but would create missing outcome data among placebo patients after they began taking the drug. The study had staggered enrollment, so that the number of monthly observations of serum creatinine for each placebo patient depended on the time of entry into the study. Figure 2 illustrates the general pattern of monotone missing data with the same length follow-up for each patient. Again, X represents baseline covariates, $Y(0)$ represents repeated measures of serum creatinine for placebo patients, and $Y(1)$ represents repeated measures of serum creatinine for Fabrazyme patients.

In order to impute the missing outcomes under placebo, a complex hierarchical Bayesian model was developed for the progression of serum creatinine in untreated Fabry patients. In this model, inverse serum creatinine varies linearly and quadratically in time, and the prior distribution for the quadratic trend in placebo patients is obtained from the posterior distribution of the quadratic trend in an analogous model fit to a historical database of untreated Fabry patients. Thus, the historical patients' data only influence the imputations of the placebo patients' data rather subtly—via the prior distribution on the quadratic trend parameters.

Although the model fitting algorithm is complex, it is straightforward to use the algorithm to obtain draws from $p(\theta|Y_{obs})$ for the placebo patients and then draw Y_{mis} conditional on the drawn value of θ, where, as earlier, θ represents all model parameters. Drawing the missing values in this way creates a sample from $p(Y_{mis}|Y_{obs})$ and thus an imputation for the missing values in the placebo group.

The primary analysis will consider the time to an event, defined as either a clinical event (e.g., kidney dialysis, stroke, death) or a substantial increase in

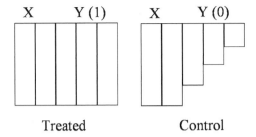

Fig. 2. Pattern of missing data for Genzyme trial.

serum creatinine relative to baseline. The analysis will be conducted on each imputed data set and the results combined (as outlined earlier in Section 5.1.) to form a single inference. Although Fabrazyme is not a medical device, the missing data mechanism in this example may be similar to those in medical device trials. Also, the availability of potentially relevant historical data is common in medical device trials.

6.3. National Medical Expenditure Survey

The National Medical Expenditure Survey (NMES) collects data, including hundreds of measurements of medical expenditures, background information, and demographic information on a random sample of approximately 30,000 members of the US population. Again, although NMES does not explicitly deal with medical devices, the general pattern of missing data and corresponding issues arising in this medical database may also arise in medical device databases.

Multiple imputation for NMES was more complicated than in the previous two examples because the missing data pattern was not monotone. Figure 3 shows a tremendous simplification of the missing data pattern for NMES, where, if Y_1 were fully observed, the missing data pattern would be monotone.

Rubin[19] imputed the missing data in NMES by capitalizing on the simplicity of imputat1:50 PMion for monotone missing data by first imputing the missing values that destroyed the monotone pattern (the "nonmonotone missing values"), and then proceeding as if the missing data pattern were in fact monotone, and then iterating this process. More specifically, after choosing starting values for the missing data, iterate between the following two steps. (1) Regress each variable with any nonmonotone missing values (i.e., Y_1), on all the other variables (i.e., Y_0, Y_2, Y_3), treating the current imputations as true values, but use this regression to impute only the nonmonotone missing values. (2) Impute the remaining missing values in the monotone pattern; first impute the

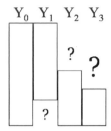

Fig. 3. Illustrative display for type of pattern of missing data in National Medical Expenditure Survey.

variable with the fewest missing values (Y_2 in Figure 3), then the variable with the second fewest missing values (Y_3 in Figure 3), and so on, treating the nonmonotone missing values filled in in Step 1 as known. This process was repeated five times to create five sets of imputations in the NMES example.

7. Summary

MI is a flexible tool for handling incomplete data sets. MIs are often straightforward to create using computational procedures such as data augmentation or using special MI software now widely available. Moreover, the results from imputed data sets are easy to combine into a single MI inference. Although MI is Bayesianly motivated, many MI procedures have been shown to have excellent frequentist properties.[21] In small samples, the impact of the prior distribution on conclusions can be assessed by creating MIs using several different prior specifications, and more generally, the impact of different models on conclusions can be analogously assessed. Furthermore, although only MAR procedures have been considered here, missing data arising from an NMAR mechanism may be multiply imputed by jointly modeling the data and the missingness mechanism; in some cases, results are insensitive to reasonable missingness models and the missing data can then be effectively treated as being MAR.[11] Rubin, Schafer, and Little and Rubin[1,11,12] are excellent sources for more detail on the ideas presented here, the last two being less technical and more accessible than the first.

References

1. Roderick, J., Little A., and Rubin, D. B. 2002. *Statistical Analysis With Missing Data.* 2nd ed. Wiley Interscience, New Jersey.
2. Zhang, J. L. and Rubin, D. B. 2003. Estimation of causal effects via principal stratification when some outcomes are truncated by "death." *J. Educ. Behav. Statist.* 28:353–368.

3. Rubin, D. B. 1978. Multiple imputations in sample surveys: a phenomenological Bayesian approach to nonresponse (with discussion). *ASA Proceedings of the Section on Survey Research Methods.* 20–34.

4. Rubin, D. B. 1976. Inference and missing data (with discussion). *Biometrika.* 63:581–592.

5. Tanner, M. A. and Wong, W. H. 1987. The calculation of posterior distributions by data augmentation (with discussion). *J. Am. Statist. Assoc.* 82:528–550.

6. Dempster, A. P., Laird, N. M., and Rubin, D. B. 1977. Maximum likelihood estimation from incomplete data via the EM algorithm (with discussion). *J. Royal Statist. Soc., Series B.* 39:1–38.

7. Geman, S. and Geman, D. 1984. Stochastic relaxation, Gibbs distributions, and the Bayesian restoration of images. *IEEE Trans. Pattern Anal. Machine Intel.* 6:721–741.

8. Metropolis, N. and Ulam, S. 1949. The Monte Carlo method. *J. Am. Statist. Assoc.* 49:335–341.

9. Hastings, W. K. 1970. Monte Carlo sampling methods using Markov chains and their applications. *Biometrika.* 57:97–109.

10. Gelman, A., Carlin, J. B., Stern, H. S., and Rubin, D. B. *Bayesian Data Analysis.* 2nd ed. Chapman & Hall, London, 2003.

11. Rubin, D. B. 2004. *Multiple Imputation for Nonresponse in Surveys.* 2nd ed. John Wiley & Sons, New York.

12. Schafer, J. L. 1997. *Analysis of Incomplete Data.* Chapman and Hall, London.

13. Rubin, D. B. and Schenker, N. 1991. Multiple imputation in health-care databases: an overview and some applications. *Statist. Med.* 10:585–598.

14. Barnard, J. and Rubin, D. B. 1999. Small-sample degrees of freedom with multiple imputation. *Biometrika.* 86:948–955.

15. van Buuren, S., Boshuizen, H. C., and Knook, D. L. 1999. Multiple imputation of missing blood pressure covariates in survival analysis. *Statist. Med.* 18:681–694.

16. Raghunathan, T. E., Lepkowski, J. M., Van Hoewyk, J., and Solenberger, P. 2001. A multivariate technique for multiply imputing missing values using a sequence of regression models. *Survey Methodology*, 27:85–95.

17. Spiegelhalter, D. J., Thomas, A., Best, N. G., and Gilks, W. R. 2002. *WinBUGS (Version 1.4).* Medical Research Council, Biostatistics Unit, Cambridge, England. Available from: www.mrc-bsu.cam.ac.uk/bugs/winbugs/contents.shtml.

18. Horton, N. J., and Lipsitz, S. R. 2001. Multiple imputation in practice: comparison of software packages for regression models with missing variables. *Am. Statist.* 55:244–254.

19. Colton, T. Piantadosi, S., and Rubin, D. B. 2001. Multiple imputation for second-look variables based on intergel pivotal trial data. Report submitted to FDA.

20. Rubin, D. B. 2003. Nested multiple imputation of NMES via partially incompatible MCMC. *Statist. Neerlandica*, 57:3–18.

21. Rubin, D. B. 1996. Multiple imputation after 18+ years (with discussion). *J. Am. Statist. Assoc.* 91:473–520.

15

Case Studies on the Local Coverage Process*

Mitchell I. Burken

1. Introduction

As providers, beneficiaries, device manufacturers, and other stakeholders strive to more fully understand the working parameters of the Medicare local coverage process, there is considerable value in presenting a more global, integrated approach. There are three major defining forces, which provide such a framework, and can be further exemplified by selected recent coverage case studies. These three forces are (1) specific regulatory mandates of the Medicare program, (2) the creation of stakeholder partnerships, and (3) the need to properly use medical evidence. Most coverage policies represent a combination of these forces. In fact, there is only the occasional local coverage scenario, which is characterized by the pure expression of any solitary element.

2. Local and National Coverage Decisions

The provision of services by the Centers for Medicare and Medicaid Services (CMS) to its beneficiaries is predicated on the determination of medical necessity, once a service has a proven benefit category within the statutorily defined parameters of the program. Title XVIII of the Social Security Act, Section 1862 (a)(1)(A)[1] states "No payment may be made under Part A or Part B for any expenses incurred for items or services which are not reasonable and necessary for the diagnosis or treatment of illness or injury or to improve the functioning of a malformed body member." Two pathways exist in which ser-

---

*Any opinions expressed herein are those of the author and do not necessarily reflect the views of TrailBlazer Health Enterprises, LLC[SM] or Blue Cross Blue Shield of South Carolina. Further, any statements made concerning an identifiable product are those of the author and are not intended to be construed as either a favorable or an unfavorable recommendation of the product by TrailBlazer Health Enterprises, LLC or Blue Cross Blue Shield of South Carolina.

From: *Clinical Evaluation of Medical Devices: Principles and Case Studies, Second Edition*
Edited by: K. M. Becker and J. J. Whyte © Humana Press Inc., Totowa, NJ

Table 1
Process Comparisons

	LCD	NCD
Requestors	Usually provider-based	Highly diverse
Breadth of services addressed	Variable	Usually narrow
Role of systematic literature reviews	Variable	Critically important
Advisory Committee referral	Mandatory (Carrier Advisory Committees): relatively broad review responsibilities	Elective (Medicare Coverage Advisory Committee): Focus on evidence evaluation
Key milestones and public comment opportunities	Initial draft policy 90-day notice and comment finalize policy	Tracing sheet "announcement" draft decicion memo 30-day notice and comment final decision memo

LCD, local coveage determination; NCD, national coveage ddetermination.

vices may be granted coverage under the Medicare program: (1) local coverage determinations (LCDs), formerly known as local medical review policies, and (2) national coverage determinations (NCDs). The overwhelming majority of decisions are made at the local level. Emerging, often high-impact technologies typically find themselves in the realm of the NCD evaluation process. Although linked by the overriding need to establish medical necessity before coverage may be granted, these processes differ in some fundamental ways, as summarized in Table 1. First and perhaps most importantly, LCD requests for coverage originate largely with Medicare contractor (that is, either Part A fiscal intermediaries or Part B carriers) interactions with provider stakeholders, whereas NCDs can be generated from a much broader base of requestors, including, but not restricted to, manufacturers, beneficiaries, providers, legislators, and even contractors.[2] Because local Medicare contractors tend to partner with local providers, often through local (as well as national) medical societies, requests from manufacturers are often expressed through local providers that have embraced emerging technologies (e.g., drugs, devices) within their practices, and, in turn, contact contractors about coverage.

Second, LCDs are often broad, having been designed to adjudicate claims for a diverse set of services. To illustrate, a physical therapy service's LCD typically might cover multiple restorative modalities, just as an LCD on chemotherapy would address many oncologic indications of salient therapeutic agents. Thus, the core of an LCD is the delineation of medical necessity, which can be translated into *Current Procedural Terminology/International Classification of Diseases*, 9th Revision, Clinical Modification (CPT/ICD-9-CM) code pairings, for the predominately electronic auto-adjudication of claims. Alternatively, the scope of an NCD tends to be relatively narrow in that specific diagnostic/therapeutic modalities are thoroughly evaluated through an evaluation of all available peer-reviewed literature, along with other sources (e.g., national specialty society position statements, practice guidelines). Except under limited circumstances, such as the November 2001 rule regarding laboratory NCDs,[3] the NCD process defers the actual formulation of code pairings to local contractors. Third, there is a mandated review[4] of all proposed LCDs by contractor-based Carrier Advisory Committees (CACs). In contrast, at the national level, there is an elective referral of pending requests for either contract technology assessments, via CMS' partnership with the Agency for Health Research and Quality (AHRQ) and/or deliberations by the Medicare Coverage Advisory Committee (MCAC). Thus, such referrals depend on the need for CMS to obtain additional analytical support on specific issues that is above and beyond what can be feasibly generated via internal systematic literature reviews. Whereas the local CAC is constructed mainly to accumulate provider input of varying types, the MCAC is responsible for technical perspectives on the state of the medical evidence, along with consumer and other advocacy viewpoints. Finally, publication formats differ for the LCDs and NCDs. For an LCD, draft policies are posted on contractor Web sites for 90-day review, during which time both general (public) comments are obtained, in tandem with those from CAC representatives. Following this notice-and-comment period, draft LCDs are finalized and then implemented in accordance with local systems specifications, although the term *finalization* is a misnomer because policy updates and revisions can occur at any time providing that stakeholders give sufficient justification for coverage expansion. Referrals through the CAC/ comment process are required only under the scenario of possible restricted coverage. In addition, under recent legislative mandates,[5,6] the new LCD format requires the publication of companion documentation, including coding guidelines, which must be separate from the expression of medical necessity language in the LCD itself. Both LCD reconsiderations[7] and appeals[8] may be exercised as additional pathways for policy alteration. During the evaluation period, new NCDs are posted on the CMS Web site via a tracking sheet, and the subsequently published draft Decision Memoranda are subject to a 30-day public comment period. This differs somewhat from the draft LCD counterpart, which is posted at the conclusion of its initial deliberative period. Sixty days after the conclusion of this comment period, a final Decision Memorandum is posted, and implementation instructions are concurrently available. As noted above, certain implementation steps, such as matching payable CPT codes with ICD-9-CM codes are usually reserved for development by local contractors, pursuant to the receipt of such instructions. Any subsequent national policy alterations must occur via a separate formal request for reconsideration[9] or an appeal.[10]

3. Decisions Made At a Contractor Level

Furthermore, in the case of TrailBlazer Health Enterprises, LLC[SM], policy-making at the contractor level, with respect to physician services under Part B, is described by a complex myriad of forces in which resources are accordingly allocated, such that approximately 100 million claims per year can be properly adjudicated within its five jurisdictions (Delaware, District of Columbia, Maryland, Texas and Virginia). Whereas the LCD is a working document designed to enable electronic edits to auto-adjudicate claims via the assignment of limited diagnostic codes[11] to specific procedural codes,[12] many other coverage decisions and related activities occur outside the LCD process described above. Specifically, there is a multidisciplinary group of clinicians and nonclinicians (i.e., the medical policy team) who follow various deliberative pathways on emerging issues (e.g., procedures, devices), which are typically referred from stakeholders (e.g., local providers, industry). Some of the issues the groups consider include:

1. Coding and/or pricing related issue(s) with no coverage action necessary.
2. When an LCD on a service already exists, the issue is whether to expand current coverage (issues regarding contraction of existing coverageneeds to be automatically referred back to the CAC/Web site review process).[13]
3. If no LCD on the service exists, one might be considered via referral to the triannual Selection Meetings in which the medical policy team decides which draft LCDs will be published for comment and review during the next CAC cycle (in the interim, no edits are in place to restrict coverage).
4. There is no pre-existing LCD, and medical policy team opts for noncoverage and informs the requestor(s), who may submit additional information and/or medical evidence when it becomes available.
5. No LCD on the service exists, because the NCD specifications (including any revisions) provide enough local guidance to make an LCD unnecessary.
6. Under rare circumstances, individual patient considerations can be granted via special-need (e.g., compassionate use) situations, which are outside the parameters of the policy-development process.

The policy selection process requires further elaboration because it synthesizes the different priorities as faced by the medical policy team. For example, interjurisdictional contractors such as TrailBlazer have had to consolidate all prior state-specific LCDs into single documents, which now cover all its jurisdictions. To illustrate, during calendar year 2004, TrailBlazer emphasized such consolidation of existing policies, in lieu of new local policy development. As new multistate Medicare Administrative Contractors[14] are created under the Medicare Modernization Act of 2003, this type of policy consolidation will likely become much more visible. Also, data-driven considerations, such as claims volumes, are key determinants of this selection process. Perhaps most importantly, the process of selecting such LCDs, in tandem with subsequently drafting new policies or redrafting existing policies, reflects the underlying dynamic in local policy formulation, as well as the interplay of regulatory, collaborative and evaluative forces.

Regulatory forces comprise those mandates under which Medicare contractors must craft policy. These are multifaceted and include the Program Integrity Manual and the "National Coverage Determination" publication (100-3) of the Internet Only Manual, Change Requests, and *Federal Register* notices, in tandem with management directives from CMS staff. The appropriate and timely assimilation of such diverse information provides the fundamental backdrop against which all other coverage activities must take place.

Collaborative forces, in the context of Medicare Part B, characterize the relationship that contractors (carriers) develop with their local provider communities, across all their state jurisdictions (parallels can be formulated with respect to fiscal intermediaries and their institutional providers). Although periodic CACs may symbolize the expression of this ongoing partnership, its backbone is the much wider array of local practitioners, who may be requesting new covered services, either independently or in concert with additional stakeholders (e.g., manufacturers). In fact, the Web posting of draft LCDs is ultimately intended for the medical community at large and is not restricted to CAC participants.

3.1. The Role of Evidence-Based Medicine

The evaluative forces modulate the above regulatory and collaborative forces by enabling the "reasonable and necessary" application of covered services to illnesses and injuries that are based on adequately documented support in the published medical literature. Although this paradigm of evidence-based medicine (EBM) cannot always apply to every situation in which LCDs must be made, it remains an overriding theme in this admixture of policy development.

In practice, EBM represents a spectrum of evaluative endeavor, which, in its purest form, involves the systematic search for improved health outcomes resulting from the medical device, drug, or procedure under examination. There is a hierarchy of medical evidence in which certain types of study designs (e.g., randomized controlled trials [RCTs]) are more robust in demonstrating improved outcomes than are less rigorous counterparts, such as epidemiological studies and case series analyses. This systematic review of the published literature on a given service can consequently maximize the opportunity for Medicare contractors to properly evaluate services during policy development.

Relying on EBM presents several practical considerations found at the local level. First, in the event that abundant published literature is available, such complex systematic reviews may be beyond the usual scope of contractor-based abilities. In such situations, any available systematic reviews (or technology assessments), such as those published by the Blue Cross Blue Shield Association (BCBSA) Technology Evaluation Center (TEC) or other AHRQ-designated evidence-based practice centers, can be extremely helpful.

Second, by way of contrast, there can be the persistent dearth of published evidence, particularly among new devices, which have not undergone extensive trials. Whereas large-scale RCT designs for new drugs often provide benchmarks of adequate rigor, such resource-consuming efforts often may not be feasible in the context of medical devices.

Third, there are frequently inherent shortfalls in the ability of the published literature to demonstrate that diagnostic tests have reasonable and necessary clinical use in that they improve measurable health outcomes. By contrast, median overall survival is a type of health outcome that might be used to assess a new chemotherapeutic agent, thus underscoring this particular relative convenience in considering therapeutic modalities. Although a new diagnostic test may strengthen the ability to detect a particular disease (i.e., test sensitivity), as well as minimize the occurrence of false-positive diagnoses (i.e., test specificity), such studies of test performance[15] are not configured to determine whether patients will ultimately have improved health outcomes. However, one might infer that improved diagnosis can, in pertinent situations, lead to more timely and effective subsequent treatment.

Finally, some types of EBM studies, such as cost-effectiveness or cost-benefit analyses, highlight resource limitation as a potential consideration in formulating coverage decisions, although at the present time, CMS prohibits contractors from basing Medicare reasonable and necessary determinations on cost.

3.2. Understanding the Balance of Factors

Readers should be cautioned that there is not always a clear demarcation between regulatory, evaluative, and collaborative interests. For example, although one might alternatively suggest that EBM could be included within the regulatory component, it is equally reasonable to assert that it should be deemed an independent element, given the historically ill-defined nature of the term *reasonable and necessary*; CMS has not issued specific defining criteria for it.[16] In addition, many clinical specialty societies have developed practice guidelines that may be constructed according to an EBM-type model and/ or via input from clinical leaders in the absence of published rigorous evidence to demonstrate improved health outcomes. Thus, if and when contractors use practice guidelines in their coverage deliberations, it might be considered collaborative to the extent that it expresses partnering with the clinical community and evaluative in that many guidelines are crafted according to the principles of EBM.

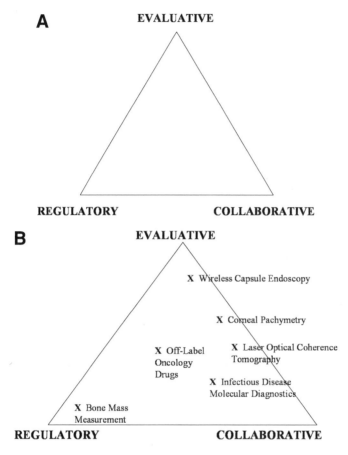

Fig. 1. (**A**) General paradigm of local coverage elements (**B**) using pertinent examples.

Thus, these factors form a triangle (Fig. 1) that is bounded by each of these three metrics. In a theoretical local policy milieu, which is purported to be optimally balanced or symmetrical, one might further suppose that policy is developed in the context of nearly equal regulatory, collaborative, and evaluative components, although this is rarely or ever the case. To illustrate the different configurations of predominate forces, we present some pertinent TrailBlazer medical policy team case studies, which, in turn, enable may help readers circumscribe the policy-development process in both a more informative and more realistic manner.

Four diagnostic device topics followed by a more abbreviated nondevice application referring to off-label coverage of oncology drugs will be presented (Fig. 1B). Although the medical policy team has considered other recent therapeutic devices, their didactic value has been somewhat diminished by the combination of limited published literature, in tandem with a relatively narrow spectrum of local provider interest. The case studies are:

4. Case 1: Osteoporosis Screening

The diagnosis of osteoporosis, a disease characterized by low bone mass, and the concomitant determination of fracture risk, has posed a continuing challenge to both radiologists and clinicians.[17] The Balanced Budget Act of 1997 provided the Medicare program with a new osteoporosis screening benefit in which multiple bone mass measurement techniques (e.g., some applied to the skeleton and others to peripheral sites such as the wrist or heel) would be covered within certain frequency parameters if a defined set of beneficiary eligibility criteria were met. In this case, the resultant regulation[18] was sufficiently complex that an LCD[19] was necessary to properly execute the legislation and ensure that appropriate limited coverage was established. Notably, only limited public commentary could be incorporated by the medical policy team, given the instrinsically prescriptive nature of the benefit. For example, during the 2004 statewide consolidation process, multiple clinicians from the different TrailBlazer CACs suggested various clinical indications for which osteoporosis screening would be necessary; however, only ICD-9-CM codes pertaining to the following five qualifying patient categories were permissible: estrogen deficiency, vertebral abnormalities attributable by X-ray to low bone mass, glucocorticoid therapy for at least 3 months, primary hyperparathyroidism, and monitoring for osteoporosis drug therapy. In addition to emphasizing such limitations at each CAC itself, the existing process of including medical policy team responses to all draft LCD recommendations on the published LCD establishes a consistent means for ensuring the transparency of the policy-development process. To respect and maintain the critically important covenant between a contractor and its various stakeholders, commentors are fully entitled to understand decisions regarding whether their suggestions were incorporated into the published LCD.

5. Case 2: Molecular Diagnostics for Infectious Diseases

Molecular diagnostic laboratory testing, which includes DNA and RNA analysis, often provides sensitive, specific, and timely (i.e., relative to conventional methods) identification of diverse biological entities, including microorganisms and tumors. The relevance of obtaining various types of molecular signatures is assuming a prominent role in diagnostic medicine.

Consequently, it is incumbent on the Medicare program to adjudicate claims for such services in a manner that is commensurate with both the clinical and scientific state of the art. Various infectious disease molecular diagnostic assay platforms, such as nucleic acid amplification testing, are relatively well delineated by the current CPT coding structure. By contrast, the characterization of tumors via gene expression assays is much more complex within the existing CPT framework; therefore, using a set of infectious disease illustrations is more informative.

There are numerous occasions when the need to craft new local policies is a result of previous unmet needs in existing policies. One such example involved the TrailBlazer Non-Covered Services policy,[20] which is a broad compendium of procedural codes pertaining to diagnostic and therapeutic services that have failed to satisfy the requisite degree of medical necessity. For instance, NCD coverage exclusions provide one component of this master list. During release of this revised LCD, as part of the first CAC cycle during calendar year 2003, multiple commentors wrote that many molecular diagnostic codes (e.g., specific to bacteria or viruses) should not have been included on this list. The primary reason for the original inclusion of such codes in the policy was the lack of claims track records; as a result, it was necessary to focus on the potential use of such codes in the future, rather than continuing in the mode of retrospective data analysis, because such diagnostic techniques had been rapidly gaining a foothold in routine patient evaluation and management.

In fact, during the 90-day comment period, a considerable amount of literature, such as letters of support, were received from various stakeholders, including local pathology practices, national specialty and trade organizations, diagnostic test manufacturers, and academic medical centers. After reviewing this material and calling stakeholders, it became abundantly clear to medical directors that the most prudent course of action would be to develop an Infectious Disease Molecular Diagnostic Testing LCD[21] specifically crafted to address these complexities. The current venue of treating such codes in "reverse fashion"—that is, simply continuing to delineate non-coverage only—was deemed unsatisfactory. Thus, this new file served as a substrate for policy development.

The necessary work had only just begun, because a host of individual coverage decisions needed to be made. The CPT manual listed 23 separate microorganisms, and under the pertinent molecular diagnostic codes, each one needed an assay-specific coverage determination. All but one of the microorganisms have three repeating procedural codes, which correspond to the following:

1. Direct probe technique, in which the nucleic acid signature of a suspected microorganism can be detected in a relatively straightforward manner.
2. Nucleic acid amplification testing, which requires that the nucleic acid content be replicated or amplified such that a sufficiently powerful signal can be gener-

ated to detect that micro-organism.

3. Quantitative technique, in which numeric data on nucleic acid content are measured (e.g., "viral load" response may be measured in patients undergoing treatment for human immunodeficiency virus [HIV] infection).

Given that CPT- and ICD-9-CM-limited coverage was specified in the laboratory NCD for both HIV-1 and HIV-2, this left approximately 60 diagnostic categories in which individual coverage determinations would be necessary. Given TrailBlazer's staffing limitations along with the lack of outcomes-based studies on the use of such emerging technologies, the contractor undertook an appropriate pathway of stakeholder collaboration and some degree of evaluation. In addition to reviewing pertinent practice guidelines, when available, and performing some relevant diagnostic tests (i.e., sensitivity and specificity), the contractor placed considerable focus on consultations with expert members of the laboratory community, most notably the Association for Molecular Pathology and the American Society for Microbiology. Furthermore, this partnership enabled the development of a working list of applicable ICD-9-CM codes—corresponding to those clinical presentations associated with current CPT-specified micro-organisms and those not yet listed (e.g., severe acute respiratory syndrome-related coronavirus) that might require molecular diagnostic testing—and allowed claim submission using *not otherwise classified* codes.

This case study illustrates an essential characteristic of the LCD process: its dynamic nature. In this process, a basic underlying assumption is that medical policy team will convert information on new specific molecular assay/microorganism pairings that have become medically necessity into timely upgrades of the policies.

6. Case 3: Ophthalmologic Diagnostic Testing

Two emerging diagnostic techniques in ophthalmology continue to illustrate the model of mixed collaboration and evaluation found in Case 2. Case 3 highlights an instance in which both LCD and non-LCD-based coverage approaches can be equally appropriate.

6.1. Laser Optical Coherence Tomography (LOCT)

LOCT is a noninvasive, noncontact imaging technique that produces high-resolution longitudinal cross-sectional tomographs of ocular structures in real time, consequently facilitating more precise diagnoses. LOCT has been well documented as an imaging diagnosis in the early detection of glaucoma.[22,23] It is also a valuable technique in the evaluation and treatment of patients with retinal disease,[24,25] particularly certain macular abnormalities such as cysts, holes, pseudoholes, and puckering. Before LOCT, more limited tools (i.e., clinical examination and fluorescence angiography) were available to evaluate such pathologies. Furthermore, LOCT can aid in making surgical management decisions, such as helping to determine the presence of vitreoretinal traction, which may influence the use of laser vs surgical approaches in diabetic macular edema.

During the routine LCD selection process in 2003, the medical policy team weighed the options of combining individual state-specific policies on LOCT and policy retirement, because contractors need to periodically review existing LCDs to determine whether they are achieving their objectives of high-quality adjudication of claims. After consultations with both general ophthalmologists and retinal specialists, the medical policy team decided to consolidate and refine the existing policies to allow this new technology to target appropriate patient populations properly.[26] Although its decision was not driven by RCT data, the medical policy team was quite satisfied that the various emerging applications of LOCT met the threshold of reasonable and necessary.

6.2. Corneal Pachymetry

The ophthalmology community's recent growth in interest in measuring central corneal thickness (CCT) via corneal pachymetry was triggered by two publications in the June 2002 *Archives of Ophthalmology*. Subsequently, the American Academy of Ophthalmology practice guidelines[27] deemed these studies as strong evidence for measuring CCT in the evaluation of primary open-angle glaucoma (POAG). Whereas Kass et al.[28] did not directly address CCT in their RCT, which determined that an ocular hypotensive treatment strategy delayed or prevented the onset of POAG, the companion epidemiological study by Gordon et al.[29] found that CCT was a significant predictive factor in the development of POAG among individuals with elevated intra-ocular pressure.

After TrailBlazer received correspondence during late 2002 and 2003 from various clinicians in its Part B jurisdictions, the medical policy team reviewed these two sentinel studies in tandem with its usual consultations. Although these studies suggested that CCT measurement might not yet be a fully understood, quantifiable entity in the overall risk assessment of glaucoma progression among patients with ocular hypertension, the combination of this evidence, coupled with provider input, ultimately supported positive coverage. When the medical policy team assigned corneal pachymetry a Category I CPT code effective January 2004, it did not implement limited coverage; therefore, no LCD was necessary. As with any new and/or existing physician service, the medical policy team periodically reviews claims data to determine if a future LCD might be warranted. Ultimately, the medical policy team did not consider decision making at the level of specific pachymetry devices. Although there is a common operational principle in the reflection of light or ultrasound from anterior and posterior corneal surfaces, it is not within the regulatory purview of a Medicare contractor to evaluate safety and effectiveness; instead, its purpose is to assess the clinical utility of the overall service encompassed by CPT code 76514©.

7. Case 4: Wireless Capsule Endoscopy

The Food and Drug Administration's decision to allow marketing of PillCam™ SB Capsule Endoscopy Given® Diagnostic System (also known as WCE) in August 2001 generated considerable literature on how small bowel pathology could be more optimally visualized. Whereas both upper and lower endoscopy (i.e., colonoscopy) enable evaluation of common disorders of the gastrointestinal system—a region that is relatively less accessible than others—diseases with lower prevalence (e.g., tumors) have been less easily identified, particularly in the absence of upper and lower endoscopic findings. With WCE, the patient digests a small (11 × 26 mm) capsule that contains a capsule, which generates images and data as it passes through the digestive system, while maintaining normal activities. During the 8-h examination, information from the camera is transmitted to a recorder device worn around the waist. Local providers brought WCE to the attention of the medical policy team, which found it necessary to incorporate this device into its ongoing working agenda.

In February 2003, the BCBSA TEC published a systematic review of the literature. Based on the mandate of the BCBSA TEC's established criteria, the group would need to demonstrate that WCE both "improves the net health outcomes; and ... [is] as beneficial as any established alternatives ..."[30] to make a favorable decisionon. This technology assessment critiqued three key published studies, involving a total of 72 patients, in which WCE was compared with two alternative modalities, push enteroscopy and a radiographic small bowel barium evaluation. Based on the positive cumulative findings from the appropriately constructed studies, WCE met the BCBSA TEC criteria "in obscure digestive tract bleeding suspected to be of small-bowel origin." This technology assessment was the critical factor in TrailBlazer supporting positive coverage for WCE.

BCBSA TEC's initial finding on WCE did not end the discussion on providing coverage for the procedure. According to the Food and Drug Administration's label,[31] WCE "may be used as a tool in the detection of abnormalities of the small bowel," but whether the technology improved health outcomes (e.g., the ability to both influence and improve patient management decisions based on WCE) regarding indications other than suspected small bowel bleeding had not been determined. Thus, BCBSA TEC published a much more extensive follow-up technology assessment in December 2003,[32] reflecting the burgeoning literature on potential applications such as Crohn's disease.

Similar to its decision on corneal pachymetry (*see* Case 3), the medical policy team has elected to track claims data for this device to later determine whether an LCD might be necessary to better direct reimbursement for WCE.

8. Case 5: Off-Label Oncology Drug Coverage

Although not within the realm of device evaluation, it is reasonable to close this presentation of various case scenarios by posing the question of whether there might be an identifiable paradigm in which all three forces can coexist in fairly even balance. It should be noted that Medicare contractors need to make coverage determinations on non-self-administered chemotherapeutic agents used off-label. Based on CMS directives,[33] decisions must include a substantive evidence-based (evaluative) component. Furthermore, because local oncologists aggressively keep abreast of the research on treatments through reading the latest studies and attending professional society meeting presentations, there exist opportunities for such a convergence of these three elements.

9. Conclusion

In summary, this chapter presents the backdrop under which Medicare carriers are chartered to make LCDs and has provided a triangular paradigm for expressing the regulatory, collaborative, and evaluative boundaries that surround this decision making. Although selected diagnostic devices have been discussed, this model should be extrapolated to the full complement of policy issues. Whereas the more regulatory or prescriptive approach applies to numerous areas, such as podiatry and ambulance services, other issues harmonize greatly with the evaluative WCE approach. In any case, all processes are intended to achieve the endpoint of timely, efficient, and the most clinically appropriate adjudication of Medicare claims.

Acknowledgments

The following individuals from both TrailBlazer Health Enterprises, LLC and its parent company, BlueCross BlueShield of South Carolina, have greatly assisted the author in the preparation of this chapter: Ruth Ledbetter and Kendall Walker, Esq.

References

1. 42 U.S.C. §1395y(a)(1)(A). 2000; *see also* Centers for Medicare and Medicaid Services, U.S. Department of Health and Human Services. *Medicare Benefit Policy Manual.* Pub. No. 100-2, chapter 16, §20. Available from: www.cms.hhs.gov/manuals/102_policy/bp102c16.pdf.
2. 68 *Fed. Reg.* 55,634; 2003.

3. Centers for Medicare and Medicaid Services, U.S. Department of Health and Human Services. *Medicare Claims Processing Manual*. Pub. No. 100-4, chapter 16, §120. Available from: www.cms.hhs.gov/manuals/104_claims/clm104c16.pdf.

4. Centers for Medicare and Medicaid Services, U.S. Department of Health and Human Services. *Medicare Program Integrity Manual*. Pub. No. 100-8, chapter 13, §13.8.1. Available from: www.cms.hhs.gov/manuals/108_pim/pim83c13.pdf.

5. Medicare, Medicaid, and SCHIP Benefits Improvement and Protection Act Of 2000. Public Law No. 106-554, §522, 114, Stat. 2763 [hereinafter "BIPA"], §522.

6. Centers for Medicare and Medicaid Services, U.S. Department of Health and Human Services. *Medicare Program Integrity Manual*. Pub. No. 100-8, chapter 13, §13.1.3. Available from: www.cms.hhs.gov/manuals/108_pim/pim83c13.pdf.

7. Centers for Medicare and Medicaid Services, US Department of Health and Human Services. *Medicare Program Integrity Manual*. Pub. No. 100-8, Chapter 13, §13.11. Available from: www.cms.hhs.gov/manuals/108_pim/pim83c13.pdf.

8. Id.; see also BIPA, §522.

9. 68 Fed. Reg. 55,634; 2003.

10. BIPA, §522, codified in part at 42 U.S.C. §1395ff(f).

11. Practice Management Information Corporation. 2004. *International Classification of Diseases*. 9th Rev. Clinical Modification. 6th Ed., Los Angeles, CA.

12. American Medical Association. 2004. *Current Procedural Terminology*, Chicago, IL.

13. Centers for Medicare and Medicaid Services, U.S. Department of Health and Human Services. Medicare Program Integrity Manual. Pub. No. 100-8, chapter 13, §13.11. Available from: www.cms.hhs.gov/manuals/108_pim/pim83c13.pdf.

14. Medicare Prescription Drug, Improvement and Modernization Act of 2003, Pub. Law No. 108-173, §911; 2003.

15. Sox, H. C., Blatt, M. A., Higgins, M. C., Marton, K. I., eds. 1988. *Medical Decision Making*. Butterworth-Heinemann, Stoneham, MA.

16. Tunis, S. R. 2004. Why Medicare has not established criteria for coverage decisions. *N. Engl. J. Med.* 350:2196–2198.

17. Genant, H. K., Guglielmi, G., Jergas, M., eds. 1998. *Bone Densitometry and Osteoporosis*. Springer-Verlag, Berlin.

18. Centers for Medicare and Medicaid Services, U.S. Department of Health and Human Services. *Medicare Claims Processing Manual*. Pub. No. 100-4, chapter 13, §140.1-140.2. Available from: www.cms.hhs.gov/manuals/104_claims/clm104c16.pdf.

19. TrailBlazer Health Enterprises. Bone mass measurement; local coverage determination. Available from: www.trailblazerhealth.com/lmrp.asp?ID=2038&lmrptype=dcde. Accessed August 2004.

20. TrailBlazer Health Enterprises. Non-covered services; local coverage determination. Available from: www.trailblazerhealth.com/lmrp.asp?ID=2070&lmrptype=dcde. Accessed August 2004.

21. TrailBlazer Health Enterprises. Infectious disease molecular diagnostic testing; local coverage determination. www.trailblazerhealth.com/lmrp.asp?ID=1777&lmrptype =dcde. Accessed August 2004.

22. Bowd, C., Zangwill, L. M., Berry, C. C., et al. 2001. Detecting early glaucoma by assessment of retinal nerve fiber layer thickness and visual function. *Invest. Ophthal. Visual Sci.* 42:1993–2003.

23. Greaney, M. J., Hoffman, D. C., Garway-Heath, D. F., et al. 2002. Comparison of optic nerve imaging methods to distinguish normal eyes from those with glaucoma. *Invest. Ophthal. Visual Sci.* 43:140–145.

24. Lattanzio, R., Brancato, R., Pierro, L., et al. 2002. Macular thickness measured by optical coherence tomography (OCT) in diabetic patients. *Eur. J. Ophthal.* 12:482–487.

25. Massin, P., Duguid, G., Erginay, A., Haouchine, B., Gaudric A. 2003. Optical coherence tomography for evaluating diabetic macular edema before and after vitrectomy. *Am. J. Ophthal.* 135:169–177.

26. Laser (optical) coherence tomography; local coverage determination. Available from: www.trailblazerhealth.com/lmrp.asp?ID=2078&lmrptype=dcde. Accessed August 2004.

27. Preferred Practice Pattern™. 2002. *Primary Open-Angle Glaucoma Suspect.* American Academy of Ophthalmology, San Francisco. Available from: www.aao. org/education/ppp/poags_new.cfm. Accessed August 15, 2004.

28. Kass, M. A., Heuer, D. K., Higginbotham, E. J., et al. 2002. The ocular hypertension treatment study: a randomized trial determines that topical hypotensive medication delays or prevents the onset of primary open-angle glaucoma. *Arch. Ophthal.* 120:701–713.

29. Gordon, M. O., Beiser, J. A., Brandt, J. D., et al. 2002. The ocular hypertension treatment study: baseline factors that predict the onset of primary open-angle glaucoma. *Arch. Ophthal.* 120:714–720.

30. Technology Evaluation Center Assessment Program. 2003. *Wireless Capsule Endoscopy.* Blue Cross and Blue Shield Association, Chicago.

31. Food and Drug Administration, Department of Health and Human Services. 2003. Indications for Use Statement for Given Diagnostic System with Suspected Blood Indicator (SBI) *in* Letter from Nancy C. Brogdon, Director, Division of Reproductive, Abdominal, and Radiological Devices, Office of Device Evaluation, Center for Devices and Radiological Health (on file with author).

32. Technology Evaluation Center Assessment Program. 2003. *Wireless Capsule Endoscopy For Small-Bowel Diseases Other Than Obscure GI Bleeding.* Blue Cross and Blue Shield Association, Chicago.

33. Centers for Medicare and Medicaid Services, U.S. Department of Health and Human Services. *Medicare Benefit Policy Manual.* Pub. No. 100-2, chapter 15, §50.4.2. Available from: www.cms.hhs.gov/manuals/102_policy/bp102c15.pdf.

16

Reimbursement Analysis

From Concept to Coverage

Robin Bostic

1. Establishing Reimbursement

Establishing reimbursement early in product development is essential to the success of a medical device company. Although reimbursement is comprised of coverage, coding, and payment procedures, coverage is the essential first step that drives subsequent coding and payment procedures. After all, if a product is not covered by insurance plans, there can be no reimbursement. Coverage occurs when a product is deemed a "reasonable and necessary" medical treatment. It is widely reported that Medicare takes an average of 2 to 5 years to create coverage for a new product. Given that a new device's product life cycle may be only 10 years, the earlier the process is initiated to obtain coverage, the sooner reimbursement will be established. Perhaps most importantly, this 2- to 5-years anticipated timeframe for coverage can be reduced if a reimbursement plan is implemented early in product development. When developing a reimbursement plan, companies should address the following questions, preferably while their product is still in the development phase:

1. Where will this product fit in the larger health care arena?
2. How will this product meet the Food and Drug Administration (FDA) "safe and effective" standards, as well as payors' "reasonable and necessary" requirements?
3. How can the reimbursement strategy support the company's overall objectives?

1.1. Where Will It Fit?

Companies must decide in which of the following three categories does the technology or procedure fit best. Is the technology or procedure (1) similar to another product already on the market (a "me too" device); (2) is it an expansion or different use of an existing technology; or (3) is it truly new and innovative? Table 1 illustrates the time (and therefore profit) implications for reimbursement categories.

From: *Clinical Evaluation of Medical Devices: Principles and Case Studies, Second Edition*
Edited by: K. M. Becker and J. J. Whyte © Humana Press Inc., Totowa, NJ

Table 1
Reimbursement Analysis: Implications of Where a Product Fits

	Similar to another product	Expansion of existing technology	Truly new and innovative
Reimbursement components that must be developed	Confirm existing code and inclusion for coverage of this product	Alter coverage, coding, and payment to include this product	Create new coverage, coding, and payment for this product
Science required	Usually FDA approval for same indications suffice for inclusion in existing coverage	Publication of controlled studies (usually 1–2)	Publication of controlled studies (usually 2–4) and cost-effectiveness data (publications and/or registry data)
Typical timeline for these components post-FDA approval	6 months to 1 year	1 to 2 years	2 to 5 years

FDA, Food and Drug Administration.

Reimbursement is easiest to obtain with "me too" products, because coverage, coding, and payment has been defined for a similar product. If this is where the product fits, the primary task is to ensure that it is identified under the existing technology and grouped with the existing code to trigger appropriate payment.

Indication expansion of an existing technology often requires altering coverage, coding, and payment to address the new indication. Published studies supporting the proposed additional indication and revisions of established medical policy will be necessary to create coverage. Codes may need to be revised, including new code descriptions, which can trigger different payment rates.

Finally, if the product is a new and innovative technology, a new reimbursement structure will need to be constructed and implemented to address coverage, coding, and payment.

Regardless of the product's category, a reasonable timeline for developing and implementing the reimbursement plan must be anticipated. If coverage, coding, and payment already exist for a similar product, 6 to 12 months is typical for getting a product positioned within the existing category. To obtain or modify a new Medicare code, whether related to physician procedures under the *Current Procedural Terminology* or the device itself under the Healthcare Common Procedure Coding System (HCPCS), the process usually takes approximately 1 to 2 years from the product's launch into the health care market. For new technologies, Medicare historically has taken 2 to 5 years to create national medical coverage or to substantially expand existing guidelines. Given these timeframes, it is advantageous, as well as economically beneficial, to address these reimbursement issues during product development.

2. Meeting Safe and Effective and Reasonable and Necessary Payor Criteria

2.1. Case Study: The Exogen Sonic Accelerated Fracture Healing System

The Exogen Sonic Accelerated Fracture Healing System® (SAFHS) shown in Fig. 1 was introduced into the market in 1994. At that time, the FDA indication was limited to the acceleration of the time to heal fresh fractures of the distal radius and tibia. These fractures had to be orthopedically managed by closed reduction and cast immobilization.

The SAFHS accelerated bone healing in fresh fractures (fractures less than 3 months from date of fracture) of the tibia and radius. The usual and customary price of the device to insurers was approximately $3500 per patient. Because the SAFHS device was the first of its kind, no coverage, coding, or payment had been established at the time of FDA approval. The FDA approval was based on two randomized, double-blind, placebo-controlled studies that

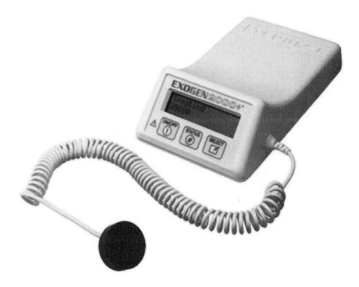

Fig. 1. Exogen SAFHS.

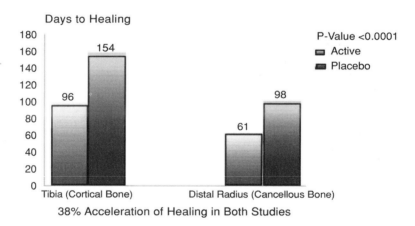

Fig. 2. Exogen key study results.

showed 35% acceleration of bone healing. Both studies had been published in the *Journal of Bone and Joint*, a peer-reviewed publication. Figure 2 illustrates the key findings from these studies.

Most payors, private (e.g., Aetna, Blue Cross Blue Shield [BCBS]) or public (e.g., Medicare, Medicaid), require treatment and procedures to be reasonable and necessary, although they often vary in how they define the term. Thus, a first step in a reimbursement plan is to identify and target the anticipated primary payors for the product before initiating studies in order to build the

Table 2
Initial Questions for Meetings With Medical Directors

1. What are your criteria for establishing that a device is "reasonable and necessary"?
2. Have you ever evaluated a product similar to this one? If yes, what aspects proved important in your decision on that product?
3. Is there a national or larger regional carrier whose decisions you follow in making your policies?
4. Are there particular technology assessment reviews (such as Blue Cross Blue Shield Association's) that you use in making your determination?
5. Do you require a particular determination before you will reimburse for a product? Do you customarily review a product only when it reaches a certain level of usage among your population?
6. How long does it ordinarily take for you to make this determination on a product?
7. Are there any indications that your criteria for this might change while we are completing our studies? Do we need to add/alter anything to accommodate evolving criteria?
8. Is there a separate utilization review or case management organization, which may be separately reviewing each case for your company? Do we know how its criteria fit with your company's criteria?

reasonable and necessary aspect into the clinical protocols. This is best accomplished by meeting face-to-face with the medical directors of the identified payors, and clarifying their expectations and requirements for a product to meet reasonable and necessary criteria. Initial questions for medical directors are provided in Table 2. If clinical studies are already underway, it is still appropriate to meet with key medical directors to either add this reasonable and necessary component to the protocols or to establish separate procedures to ensure reasonable and necessary conditions can be demonstrated.

In the example of the Exogen product, the initial strategy for coverage, coding, and payment included targeting payors that were interested in returning a patient to work more promptly or avoiding nonunion fractures. Therefore, workers' compensation carriers and private insurers that paid disability insurance to patients were initially approached with the clinical evidence as well as a cost-effectiveness study showing clinical benefit with cost savings. Larger workers' compensation carriers such as Liberty Mutual, Travelers Insurance, Kemper Insurance Company, and Wausau Insurance Company were targeted based on the number of workers insured and the opportunity to educate case managers and claims adjustors on the clinical benefit of the device through manufacturer-sponsored continuing education programs. Because most workers' compensation plans do not create medical policy or perform their own technology assessments, these two groups provide approvals on a case-by-case determination.

Private payors will sometimes reimburse for a new therapy until the requests for pre-authorization of approval increase to a certain volume or economic threshold. Private payors may then begin to deny coverage, placing the technology under an *experimental and investigational* category until a medical policy is created to establish or deny coverage. Accordingly, manufacturers may erroneously conclude that initial coverage suggests their device has been approved by a carrier, when in fact, the payor may initially reimburse because it did not identify the product as new or having expanded indications. The product falls below the reimbursement radar.

Using the approach of creating plan-by-plan coverage, local medical coverage gives manufacturers the opportunity to gradually develop a reimbursement plan. Reasonable and necessary requirements vary by plan, and coverage by one plan can influence other plans. For example, larger insurance plans, such as Anthem BCBS, are more influential than others owing to their market size. A positive coverage decision by Anthem BCBS can influence a favorable decision by BCBS of Texas. With other private payors, the local medical director reviews clinical evidence to make local coverage determinations (LCDs) and recommendations when a national medical policy is made. As the technology becomes more broadly accepted, many national private insurers will establish national medical policy. Groups such as Aetna, Cigna, United Health Care, and Humana have in-house technology assessment groups that review technologies and establish national medical policies for their local plans. Local health maintenance organizations (HMOs) may use their own assessment processes as they create their specific medical policy.

If the product becomes well used, some payors such as HMOs will request technology assessments of the therapy to provide guidance in the coverage determination process. A technology assessment is a systematic evaluation of scientific evidence used to form conclusions on the benefits and risks of a particular device, often in relation to its potential clinical use for a defined group of individuals. Thus, technologies now need to obtain positive technology assessments from such groups as BCBS Association (BCBSA) Technology Evaluation Center (TEC), Hayes Group, or ECRI (formerly the Emergency Care Research Institute). Many of the BCBS plans will consider BCBSA TEC's assessment as they draft their own specific medical policies. Other payors will purchase analyses completed by Hayes or ECRI to determine whether they should provide coverage for a new therapy. It is important that these organizations are approached when the body of clinical evidence is strong; however, if the therapy is truly innovative or expensive and has the possibility of high volume, these assessments are typically initiated at the request of the local plans without manufacturer contact or input. It is important to determine when these technology groups should be approached as the reimbursement plan is

created. The process and criteria for BCBSA TEC is available at www. bcbs.com/
tec/, and for ERCI at www.ecri.org/Products_and_Services/Membership _Pro-
grams/Health_Technology_Assessment_Information_Service/Evidence_
Based_Practice_Center.aspx.

2.2. Blue Cross Blue Shield Coverage

Owing to the potential wide-scale use and appreciable expense, the BCBSA
TEC completed a review of the SAFHS within 1 year of FDA approval. Based
on the two published studies showing the device accelerated tibia and radius
fracture healing, BCBSA TEC made a positive report in 1995. Based on this
recommendation, approximately 33% of the BCBS plans developed medical
policies for the Exogen device.

Two complications emerged after the BCBSA TEC made its assessment.
First, approximately 67% of the BCBS plans did not establish medical policy.
Second, the coverage they developed was narrow, as only patients with
nonunions demonstrated by two sets of X-rays obtained at least 90 days after
initial surgery were eligible for this product. With such limited medical policy,
approximately 50% of physician-ordered prescriptions for SAFHS were preauthorized by individual BCBS plans across the United States.

2.3. Medicare

Following FDA approval in 1995, Exogen contacted four durable medical
equipment regional carriers (DMERCs) to review a possible LCD as dictated
by the Centers for Medicare and Medicaid Services (CMS). DMERCs' unani-
mous decisions amount to Medicare national coverage decisions (NCDs). If
the four DMERCs do not feel comfortable making a decision regarding a technology, they typically ask for the CMS Coverage and Analysis Group (CAG)
to issue an NCD.

After another 12 months of reviewing the technology, no consensus on coverage emerged from the four DMERCs, so the DMERC medical directors asked
the CAG to decide if Medicare should cover the Exogen device for the acceleration of bone healing in the tibia and radius fractures and/or nonunion.

During the time before the Medicare decision was made, the manufacturer
elected to provide the device to Medicare patients with nonunion fractures that
had not healed in the previous 6 months. Exogen selected these patients to
prove to Medicare that the noninvasive SAFHS device could heal complex
nonunion fractures without revision surgery. The manufacturer's strategy was
developed to address several needs, including:

1. Showing the benefit of the technology on the Medicare population. The DMERC
   medical directors had complained that the initial clinical study was done on a
   younger population (average patient age = 55), so the study outcomes could not

be assumed with a population older than 65 years. In addition, the study design excluded patients on medications that Medicare patients typically use. Putting the SAFHS on Medicare patients would allow the manufacturer to demonstrate that outcomes were in fact similar or to identify relevant circumstances that should be considered for the Medicare population.

2. Providing patients with treatment that they truly needed, and supporting physician decisions. Doctors prescribe treatments but quickly lose interest in them if the insurance carrier denies the prescribed treatment and patients have to pay for it or when hospitals have to absorb the usage costs.

A review of 150 claims by the DMERC medical directors continued to support the notion that clinicians saw the therapy as a medical standard in treating nonunion fractures of the elderly and Medicare coverage was necessary. It took 3 years to go through three levels of appeals, which included DMERC telephonic hearings, in-person reviews, and adjudicated law judge (ALJ) proceedings, with the ALJ judgment reviewed by the DMERC medical directors before payment. This process took 3 years because Medicare Regional Carriers (RCs) sometimes lost requests for hearings or documentation and the manufacturer created an internal administrative structure late in the process to meet appeal deadlines, confirm hearing requests had been received, and obtain additional medical reports and outcome data. Claims were lost—some forever—and the resurrection of others took valuable time. By providing administrative support and documenting when information was received, RCs are required to meet the CMS' appeal timeframes. If they do not meet these deadlines, RCs can be required to pay interest on the total amount owed. All appeals were denied until the claims were reviewed at the ALJ level. Expert medical testimony confirming medical necessity was essential, as was the manufacturer's positive clinical outcome data for each appealed case. The ALJ reversed the earlier denials by the DMERC and ordered payment 100% of the time, resulting in the DMERCs paying more than 100 claims. Costs to adjudicate appeals including administrative and medical expert time, materials, and travel expense were estimated at $150 per claim. Devices had been provided to the Medicare population since 1996, but Exogen did not receive the first payment from the DMERCs until 1999. Although the appeal process cost approximately $15,000, it ultimately generated more than $210,000 in revenue.

Important factors in appealing Medicare claims include:

1. Establishing a company internal structure to administer, move, and monitor claims through Medicare appeal system.
2. Meeting Medicare timeframes for filing all levels of appeal. If a deadline is missed, the claim will be eliminated based on filing requirements alone. These Medicare deadlines can also be used to help expedite the process and keep the Medicare reviewers on schedule and accountable.

3. Maintaining complete medical documentation detailing clear medical necessity and providing individual patient outcomes. The manufacturer appealed only those claims in which the patient's fracture healed with the use of the Exogen device.
4. Identifying and using independent expert medical witnesses (e.g., members of medical societies who prescribe the product, physicians involved in the early clinical trials) at the ALJ hearing to establish reasonable and necessary medical care, thereby meeting Medicare's criteria for coverage.

If Exogen's intent in providing Medicare beneficiaries with the device had been to increase revenue, the appeal process would have proven a poor tactic. Instead, Exogen provided the SAFHS to this patient population to meet physicians' requests and expectations. How the appeal process is employed is critical to building a successful proactive reimbursement strategy. Companies must balance what can realistically be gained and the resources required, as well as clarify what must be accomplished for this strategy to merit success.

Two years after FDA approval, the DMERC medical directors requested that CMS CAG review the technology. During the 2-years interval, the manufacturer initiated a postmarket registry for outcomes, including off-label use, for all patients. This data, which included positive outcomes for more than 500 patients who had used the device for the treatment of nonunion fractures, was presented at medical meetings and published. Although CMS reviewed the new data supporting clinical efficacy in the noninvasive treatment of nonunion fractures, the agency issued a negative NCD in 1997, because it lacked FDA indications for this use and had limited FDA labeling for only "acceleration of bone healing." This decision was made primarily on the basis that the data did not appear generalizable to the Medicare population and the outcomes did not seem clinically relevant (i.e., how did radiographic healing translate to clinical outcomes such as reduction in comorbidity?). This demonstrates how outcomes for safe and effective may differ from medically necessary and reasonable and emphasizes why manufacturers need to talk to payors early on in trial development.

In anticipation of a favorable NCD and to hasten Medicare and private payor payment, a code had been requested for the Exogen device in 1996. In general, because a payment code is unrelated to product approval, it can be obtained while coverage is being developed, rather than after coverage is obtained—saving manufacturers yet another year in the reimbursement process. Unfortunately, in this scenario, Medicare created a HCPCS code that identified the product as not having Medicare coverage.

This decision not only affected access to treat the Medicare population but also created a domino effect with the private payors. Those who had not created medical policy or who administered Medicare plans now adopted Medicare's noncoverage decision.

With limited resources, Exogen continued to invest in its postmarket registry. Exogen augmented its registry data with smaller European studies focusing on the effects of ultrasound in the healing of nonunion fractures. Relying on these small studies and the 1000-patient registry, the manufacturer filed for an FDA Postmarket Approval supplement, which included indications for the noninvasive treatment of established nonunions excluding skull and vertebra. This would allow coverage for the SAFHS beyond the narrow tibia and radius fractures. In response, FDA indications were expanded to include noninvasive treatment of established nonunions 6 years after the first indications were established.

With the new nonunion labeling, additional studies, and outcome analysis, payors were willing to review their medical policy and began to cover the device for patients with nonunion fractures within 6 months of the labeling change. Before the device-labeling expansion, only 60% of the orders had been preauthorized by payors. After Medicare coverage was established, 80% of prescriptions were authorized.

By 2000, the CMS coverage process required a technology review to be completed within 180 d. Meetings were held in February 2000 with a CMS coverage group to review the clinical efficacy of the technology before submitting a request for reconsideration of the no-coverage decision. In May 2000, Exogen asked CMS to reconsider the Exogen technology as a noninvasive treatment for nonunion fractures. Relying on the new device indications and the clinical evidence showing benefit to the Medicare population, CMS issued a positive NCD for the use of the SAFHS in the treatment of nonunion fractures that had undergone surgical intervention for stabilizing the fracture. Of the 14 retrospective reviews and case studies submitted, eight were actually utilized in establishing coverage, because they were either published, pending publication, or included outcome analysis on more than 1000 patients. Table 3 summarizes the clinical evidence considered by CMS during the coverage analysis.

The positive CMS NCD went into effect January 2001, 7 years after the initial FDA approval for acceleration of fresh fractures of the tibia and radius and 1 year after expanded labeling to include treatment of nonunion. The indication of healing nonunion fractures without the need for secondary surgery was beneficial to the Medicare population. Futhermore, studies showed that 80 to 100% of patients experienced healing; even under the more stringent "intent to treat" (i.e., including anyone who receives the device for even 1 d) analyses, 64 to 82% of patients healed with this therapy. Figure 3 illustrates the timeline from FDA indication to Medicare coverage for the SAFHS.

The body of clinical evidence and its use significantly affect CMS coverage determinations. The hierarchy of evidence favors randomized double-blind, placebo-controlled studies published in a peer-reviewed journal, preferably

Table 3
Nonunion Results

Author	Overall heal rate	Heal rate (intention to treat)	*p* Value
Gebauer D, Mayr E, Orthner E, Heppenstall RB, McCabe JM	85% (57/67)	82% (70/85)	$p < 0.00001$
Nolte PA, van der Krans A, Patka P, et al.	86% (25/29)	80% (33/41)	$p < 0.00001$
Mayr E, Frankel V, Ruter A	86% (314/366) 94% (15/16)	Not performed	Not performed
Moyen B, Mainard D, Azoulai J, et al.	89% (39/44)	75% (39/52)	$p = 0.00001$
Heppenstall RB, Frey JJ, Ryaby JP, McCabe J	80% (249/313)	64% (351/551)	$p = 0.00001$
Frankel VH, Koval KJ, Kummer FJ	84% (146/174)	Not performed	Not performed
Mayr E, Wagner S, Ruter A	83% (201/241) 93%	Not performed	Not performed
Choffie M, Duarte L	100% (26/26)	Not performed	Not performed
Exogen registry data (June 15, 2000)	83% (1283/1546)	Not performed	Not performed

published in the United States (*see* Fig. 4). Less attractive are open trials in which both patients and physicians know that the patient is receiving the product or retrospective studies in which researchers review charts after a group of patients have received a treatment. Case studies are rarely considered in a technology coverage decision.

3. Lessons Learned

In a perfect world, steps to reimbursement would be simple. Manufacturers would initiate randomized double-blinded, placebo-controlled studies that report clear statistically significant long-term clinical benefits and simultaneously demonstrate cost effectiveness to payors. For some technologies, an independent patient registry would follow, be monitored by a medical group with no connections to the manufacturer, and lead to ongoing peer-reviewed publications.

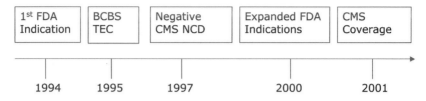

Fig. 3. Timeline from indication to national coverage for SAFHS.

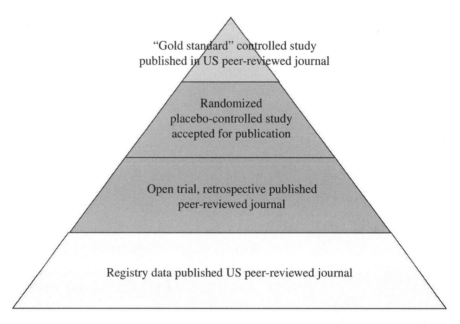

Fig. 4. Hierarchy of clinical evidence considered by payors.

Many companies designing innovative devices are start-ups that make one technology and have limited financial resources. Meetings with investors seem continuous—with reimbursement an important issue in an investor's decision to provide funding. The need to show some clinical effect early often outweighs the need to construct a comprehensive body of clinical evidence that will satisfy payors' requirements. As shown above, despite early positive trials, payors required additional studies specific to their populations or simply additional studies showing substantial evidence of benefit from this device. Although such requirements sometimes seem arbitrary at best, identifying the likely payors for a product—and even partnering with them in planning studies so that the results address their needs—can save years in the reimbursement process. These years translate to financial viability, which can mean life or death to a manufacturer. During product development, attention to clinical, regulatory, sales, and marketing should be coupled with attention to reimbursement. Almost immediately, a company must start defining who will primarily receive the product and who will pay for it. In addition, other issues, including who will actually own the product (e.g., the patient, the payor, the hospital, the physician) and who will perform the procedure and in what setting, also become important in defining a reimbursement plan.

Ideally, reimbursement issues should be considered before the creation of the first clinical protocol. Manufacturers should direct attention to what potential outcome measures are actually being studied and what impact this product will have on the quality of life for particular populations, such as Medicare recipients and private payors. Once that has been determined, meetings with medical directors of the payors that will potentially reimburse for the technology are helpful, if not essential. In preparation of these meetings, the company should consider how resulting indications benefit this payor's population. For example, Medicare would not be interested in a device that accelerated healing in new fractures but was not associated with clinically useful outcomes. Medicare rarely reimburses for prophylactic indications, because its population is not usually employed full time and the agency does not reimburse for missed work.

To Medicare, accelerated radiographic healing alone would not be considered reasonable and necessary medical treatment, although an indication for salvage of a nonunion fracture without repetitive surgeries might appeal to it. In this scenario, Medicare can avoid paying for secondary surgical interventions, and if the clinical benefits were similar to surgery, the technology could be deemed reasonable and necessary. If a primary payor target were workers' compensation boards instead of Medicare, accelerated healing to return an injured worker to work would be deemed a benefit. Accordingly, clinical protocols should include measures to demonstrate this reasonable and necessary component consistent with the requirements of anticipated and targeted primary payors.

Even if the Medicare population is not the primary population, companies should realize many private health care plans use Medicare medical policies, codes, and payments as a guide for determining their own policies. If CMS issues a negative NCD, managed care plans may also decide to deny coverage. Even if Medicare is not anticipated to be a primary payor, companies may still find it prudent to meet with local Medicare carriers to clarify relevant regional coverage decisions, as well as potential codes that can be used to identify the product and procedure. Likewise, positive local Medicare carrier decisions can be used to gain coverage with local private payors. Thus, modifying any clinical study designs to benefit the Medicare population will often expand opportunities for broader coverage.

If the initial study of the Exogen technology had considered nonunion fractures and been given that indication by the FDA in 1994, Medicare most likely would have issued a favorable coverage decision in 1995 instead of in 2001. The impact of being reimbursed for 80% of the devices instead of 50% would be substantial for any company not only from a revenue impact but also in terms of physician satisfaction.

A positive coverage decision also drives the need for a code to identify the product and or procedures. These codes help to establish payment by Medicare and private payors. However, confirmation of existing codes that can be used for the product can be determined before FDA approval. Applications for new codes usually require 6 months of FDA postmarket approval volume to demonstrate a need for a new code.

Companies must also anticipate costs to the health care system, because it is increasingly necessary to prove a product is not only clinically beneficial, but also cost-effective. Access to billing charges, including existing codes, can be collected with clinical data during the clinical trial phase by designated data collectors and help clarify whether a modification of a code or creation of a new code will be necessary.

If studies do not support reimbursement, payors may be able to describe what information would prompt them to re-evaluate the device. A small retrospective review study or specific outcome analyses might impact their initial decision. In addition, a nationally audited registry maintained by an outside source may be acceptable. If they insist on another study, ensuring the protocol will provide them information that is essential. The cost and time of additional studies can be weighed against the potential benefits on reimbursement and revenue.

3.1. Establish a Reimbursement Strategy to Achieve Company Objectives

Reimbursement should be established as a core competency in an organization. With a strong reimbursement infrastructure, companies are poised to develop, obtain, or launch new technologies to achieve their revenue objectives. Executives should understand the effects of reimbursement on the overall company objectives, as well as how departments such as clinical, regulatory, marketing, sales, and reimbursement should collaborate to understand how payment of a product effects their overall goals. These department managers should be apprised early of reimbursement issues, work with reimbursement specialists to resolve these issues, and adjust product launches to include timeframes required to establish coverage, coding, and payment. Internal reimbursement should become an early part of a company's overall strategic plan, with the above departments all understanding their roles in the reimbursement process, and contributing and then investing in the reimbursement strategy.

4. Summary

The primary components of a reimbursement strategy are obtaining medical coverage through well-designed studies with appropriate outcome measures, defining coding, and establishing adequate payment. Initial reimbursement efforts focus on coverage by determining where a technology fits, what indications are most appropriate or potentially limit coverage, and what coding modifications will be needed to obtain payment given the coverage anticipated. After reviewing the clinical study protocols with targeted payors, revisions are often needed to demonstrate that the product is also reasonable and necessary under that payor's criteria, as well as cost effective. If these topics are addressed, a reimbursement strategy sensitive to the needs of the customer, the payor, and the company can be developed and successfully implemented.

17

Polyurethane Pacemaker Leads

The Contribution of Clinical Expertise to the Elucidation of Failure Modes and Biodegradation Mechanisms

Ken Stokes

1. Introduction

The term *clinical study* can mean many things. Premarket clinical studies are necessary to verify the safety and efficacy of a new device, but they may not be able to detect low-level or long-term complications. No matter how much premarket work one does and no matter how sophisticated the protocols, the only valid proof of long-term reliability is performance in the field through postmarketing surveillance. Even studies on the long-term performance of marketed products may be misleading if not done appropriately. Clinically based postmarket surveillance can reveal the true actuarial survival of a device and the clinical mode of failure. Analysis of returned products may be required to understand the details of failure mechanisms *per se*.

In the case of polyurethane-insulated cardiac pacing leads, we discovered three previously unknown failure mechanisms in marketed products that occurred in spite of state-of-the-art premarket engineering tests and thorough premarket clinical studies. To understand how these failures could have occurred, we briefly review the development and clinical history of the first polyurethane-insulated cardiac pacemaker leads. We also review the discovery of these failure mechanisms and how those discoveries changed the way we measure chronic reliability. We look at the state-of-the-art interactions between clinical postmarket surveillance and analysis of returned products in the development and monitoring of increasingly reliable implantable cardiac pacemaker leads. Before considering these factors, we must first provide some background about the device itself.

From: *Clinical Evaluation of Medical Devices: Principles and Case Studies, Second Edition*
Edited by: K. M. Becker and J. J. Whyte © Humana Press Inc., Totowa, NJ

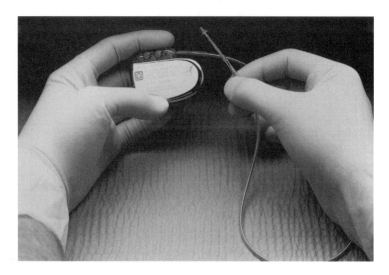

Fig. 1. A typical unipolar single-chamber pulse generator has its electrical circuitry and battery hermetically sealed in as titanium pluse-generator can. A unipolar lead is attached to the pulse generator. The distal tip of the polyurethane-insulated lead has four pliable tines designed to anchor the ring-shaped electrode within the endocardial structures of the heart. Many pacemakers sense and stimulate both atrial and ventricular chamber, requiring two leads.

2. The Implantable Cardiac Pacemaker

2.1. The Device

Cardiac pacemakers have two components: a pulse generator and a lead (*see* Fig. 1). The pulse generator includes hermetically sealed circuitry and a battery, with an external connector module. The hermetic container or "can" is usually composed of titanium, whereas the connector modules are typically polyurethane or epoxy. The leads contain one or more metallic conductor coils. The conductors are insulted with Pellethane 2363-80A, 90A, or 55D polyether polyurethane (Dow Chemical), or silicone rubber. The electrodes are usually composed of vitreous carbon, titanium, or platinum/iridium alloy. Unipolar leads have only one insulated conductor (as shown in Fig. 2) and one electrode at the distal end (*see* Fig. 1). The metallic pulse-generator can is the second electrode in a unipolar pacemaker. Bipolar leads have two conductors (as shown in Fig. 2) and a second electrode 10 to 28 mm proximal from the tip electrode. The pulse-generator can is not electrically active in a bipolar pacemaker. The distal ends of the leads usually have fixation mechanisms to assure stable contact of the electrode with the endo- or myocardium. These are either

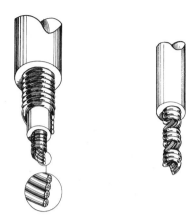

Fig. 2. A unipolar conductor coil is shown on the right. This uses a single, multifilar conductor coil with one layer of insulation. One end of the conductor is connected to a distal electrode, the other to a terminal assembly. A coaxial bipolar conductor is shown on the left. Here the inner insulated conductor coil connects with a more proximal second electrode and a ring in the terminal assembly.

"passive" fixation devices, such as the tines shown in Figs. 1 and 3, or "active" fixation corkscrews (*see* Fig. 3).

2.2. The Implant Procedure

Transvenous leads are threaded through the venous system to place the electrodes in the heart chamber. This is accomplished either by a venotomy (usually in the cephalic vein) or a subclavian "stick" using a percutaneous introducer.[1,2] Once the lead tip is in the vein, it is advanced to place the distal tip either in the right ventricle or right atrium. For dual-chamber pacemakers, both an atrial and ventricular lead are placed. After the lead is positioned and tested, it is fixed at the venous insertion site or to the muscle with a suture, with or without an anchoring sleeve, as will be discussed later. The lead terminal(s) are inserted into the pulse-generator connector and are secured with set screws. Then the excess lead is coiled around the pulse generator, which is placed in a subcutaneous or intermuscular pocket.

3. The Polyurethane Lead Story
3.1. Why Polyurethane Leads?

Through the 1970s, the vast majority of transvenous leads were insulated with silicone rubber. Silicone rubber is an excellent implantable material, but its physical properties are limited. It is relatively weak, with low tear strength. To provide reasonable protection against mechanical damage (e.g., wear, creep,

Electrode Retracted

Electrode Extended

Fig. 3. A "corkscrew" electrode in its retracted position is shown on the right. During venous insertion and passage, the corkscrew is retracted to protect intravascular tissues. When the lead tip in in position, the conductor coil is rotated to extend the helix, as shown in the center. The corkscrew penetrates the myocardium to provide secure "active" fixation as well as electrical stimulation. A bipolar tine lead tip is shown on the left.

ligature cut-through) silicone rubber must be used in relatively thick cross-sections. This was acceptable as long as a single lead was to be threaded through the veins; however, with the advent of dual-chamber pacemakers, the use of two leads (preferably in one vein) became necessary. Because of a high coefficient of friction (the ratio of frictional force to the perpendicular force pressing two surfaces together) in blood (about 0.7), it was difficult to implant two relatively large silicone rubber leads in one vein. The use of two veins tended to increase postsurgical morbidity and did not necessarily lessen the problem of one lead dislodging the other when they were manipulated for positioning. Polyester polyurethanes had superior mechanical properties but were known to be hydrolytically unstable. On the other hand, polyether polyurethanes were known to be hydrolytically stable and were much more durable than silicone rubber.[3] The tear strength of Pellethane 2363-80A (P80A), for example, is about 85 kg/linear cm compared to standard silicone rubber at about 8 kg/linear cm, or the high-performance silicones at about 35 kg/linear cm. The mechanical properties of Pellethane 2363 elastomers allowed us to develop leads that were significantly smaller in diameter yet less prone to mechanical damage. Because the coefficient of friction of polyurethane in blood is low (<0.1), implanting two polyurethane-insulated leads in one vein was a relatively easy procedure. Thus, the development of polyurethane insulated leads facilitated the use of the more physiologic dual-chamber pacemakers. In addition, the physical properties of polyurethane allowed us to develop new lead designs that were not possible with silicone rubber. A good example of this is a lead with a rotatable terminal pin and conductor coil assembly (Fig. 3). The lead can be passed easily through the vasculature with the distal helical "corkscrew" electrode retracted. Then, when the terminal pin is rotated, the corkscrew electrode emerges from the distal tip to penetrate the myocardium, holding the electrode firmly within the tissue. The lubricity and higher stiffness (or elastic modulus) of polyurethane permits rotation of the conductor coil against the insulation, whereas in silicone rubber designs, binding prevents extension and retraction of the helix. The extendible/retractable corkscrew design is popular today, especially for atrial pacing in dual-chamber pacemakers.

3.2. The Development and Market Release of the First Polyurethane Cardiac Leads

In 1975, the first polyether polyurethane-insulated lead was implanted in a human as part of a neurological stimulator. The development of cardiac leads progressed at a slower rate. Several iterations were tested on the bench and in animals before optimized designs were settled on. It was accepted that in canines, cardiac leads became chronically stable well within 12 weeks. Longer-term testing showed no further significant changes. However, although these data seemed acceptable for evaluating device performance, we were uncomfortable using them for material stability. At that time, there was no history of long-term in vivo material testing in the literature except for some 8-month silicone rubber tests by Swanson and Lebeau.[4] Thus, we took the unusual step of conducting a 2-year evaluation of Pellethane 2363-80A and 55D in rats.[5] Although reversible changes in mechanical properties occurred because of moisture absorption, there was no evidence of instability. The first clinical studies of polyether polyurethane-insulated cardiac pacemaker leads began in Europe in 1977 and in the United States in 1978. Based on excellent animal- and bench-test results, as well as superior premarket clinical performance, the first polyurethane-insulated cardiac pacemaker lead products were released to US markets in April 1980.[6,7] As shown in Table 1, the clinical evaluation of the new leads demonstrated significantly improved reoperation rates resulting from lower acute complications.[8] Sales went well, with positive comments about the new lead models excellent performance and ease of implantation.

Table 1
Clinically Determined Reoperation Rates for Transvenous Leads

Medtronic model no.	Market date	Description	Reoperation rate (%)
Ventricular leads			
5818	Early 1960s	Straight lead, no fixation	40
6901	Late 1960s	Flanged silicone tip	20
6950	1976	Long silicone tines	12–15
6962	1978	Short silicone tines	5
6972	1980	Polyurethane tines	1
Atrial leads			
6994	Early 1970s	"J"-shaped silicone	20
6991	1976	Long-tined silicone "J"	7
6991U	1980	Polyurethane-tined "J"	2
6957	1980[a]	Transvenous screw-in	1–3

[a]Europe only.

3.3. The Discovery of a New Failure Mechanism, Environmental Stress Cracking

On May 15, 1981, we received a Model 6991U arterial lead that had been removed from a human after only 5 months of implantation. The lead had about a 1in. gap in the insulation at the base of the "J," similar to that shown in Fig. 4. Thorough analysis produced no evidence of chemical degradation of the device.[9,10] For example, we found no changes in the surface or bulk infrared spectra. There was no change in the molecular weight of the sample. However, optical microscopic analysis revealed an interesting occurrence. The edges of the breach, even though separated by about an inch, matched perfectly (*see* Fig. 4). In fact, it was clear that no material was missing and that the polymer had somehow pulled apart. This led us to suspect that the insulation had failed because of some kind of stress-cracking mechanism.

The insulation failure was completely unanticipated. We had seen no such cracking in 12-weeks canine implants, in 2-year rat implants, or during premarket clinical studies. Stress-cracking mechanisms were not among the known possible degradation mechanisms for polyether polyurethane elastomers. No insulation failures of the first neurologic leads were known, even after 6 years in service. A thorough review of the literature by both us and independent sources found no explanation for this. Thus, we had discovered a previous unknown failure mechanism.

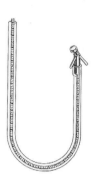 

Fig. 4. An illustration of the first polyurethane insulated atrial "J" lead marketed in 1980 is shown on the left. An example of environmental stress cracking (ESC) failure at the base of the "J" is shown on the right at about ×30. Note that the edges of the breach match closely. The insulation has cracked and pulled apart as a result of ESC in the presence of unusual tension. No material is missing.

Stress cracking is defined as cracking or crazing of a material in the presence of stress (strain) and a chemical environment.[11] Many forms of stress cracking are known in rigid plastics. For example, polyethylene will crack when bent and exposed to a detergent. Polycarbonate can stress crack in ethylene oxide because of the residual molded-in stresses. Oxidative stress cracking is known for many rigid plastics but, with the exception of natural rubber in the presence of ozone, was not known for elastomers. Indeed, it did not appear to be possible in vivo based on current knowledge. Although we conducted tests on strained polyurethane and many different chemical agents, we could not duplicate the mechanism in vitro. We could not address any chemical component of the mechanism, except that it required exposure to the in vivo mammalian environment. Therefore, the mechanism was labeled *environmental stress cracking* (ESC). We did discover that during manufacture the insulation of the Model 6991U lead was occasionally and inadvertently stretched at the approximate point where the failure had occurred in the returned lead. Although we could not control the environmental portion of the mechanism, we could control residual strain in the manufacturing process. Manufacturing techniques were changed to assure that no residual stresses remained in the device as shipped.[12] These changes appeared to be completely effective. Some patches of shallow cracks were still found in the tissue-exposed surfaces of some explanted and returned leads, although cracks through the insulation causing

Fig. 5. An electron microscopic view of environmental stress cracking at a tight ligature at about ×500. The cracks become deeper and wider closer to the ligature (to the right), and decrease in depth and width away from the source of stress (to the left).

clinical failure no longer occurred in the Model 6991U manufactured after the change date. It appeared that the problem had been identified and corrected.

Later in 1981, a few explanted bipolar ventricular leads (Model 6972) were returned with cracks in the insulation around the fixating ligature (*see* Fig. 5). Until this time, anchoring sleeves had always been supplied separately in the lead packages. The instruction manual indicated that the sleeve should be placed on the lead before ligation to prevent damage to the device; however, it was common clinical practice to ignore the anchoring sleeve and simply ligate the lead directly in the vein. Now that we had identified that the polyurethane insulation was susceptible to a form of stress cracking, it became apparent that ligating the lead directly was no longer acceptable. Therefore, in February 1982, the factory began placing an anchoring sleeve on each lead. The sleeves could not be ignored, because they would have to be cut off to be removed. The instruction manual was also changed to state that the use of the anchoring sleeve was mandatory. Based on the analysis of returned products, only a small fraction of a percentage of the leads that were sold had failed by this or any other mechanism, whereas the next best silicone rubber bipolar ventricular lead had a 5% reoperation rate (*see* Table 1). We were satisfied that the aforementioned changes had solved a problem that had affected very few leads with otherwise superior clinical performance.[13]

3.4. The Development of Accelerated Test Methods for Environmental Stress Cracking

A more global question was how to determine that design or process changes applied to new products would not cause ESC. How could we prove that the mechanism really was ESC? In our investigation of the stress-cracking mechanism, we recognized several things; ESC cannot occur without strain and the mechanism could not be duplicated in vitro. Obviously, if ESC requires a residual strain, then it should depend on processes that increased or reduced residual strain. It was also known that ESC processes required induction periods and critical strains; however, one cannot accelerate ESC in a thermoplastic by elevating temperature. These polymers are viscoelastic, which means that they flow or "creep" under load. As temperature increases, the creep rate increases and stresses are relieved. Therefore, we strained a number of samples over mandrels and implanted them in the subcutis of rabbits. We found that the time to failure (induction period) varied as a function of the magnitude of applied strain and the polymer's thermal history. Extrusion conditions and poststrain annealing (a thermal treatment to reduce stress) were found to have significant effects. Indeed, these properties fit all of the hallmarks of a stress-cracking phenomenon.[14] The animal results matched exactly with the findings on explanted and returned cracked leads.

We settled on a set of standard conditions and developed an accelerated, in vivo test that could be used to evaluate new processes and new materials. We learned how to optimally stress-relieve the devices by annealing to prevent ESC failure and incorporated those processes in evolving next-generation devices.

3.5. The Discovery of Metal Ion Oxidation

In late 1982/early 1983, we received a returned Model 6972 lead with cracks in the inner insulation. Analysis showed that the polymer had underdone auto-oxidative degradation and not ESC. Infrared spectra showed significant chemical changes, including loss of the aliphatic ether linkages and other changes.[15] Substantial molecular weight changes were found, which were again unforeseen. Auto-oxidation of polyurethanes in the environment is well known, but this process required several factors not believed to be present in vivo. *Auto-oxidation* is defined by Hawkins as the reaction with oxygen that occurs between room temperature and about 150°C.[16] It is a free radical chain reaction that requires the presence of oxygen in reasonable quantities. However, cardiac pacing leads are implanted in tissues with very low oxygen tension. In addition, auto-oxidation requires a catalyst to proceed at clinically significant rates. Photo-oxidation, for example, requires the presence of both oxygen and certain wavelengths of light to initiate and propagate the reaction. The requirement for light is not fulfilled in the case of implanted cardiac pacing leads. Thermo-oxidation requires oxygen and a relatively high temperature to initiate and propagate the reaction. The body produces heat, but the temperature remains a relatively benign $37 \pm 3°C$. Once again, no such phenomena had been seen in preclinical animal implants and were not expected based on the literature. Therefore, a second new and previously unknown phenomenon was presented to us.

3.6. Accelerated Test Methods for Metal Ion Oxidation

As we analyzed the degraded polymer, we began to suspect that metal ions released from the conductor coils may somehow be involved as catalysts.[17] In addition, the medical literature contained new reports revealing the discovery that oxygen free radicals actually could be produced in vivo. We discovered from studying the autoimmune disease and pathology literature that the mechanism by which implanted devices become encapsulated in fibrous tissue (known as the *foreign body response*) involved the release of oxidants. These include hydrogen peroxide (H_2O_2) and oxygen free radicals, such as super oxide anion ($^•-O_2$) and hydroxyl radical ($^•OH$).[18,19] These oxygen free radicals could not possibly affect the inside of the device because of their extreme reactivity. However, H_2O_2 permeates the polyurethane even more rapidly than water to decompose on the metallic conductor wire. A hypothesis evolved that metal ions released from the conductor coils as a result of interactions with H_2O_2 could catalyze auto-oxidative degradation of the polyether portion of the polymer inside the lead.[16] We immersed leads in 3% H_2O_2 and duplicated the mechanism in vitro. Thus, the mechanism was termed *metal ion oxidation* (MIO).

The first polyurethane bipolar ventricular lead, Model 6972, used a conductor wire made from a composite of silver and MP35N called *drawn brazed strand* (DBS). MP35N alloy wire (Dupont) has excellent mechanical properties. It is a "super alloy" composed primarily of nickel, cobalt, chromium, and molybdenum. The DBS wire had excellent low electrical resistance and, when coiled, unmatched fatigue-fracture resistence.[20] It had an excellent history when used with earlier silicone-rubber-insulated leads. Based on in vitro testing and analysis of returned products, we determined that DBS wire significantly accelerated MIO in Model 6972, if it was not the cause *per se*. Therefore, that conductor material was removed from the next generation Model 4012, and all future lead designs were to be replaced with solid MP35N wire. The next generation product (Model 4012) was expected to have significantly improved performance in all respects, including greatly less—if not complete elimination of—ESC and MIO. Unfortunately, this is still not the end of the story.

4. Methods of Postmarket Surveillance Used for Cardiac Pacemaker Leads

4.1. Returned Products and Lead Removablility

Implanted cardiac pacemakers are often referred to as "permanent" implants. This is a serious misnomer. Pulse generators are battery-operated devices. They have a limited functional longevity and must be replaced when the battery expires. Although this requires a surgical procedure, it is a relatively simple and risk-free operation performed under local anesthesia. The explanted pulse generators are routinely returned to the original manufacturer for analysis. Compliance is so routine and so good that analysis of returned products provides an excellent means of postmarket surveillance for pulse generators. One can accurately track failure rates and analyze returned devices to learn about failure mechanisms. However, the lead presents other difficulties.

It was well established in the canine model that silicone-rubber leads become encapsulated fully in the heart and vasculature within 12 weeks of implant (Fig. 6). Conversely, the polyurethane-insulated leads did not. They typically developed a thin translucent sheath over the distal tip (including the tines and at the ligature site). Encapsulation between the ligature and the distal tip was rare. We found that Model 6972 pulled free from the heart with about 750g once the ligature site was dissected free. Thus, we expected Model 6972 (and its successors) to be chronically removable devices. We were somewhat perplexed to discover that complete removal of a chronic lead from a human often required open-heart surgery. Not all leads were being explanted and returned. Indeed, most of what was being returned was the more readily accessible proximal portion, leaving the distal part of the device in the patient. If the lead failure was distal to the point of separation, then failure could not be verified. In addition, in the mid-1980s we discovered that it was common practice for clinicians to discard the lead, even if it was explanted. We now know that in canines and humans, the degree of encapsulation seen at 12 weeks persists, even after 2 years of implantation. It typically takes approximately 3 to 4 years for the degree of encapsulation to increase, producing a discontinuous sheath at various points along the lead body. These sheath segments can range from thin transparent collagen, thick opaque white collagen, cartilage, mineralized collagen, and can even contain bone.[3] This clinical finding could not have been predicted even after longer-term animal tests, certainly not within a 2-year experiment. We needed to validate our perception of chronic lead performance.

4.2. The Development of a Chronic Postmarket Surveillance Study

We contracted three large implanting centers and sent a clinical specialist to visit them. The specialist went through patient records to determine the actual

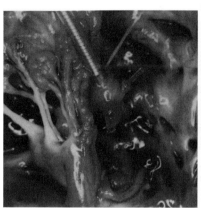

Fig. 6. A unipolar polyurethane-tined lead after 12 weeks implant in a canine is shown on the left. Note the relative freedom from encapsulation except at the distal tip. A typical silicone-rubber-insulated lead after 12 weeks implant in a canine is shown on the right. Note the thick encapsulation.

clinical failure rates of various lead models. It was not possible to determine with certainty what caused a clinical failure by this method. For example, oversensing is the situation that occurs when the pulse generator detects something other than the heart's signals but interprets the artifacts as an R-wave and inhibits. If the generator inhibits, this means that it will not emit a stimulus when it should. Such electrical noise can result from various phenomena, including an unstable electrode rubbing against tissue, a loose-set screw in the connector, or metal-to-metal contact within the device. Based on our analysis of returned products, we believed that MIO produced holes in the inner insulation that could allow such metal-to-metal contact. This meant that MIO could be one of several possible causes of oversensing. Thus, the three-center study produced data on clinical failure rates without necessarily identifying the root cause of failure. A preliminary report was issued in the October 1983 issue of *Medtronic News*, which had a circulation of about 38,000 in the medical community.[21] By February 1984, it was clear that the apparent clinical-failure rate was higher than expected, based on the analysis of returned products; it was 7 to 10% after 3 years of implantation, not less than 1%. It must be remembered that there was no industry standard on lead failure rates. We had no knowledge of—nor any way to determine—what our competitor's failure rates were. Nevertheless, this failure rate was not acceptable to us. As a result, we initiated a voluntary advisory on Model 6972.

The only completely acceptable failure rate is 0%, but the reality is that a 0% failure rate for 100% of patients for the remainder of their lifetimes is not possible for implanted cardiac pacemaker leads. As is true of all implanted devices, leads eventually wear out. Therefore, we needed to determine an acceptable failure rate for cardiac pacemaker leads so a trigger point for action could be developed in the unlikely event that one was needed. We asked our customers what would be acceptable to them. When we asked how long pacemaker leads should last, almost all initially responded "for the life of the patient." We explained that although this was our goal, it was probably not achievable. Our customers typically stated that they wanted the lead to last at least through two pulse generators. At that time, the longevity of a dual-chamber pacemaker was commonly about 5 years or less, which set the timeframe at about 10 years. When we asked customers what the the acceptable survival rate was for implanted cardiac pacemaker leads after 10 years, most typically responded, "at least 90%." Because 10 years is a long time to follow these patients and a long time to wait for a result, we set an interim trigger point at 95% survival at 5 years. The next question was how to obtain valid data to determine lead survival statistics.

Our initial experience with the three-hospital postmarket clinical study was expanded to presently include 14 large implanting centers. A medical advisory board was set up to review methods, results, and actions. The advisory board meets annually regardless of the study results but can meet at shorter intervals, if necessary. Thus began the Medtronic Chronic Lead Study (CLS), which continues to this day. The results of that study, reviewed by the medical advisory board, are published twice a year.[22]

The CLS requires that each center inform Medtronic whenever a lead complication, patient death, or loss to follow-up occurs. The data analysis assumes that there are no such events at the time of data update unless specifically reported by the center or determined by correlation with returned product analysis. A lead complication is defined as loss of capture (stimulation), loss of sensing, oversensing, skeletal muscle stimulation, conductor fracture, insulation breach, or impedance of less than 200Ω. Because this is a chronic study, these complications must be observed at or beyond 1 month postimplantation. The complication must be resolved by physically modifying, revising (excluding repositioning), replacing, or abandoning the lead. The criteria for a lead complication are summarized in Table 2. These criteria do not enable a lead hardware failure to be differentiated from other clinical events, such as elec-

Table 2
Criteria for Lead Complications in the Medtronic Chronic Lead Study[a]

A complication is considered to have occurred in the Chronic Lead Study if both of the following conditions are met.

Condition 1: One or more of the following clinical observations beyond 30-days post-implant is reported:

- Failure to capture (stimulate)
- Failure to sense
- Cardiac perforation
- Dislodgment
- Oversensing
- Extracardiac stimulation
- Conductor fracture (observed visually or radiographically)
- Insulation breach exposing the conductor (observed visually)
- Pacing impedance of 200Ω or less or 3000Ω or greater

Condition 2: One or more of the following clinical actions directly results and is reported:

- Lead abandoned
- Lead explanted
- Lead replaced
- New lead implanted
- Other lead related surgery performed
- Pacemaker mode or polarity reprogrammed to circumvent the problem (i.e., electrical abandonment)
- Lead use continued, based on medical judgment

[a]Lead positioning is not qualifying action.

trode dislodgment, exit block, or concurrent pulse-generator failure presenting as sensing or capture problems. Because the protocol reports clinical and hardware complications and cannot differentiate between them, there is a likelihood that hardware failures will be overreported. The centers are audited onsite annually to monitor overall compliance with the protocol. In the history of the study, one center has been replaced for noncompliance with the protocol.

The data are analyzed by the Cutler and Eder Life Table method.[23] The actuarial survival curves are reported with standard errors at the leading 3-month interval. *Survival probability* refers to proper functioning of the device not the survival of the patient. For example, a survival probability of 98% is a statistical assessment that at the time interval indicated, each patient has a 2% risk of incurring a device malfunction or complication. An example of the survival curves from the report is shown in Fig. 7.

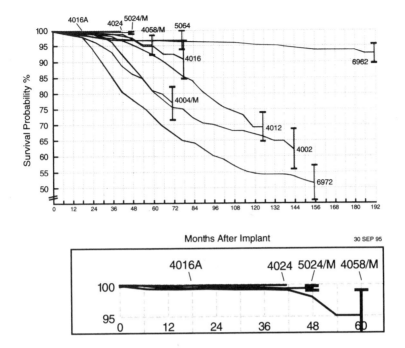

Fig. 7. Actuarial survival curves for the chronic human performance of Medtronic's present and past bipolar ventricular transvenous leads taken from the September 1995 Product Performance Report.

5. Results From a Clinically Based Postmarket Surveillance Study

5.1. Discovering the Undiscoverable

The CLS has shown that the Model 4012 lead that succeeded Model 6972 had an actuarial survival rate of about 96% 5 years after market release. Because our trigger point was 95% in 5 years, this lead was performing acceptably. Although this reflected clinical reality, it did not necessarily tell us what failure mechanisms were affecting 4% of the devices. Looking at returned products that had failed, we discovered that about 25% had ESC breaches in the outer insulation; another 25% had MIO cracks in the inner insulation. About 50% of the returned failed leads had conductor fracture. Analysis of the returned leads themselves showed that the conductor coils had failed in a crushing mode, which had not been seen previously.[24] Thus, we projected from analysis of returned products that the CLS was showing us 1% ESC failure, 1% MIO failure, and 2% crush failures with 96% failure-free performance after 5 years of implantation. In comparison, Model 6972 had about 30% ESC and MIO failure in the CLS at 5 years. Therefore, the improvements made to Model 4012 had been highly effective in reducing ESC and MIO as predicted, although still not to the ultimate goal of 100% survival. We did not anticipate crush failure in the design of Model 4012, because (once again) it was a previously unknown failure mechanism.

5.2. The Crush Fracture Mechanism

Before the introduction of Model 6972, leads were typically inserted through the cephalic or jugular vein. A new implant procedure via the subclavian vein had been introduced about the same time that the Model 6972 lead had been in its premarket clinical study. In fact, about 30% of the leads in the premarket clinical study had been implanted by the new method. In this procedure, a needle is inserted percutaneously into the subclavian vein, and then a guidewire is inserted into the venous system through the needle. The needle is removed and an introducer is inserted over the guidewire. The dilator is removed, and the lead is inserted into the vein. The introducer and guidewire are removed and discarded. Depending on how the stick was performed, the lead could be positioned between the first rib and clavicle before it entered the vein. In retrospect, we learned that the Model 6972 lead's DBS wire conductor was exceptionally resistant to crush fractures. It would flatten but not break. Extremely few had been returned for conductor fracture. It was thought that the flat spots found on some returned leads had been caused by instruments at explantation. This conclusion was supported by the presence of marks on the insulation that looked like those made by instruments. Because DBS wire was found to exacerbate MIO in Model 6972, it was replaced in Model 4012 with solid MP35N wire that had much better corrosion resistance. Conductor fracture by the well-known mechanism of flex fatigue was relatively rare with multifilar solid wire conductor coils and occurred far less often than did MIO failures in Model 6972.

Analysis of returned products showed us that the fractures occurred at a specific site, 27 ± 5 cm from the terminal pin. Studies on cadavers demonstrated that the conductor coil could be pressed "out of round" or crushed between the first rib and clavicle if the lead was inserted between the bones before it entered the vein.[25] Tests were done with copper coils and balloon pressure devices.[26] It was found that when the leads were placed between bones before entering the vein, pressures as high as 126 ± 26 psi could be obtained when the arm was placed in caudal traction. Pressures never exceeded 52 ± 30 psi when the leads were inserted in the cephalic vein (90° flexion). A survey of all explanted and returned lead records found that all crushed leads had been implanted by the subclavian stick method. By contrast, no returned leads implanted via the cephalic vein had experienced crush fracture. Why were crush fractures not found in preclinical animal studies? Canines, the generally accepted model for lead studies, have no clavicles. Therefore, it was impossible to discover crush failure in the animal model. Why was it not discovered in premarket clinical tests? The incidence was too low for a premarket clinical study of reasonable length and numbers. We discovered that the marks on the polyurethane insulation were not caused by instruments but by the pressure of the bones clamping on the device, mimicking instrument damage.

Once the mechanism was understood, modifications to the subclavian stick procedure were made and published.[27] It was recognized that subclavian crush is the result of a surgical procedure and not necessarily a lead design problem. In vitro test methods were developed to assess crush fracture in new conductor designs. Today, the results of the chronic lead study and analysis of returned products tells us that these have been effective measures.

6. Postmarket Surveillance on Leads Today

So far, the CLS has demonstrated that what we have learned and applied to present-day products has been highly effective. For example, the polyurethane bipolar ventricular transvenous lead, Model 4024, has 100% actuarial survival after 57 months in the CLS. Its silicone rubber counterpart, Model 5024, has $99.3 \pm 0.4\%$ survival after 66 months. A comparison with Model 4012 at 60 months shows $92.5 \pm 1.4\%$ survival, whereas Model 6972 is at 72.2 $\pm 3.6\%$

A comparison of how one might draw conclusions about chronic lead survival based on two methods is shown in Table 3. Twelve Medtronic lead models are presented based on the March 1997 issue of the Medtronic Product Performance Report. The report contains a table presenting the number of devices sold in the United States since market release, as well as the number of devices explanted and returned with failure to perform as intended. Data derived from this information are compared in Table 3 to the same lead models followed in the CLS. Based on analysis of returned products, all 12 lead models are performing well over the 95% trigger point we set for ourselves. One could even claim that Model 6972 has an acceptable 96.4% survival after 13.5 years, although the column representing the results of the CLS shows a very different story. Model 6972 really has only 48.7% survival in 13.5 years; not 96.4%. Model 4012 has 69.3% survival in 11.5 years; not 99.2%, as would be surmised from the analysis of returned products. It has now been mandated that all manufacturers must report postmarket surveillance on cardiac pacemaker leads by clinically based processes to the Food and Drug Administration. So far, however, we see no evidence of clinically based postmarket surveillance being reported to the medical community with the exception of the CLS study.

Table 3
Comparison of Survival Conclusions Based on Analysis of Returned Products vs Clinical Postmarket Surveillance

Model no.	Years service	Number sold	Returned products % survival	Number in CLS	% Acturial survival, CLS
6962	16	70,560	99.7	1418	92.6 ± 2.7
6961	14	44,673	99.7	608	89.7 ± 5.5
6972	13.5	43,198	96.4	1253	48.7 ± 5.9
6971	14.3	56,261	99.2	1317	83.9 ± 4.0
4012	11.5	96,901	99.2	2500	69.3 ± 3.8
4011	11.8	64,083	99.8	827	93.7 ± 3.4
4004	7.0	74,481	99.3	1638	68.5 ± 4.5
4003	5.5	39,681	99.9	441	99.5 ± 0.8
4024	4.8	121,647	99.8	721	100
5024	5.5	132,395	99.6	5236	99.3 ± 0.4
4057	5.5	11,549	99.6	259	96.4 ± 4.4
4058	6.3	104,540	99.6	1581	95.5 ± 3.2

CLS, Chronic Lead Study.

7. Conclusion

Polyether polyurethane-insulated cardiac pacemaker leads were introduced to the market with high expectations. They made a significant impact on cardiac pacing, serving many hundreds of thousands of patients well. These leads made dual-chamber pacemakers practical to use and facilitated the development of important new designs. Before their market release, the first generation of devices was subjected to animal and bench testing that was considered state of the art at the time. Additional testing that exceeded this standard was also done, proving that the polyurethane *per se* was biostable. Premarket clinical studies confirmed that these devices were easier to use and had a significantly lower complication rate than their silicone-rubber predecessors. Nonetheless, two previously unknown failure mechanisms, ESC and MIO, were discovered after clinical use, as a result of analysis of returned products (which was the generally accepted means of postmarket surveillance in the pacemaker industry at that time). Improvements were made to existing and new models to reduce ESC and MIO. Even without these improvements, the known chronic clinical failure rates based on analysis of returned products was very low. A clinically based three-center chronic lead study* revealed that the actual failure rate was much higher than was believed because chronic leads

*The first CLS study included three centers. Later, the study was expanded to 11 and then 14 centers. Additional centers are being identified in Europe and Asia.

were difficult to remove and were not being returned to the manufacturer. The postmarket surveillance results based on this study have been reported to the medical community at least twice a year since 1984.

A third previously unknown lead failure mechanism was discovered in the chronic lead study after escaping detection in animal and premarket clinical testing. Subcalvian crush fracture of transvenous leads escaped detection in animal tests because the mechanism requires bone structures not present in canines or other suitable models. It escaped detection in premarket clinical trials because of its very low incidence (estimated to be about 2% in 5 years) and appearance (similar to misuse with surgical instruments).

Cardiac-pacemaker-lead postmarket surveillance reporting based on analysis of returned product alone is unacceptable. The Food and Drug Administration has mandated that all manufacturers report on the basis of clinically based chronic lead studies. This may or may not be happening, but so far, only one company reports clinically derived actuarial survival data on its products to the medical community.

References

1. Littleford, P. O., Parsonnet, V., and Spector, D. S. 1979. A subclavian introducer for endocardial electrodes. In Meere, C., ed. *Proceedings of the VIth World Symposium on Cardiac Pacing*. PACESYMP, Montreal, Ch. 14–21.
2. Belott, P. H. 1981. A variation on the introducer technique for unlimited access to the subclavian vein. *PACE Pacing Clin. Electrophysiol.* 4:43–48.
3. Stokes, K. B., Cobian, K., and Lathrop, T. 1979. Polyurethane insulators, a design approach to small pacing leads. In Meere, C., ed. *Proceedings of the VIth World Symposium on Cardiac Pacing*. PACESYMP, Montreal, Ch. 28–32.
4. Swanson, J. W., and Lebeau, J. E. 1974. The effect of implantation on the physical properties of silicone rubber. *J. Biomed. Mater. Res.* 8:357–367.
5. Stokes, K. B. and Cobian, K. 1982. Polyether polyurethanes for implantable pacemaker leads. *Biomaterials.* 3:225–231.
6. Stephenson, N. L. 1980. Synopsis of clinical report on the Spectraflex models 6971/71 transvenous leads. *Medtronic News* X(3):16.
7. Stephenson, N. L. 1980. Synopsis of clinical report on models 6990U/6991U atrial J leads. *Medtronic News* X(2):10.
8. Stokes, K. B. and Stephenson, N. L. 1982. The implantable cardiac pacing lead—just a simple wire? In Barold, S. and Mugica, J., eds. *Modern Cardiac Pacing*. Futura, Mount Kisco, 365–416.
9. Stokes, K. B. 1982. The long-term biostability of polyurethane leads. *Stimucoeur.* 10:205–212.
10. Timmis, G. C., Gordon, S., Westveer, D., Martin, R. O., and Stokes, K. 1983. Polyurethane as a pacemaker lead insulator. In Steinbach, K., ed. *Cardiac Pacing*. Steinkopff Verlag, Darmstadt, 303–310.

11. Whittington, L. R. 1968. *Whittingtons Dictionary of Plastics.* Technomic, Stanford, 90.

12. Stokes, K. B. 1984. The biostability of polyurethane leads. In Barold, S., ed. *Modern Cardiac Pacing.* Futura, Mount Kisco, 173–198.

13. Stokes, K. B. 1984. Environmental stress cracking in implanted polyether polyurethanes. In Planck H., Egbers G., and Syré, I., eds. *Polyurethanes in Biomedical Engineering.* Elsevier, Amsterdam, 243–255.

14. Stokes, K. B., Urbanski, P., and Cobian, K. 1987. New test methods for the evaluation of stress cracking and metal catalyzed oxidation in implanted polymers. In Planck H., Egbers G., and Syré, I., eds. *Polyurethanes in Biomedical Engineering II.* Elsevier, Amsterdam, 109–128.

15. Stokes, K. B., Berthelsen, W. A., and Davis, M. W. 1985. Metal catalyzed oxidative degradation of implanted polyurethane devices. Proceedings of the ACS, Division of Polymeric Materials Science and Engineering. 53:6–10.

16. Hawkins, W. L. 1972. *Polymer Stabilization.* Wiley-Interscience, New York.

17. Stokes, K. Urbanski, P., and Upton, J. 1990. The in vivo auto-oxidation of polyether polyurethanes by metal ions. *J. Biomater. Sci. Polym.* Ed. 1:207–230.

18. Anderson, J. A. 1988. Inflammatory responses to implants. *ASAIO.* 34:101–107.

19. Sybille, Y. and Reynolds, H. V. 1990. Macrophages and polymorphonuclear neutrophils in lung defense and injury. *Am. Rev. Respire Dis.* 141:471–501.

20. Upton, J. E. 1979. New pacing lead conductors. In Meere, C., ed. *Proceedings of the VIth World Symposium on Cardiac Pacing.* PACESYMP, Montreal, Ch. 29–36.

21. Helland, J. 1983. Pacemaker lead complications: clinical significance and patient management. *Medtronic News* XIII:8.

22. Medtronic, Inc. 1995. *Medtronic Product Performance Report* UC-9300222eEN, September, Minneapolis, MN.

23. Cutler, F. and Eder, F. 1958. Maximum utilization of the life table method in analysis of survival. *J. Chronic Dis.* 8:699–712.

24. Stokes, K., Staffanson, D., Lessar, J., and Sahni, A. 1987. A possible new complication of subclavian stick: conductor fracture. In *VIII World Symposium on Cardiac Pacing and Electrophysiology.* Jerusalem: PACESYMP, 10(3), Pt.II, 748 (Abst. 476).

25. Fink, A., Jacobs, D. M., Miller, R. P., Anderson, W. R., and Bubrick, M. P. 1992. Anatomic evaluation of pacemaker lead compression. *PACE Pacing Clin. Electophysiol.* 15:510.

26. Jacobs, D. M., Fink, A. S., Miller, R. P., et al. 1993. Anatomical and morphological evaluation of pacemaker lead compression. *PACE Pacing Clin. Electrophysiol.* 16:434–444.

27. Byrd, C. L. 1992. Safe introducer technique for pacemaker lead implantation. *PACE Pacing Clin. Electrophysiol.* 15:262–267.

18

Role of Device Retrieval and Analysis in the Evaluation of Substitute Heart Valves*

Frederick J. Schoen

1. Introduction

Surgical replacement of diseased valves with functional substitutes is the dominant therapeutic modality in patients with symptomatic valvular heart disease and generally improves survival and enhances quality of life.[1,2] Nevertheless, problems associated with the valve replacement devices remain a major impediment to the long-term success of this procedure. Despite considerable improvement in the technology of heart valve prostheses since their first successful use approximately 45 years ago, both mechanical and tissue heart valve substitutes (illustrated in Fig. 1) remain imperfect, and prosthesis-associated complications have considerable impact on the long-term outlook for those who have had valve replacement surgery.

Information derived from pathological evaluation of substitute heart valves has (1) contributed to the care of individual valve replacement patients; (2) established the rates, morphology, and mechanisms of prosthesis-associated complications; (3) elucidated the structural basis of favorable valve performance; (4) predicted the effects of developmental modifications on safety and efficacy; and (5) enhanced our understanding of patient–prosthesis and blood–tissue–biomaterials interactions. The technical approaches and procedures useful in both preclinical and clinical heart valve retrieval and evaluation have been widely documented.[3–8] In this chapter, we summarize the rationale for and con-

---

*Disclosure: In the course of consultations on preclinical and/or clinical studies during approximately the past 5 years, Dr. Schoen has received or may receive something of value from the following organizations whose work is germane to the subject of this presentation: Edwards LifeSciences, Inc.; Sulzer CarboMedics, Inc.; CryoLife Cardiovascular, Inc.; Medtronic, Inc.; and St. Jude Medical, Inc.

From: *Clinical Evaluation of Medical Devices: Principles and Case Studies, Second Edition*
Edited by: K. M. Becker and J. J. Whyte © Humana Press Inc., Totowa, NJ

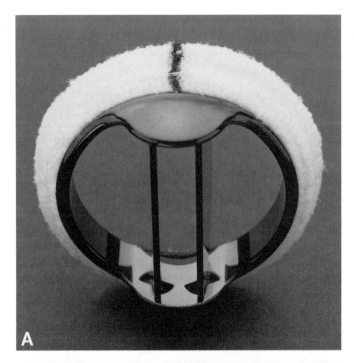

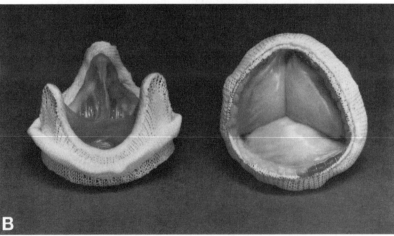

Fig. 1. Mechanical prosthetic and tissue bioprosthetic heart valve replacement devices. (**A**) St. Jude Medical carbon bileaflet tilting disk prosthesis, the most widely
used mechanical heart valve prosthesis (courtesy of St. Jude Medical Co., St. Paul,
MN). (**B**) Carpentier-Edwards porcine aortic bioprosthetic valve, a widely used type
of tissue heart valve prosthesis (courtesy of Edwards Division, Baxter HealthCare
Corp., Santa Ana, CA).

tributions of analysis of explanted valve substitutes. Other discussions of concepts, strategies, and results of cardiovascular device analysis are available, pertaining to cardiac assist devices, vascular grafts, and endovascular stents and stent grafts.[9-21]

2. General Considerations

The goals of routine hospital surgical pathology or autopsy examination of an artificial valve are generally restricted to documentation of the specific valve that has either been removed at reoperation or after the patient's death and diagnosis of any prosthesis-related abnormality such as thrombotic deposits or embolism, valve-related infection (prosthetic valve endocarditis), abnormal healing, or structural dysfunction. Detailed correlation of morphologic features with clinical signs, symptoms, and dysfunctional physiology is usually not done in routine analysis. However, directed and informed pathological examination of cardiac valve prostheses retrieved during preclinical animal studies or at reoperation or autopsy of human patients (with appropriate clinical details) can provide valuable information. For example, preclinical studies of modified designs and materials that usually include implantation of functional devices in the intended location in an appropriate large animal model and noninvasive and invasive monitoring at specified intervals, followed by specimen explantation and detailed pathological analysis are crucial to developmental advances. In addition, demonstration of specific vulnerabilities to complications in individual patients, such as a propensity toward accelerated calcification or a genetic hypercoagulability as a cause of thrombosis would affect further management. Moreover, clinicopathological analysis of cohorts of patients who have received a new or modified valve prosthesis type evaluates the safety and efficacy of the new device type to an extent beyond that obtainable by either in vitro tests of durability and biocompatibility or preclinical investigations of valve implant configurations in large animals.

Through analysis of rates and modes of failure as well as morphologic and mechanistic characterization of specific failure modes in patients with implanted medical devices, retrieval studies can contribute to development of methods for enhanced clinical recognition and elucidation of the pathogenesis of failure mechanisms that can guide future development of improved prosthetic devices to eliminate complications. Emphasis is usually directed toward failed valves; however, careful and sophisticated analysis of removed prostheses that are functioning properly, and detailed analyses of preimplantation structural features and their evolution following implantation can yield an understanding of structural correlates of favorable performance. Moreover, such studies can identify predisposition to specific failure modes such as thrombosis or mechanical failure. Thus, implant retrieval contributes important information to preclinical studies and clinical research, at both clinical trial and postmarket surveillance phases. Representative examples will be summarized throughout this chapter.

Device-retrieval analysis also has an important regulatory role, as specified in the Safe Medical Devices Act of 1990 (PL 101-629), the first major amendment to the Federal Food, Drug, and Cosmetics Act since the Medical Device Amendments of 1976.[22,23] The user reporting requirements of this legislation require health care personnel and hospitals to report (within 10 d) to the Food and Drug Administration (FDA) or manufacturers or both (depending on the nature of the occurrence) all prosthesis-associated complications that cause death, serious illness, or injury. Such incidents are often discovered during a pathologist's diagnostic evaluation of an implant in the autopsy suite or the surgical pathology laboratory.

The contemporary emphasis on health care financing concerns may stimulate a broadening of the concept of patient–prosthesis matching to include the relative rates and varying nature of complications of and generalized patient outcomes with different devices of disparate cost. Analysis of patients and prostheses provides important data that can be used to justify third-party reimbursement or approach the controversial and ethically challenging question: "Can less expensive devices with adequate performance provide sufficient benefit in particular patient populations, to obviate the use of high-performance, but higher-cost implants, where they may be unnecessary?" For example, can an inexpensive heart valve prosthesis with an expected 20-year lifetime be used in an octogenarian, thereby reserving more costly devices with an estimated 50-year lifetime for younger, more active patients?

Extensive information is available on the general failure modes and pathological features associated with many different types and models of heart valve substitutes.[24–28] Because failure modes of prosthetic heart valves depend on the device model and type, patient factors, and the site of implantation, both experimental and clinical analyses require knowledge of established and potential failure modes of various devices in particular situations as well as the clinical data pertinent to specific cases. The major objectives of substitute heart valve retrieval and analysis are summarized in Table 1.

Examination of substitute heart valves varies according to the specific goals of the evaluation. The essential and desirable prerequisites for high-quality implant-retrieval analysis are summarized in Table 2. Moreover, evaluation techniques should be stratified. For example, one approach distinguishes level 1 and level 2 studies. Level 1 studies include routine evaluation modalities capable of being done in virtually all laboratories and that characterize the overall safety and efficacy of a device, including complications, cause of death, and critical blood–tissue–biomaterial and patient–device interactions (e.g.,

Table 1
Objectives of Substitute Heart Valve Retrieval and Analysis

Establish rates, modes, and mechanisms of failure
Enhance patient management by surveillance for and noninvasive recognition
  of complications
Identify patient influences on device function
Enhance device selection and patient–prosthesis matching criteria
Establish structural correlates of favorable performance
Eliminate complications of device development
Predict effects of prosthesis modifications on efficacy and safety
Identify subclinical patient–prosthesis interactions
Elucidate or blood–tissue–biomaterials interaction mechanisms

Table 2
Features of Successful Implant-Retrieval Studies

Driven by hypothesis
Considers known and potential failure modes of specific devices and settings
Availability of pertinent clinical data
Involvement of all concerned disciplines, including a pathologist
Stratified analyses (mandatory vs elective), with availability of expert laboratories
  for specialized/advanced analyses
Recognition that some analyses may be mutually exclusive, so that material is taken
  and prepared for all reasonable analysis possibilities
Data recorded on carefully designed, study-specific forms
Data is quantified, with appropriate statistical analyses, whenever possible

gross and dissecting microscope examination, photography, microbiological cultures, histology, and, where pertinent, a radiograph of the specimen). In contrast, level 2 studies comprise well-defined and meaningful test methods that are either difficult or expensive to perform, require special expertise, or yield more investigative or esoteric data. Such test methods might involve scanning or transmission electron microscopy, or chemical, biochemical, immunological, or molecular techniques (e.g., calcium assay, protein measurement, immunoperoxidase localization in tissue sections of a protein for which an antibody is available, and molecular studies such as *in situ* hybridization to localize messenger RNA in tissue, or expression profiling, as an indication of cellular gene expression). Because some level 2 analyses may be mutually exclusive owing to the specific preparation methodologies required for different studies (e.g., some studies are destructive, some require fresh frozen tissue or material specifically preserved in different chemicals), some material might routinely need to be set aside (during level 1 analyses) in the event that more specialized level 2 studies are later indicated.

3. Preclinical Implant Retrieval

Procedures used to evaluate cardiovascular devices and prostheses after function in animals and humans are largely the same. However, subject to humane treatment considerations enumerated in institutional and National Institutes of Health guidelines that enforce the federal Animal Welfare Act of 1992, animal studies often permit more detailed monitoring of device function and enhanced observation of morphologic detail (including tissue–biomaterials interaction). Moreover, animal studies allow *in situ* observation of implants following elective sacrifice at desired intervals.

Specimens from experimental animals are often obtained rapidly, thereby minimizing the artifactual autolytic changes that can occur when tissues are removed from their blood supply. Furthermore, advantageous technical adjuncts may be available in animal but not human investigations, including in vivo studies, such as fixation by pressure perfusion that maintains tissues and cells in their physiological configuration following removal,[29,30] injection of radiolabeled ligands for imaging platelet deposition,[31] and injection of various substances that serve as informative markers during analysis (e.g., indicators of endothelial-barrier integrity).[32] Animal studies often facilitate observation of specific complications in an accelerated timeframe, such as calcification of bioprosthetic valves, in which the equivalent of 5 to 10 years in humans is simulated in 4 to 6 months.[33,34] Moreover, in preclinical studies, experimental conditions may be maintained constant among groups of subjects with the same valves, including nutrition, activity levels, and treatment conditions. Consequently, concurrent control implants, in which only a single critical parameter varies, may be done in preclinical animal investigations but are often difficult or unavailable in human studies.

4. Clinical Implant Retrieval

Clinical and pathological studies have demonstrated that virtually all types of widely used cardiac valve substitutes suffer deficiencies and complications that have limited their success. Indeed, prosthesis-associated pathology is a major determinant of the prognosis of patients who have had valve replacement. Among patients who die following valve replacement, the immediate cause of death is device related in approximately one-half.[35] Moreover, clinical investigations of individual valve types in randomized studies show that more than 60% of valve replacement patients suffer an important adverse prosthesis-associated event within 10 years postoperatively, regardless of valve type. However, patient outcome after cardiac valve replacement is also critically dependent on both irreversible cardiac pathology secondary to the original valve disease (especially left ventricular myocardial hypertrophic and degenerative changes) and other cardiac pathology that occurs even in patients who have not had valve replacements such as those who have had coronary arterial atherosclerotic occlusions. This emphasizes the value of careful and complete postmortem examination (autopsy) in patients following valve replacement, as the nature of those complications is often not appreciated clinically. It also highlights the importance of examination of a valve or other device in the functional anatomic context (e.g., the heart and patient), if at all possible.

Valve-related complications are categorized as thromboembolism and related problems, including anticoagulation-related hemorrhage, infection, structural dysfunction, and nonstructural dysfunction, as summarized in Table 3. Although overall rates of valve-related complications are similar for mechanical prostheses and bioprostheses, the frequency and nature of specific valve-related complications vary with the prosthesis type, model, site of implantation, and patient characteristics. Studies of retrieved heart valve prostheses have elucidated the features of those complications.[36–39]

Contemporary mechanical prostheses are durable except for a few notable exceptions. However, such valves are prone to thrombosis and thromboembolism (Fig. 2), necessitating chronic anticoagulation in patients who have received them. By contrast, tissue valves have a relatively low rate of thromboembolism without anticoagulant therapy, but virtually all bioprostheses used to date have had limited durability, nearly exclusively because of cuspal degeneration (primary tissue failure with calcification and tearing), as exemplified by glutaraldehyde-pretreated porcine aortic valves (Fig. 3), which is discussed in Section 4.3. In several valve types with patterns of failure, detailed pathologic analysis coupled with clinical data has implicated specific causal mechanisms of deleterious interaction.

5. Case Studies Demonstrating Utility of Retrieval Analysis

The literature contains numerous instances in which problem-oriented implant analysis studies have yielded important insights into problems with specific cohorts of substitute heart valves; these are summarized in Table 4. Several representative instructive "cases" from our personal experience are summarized in Sections 4.1.–4.7.

5.1. Braunwald-Cutter Cloth-Covered Heart Valve

Introduced in the early 1970s, the Braunwald-Cutter cloth-covered caged-ball mechanical prosthetic heart valve had a ball fabricated from silicone and an open cage apex that was covered by polypropylene mesh struts.[40] The major innovation was the cloth-covered cage struts, intended to encourage tissue ingrowth and thereby decrease thromboembolism.

Table 3
Complications of Cardiac Valve Substitutes

Generic	Specific
Thrombotic limitations	Thrombosis
	Thromboembolism
	Anticoagulation-related hemorrhage
Infection	Prosthetic valve endocarditis
Structural dysfunction	Wear
	Fracture
	Poppet escape
	Leaflet immobility
	Cuspal tear
	Calcification
	Commissural region dehiscence
Nonstructural dysfunction	Pannus (tissue overgrowth)
	Entrapment by suture or tissue
	Paravalvular leak
	Disproportion
	Hemolytic anemia
	Noise

Modified from ref. *48*.

In preclinical studies of the Braunwald-Cutter valve concept using mitral valve implants in pigs, sheep, and calves, the cloth-covered struts were rapidly and appropriately healed by endothelium-coated fibrous tissue (a neointima).[41,42] However, subsequent clinical studies of these valves demonstrated that cloth wear was abundant in both mitral and aortic prostheses[43] (Fig. 4). Some patients had cloth wear accompanied by abrasive wear of the poppet sufficient to permit the ball to escape through the spaces between cage struts, a potentially rapidly fatal complication. With aortic valves, both cloth and ball changes were accentuated, and ball escape was more likely.

Detailed and formal analysis of clinically removed mitral and aortic Braunwald-Cutter prostheses, including semiquantitative and quantitative characterization of strut coverage and poppet wear, elucidated the mechanisms of the disparate clinicopathologic behavior between mitral and aortic sites. Tissue ingrowth was usually present on (at least) a portion of the struts of the mitral prosthesis after prolonged periods of implantation, but tissue did not cover the fabric on aortic valves.[26] Importantly, these data suggested that

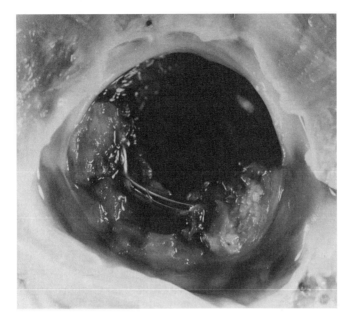

Fig. 2. Thrombotic occlusion of a mechanical heart valve prosthesis viewed from distal (outflow) aspect. (Reproduced with permission from ref. *48.*)

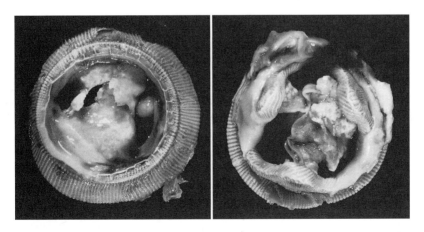

Fig. 3. Porcine bioprosthetic valve, failing by calcification with secondary cuspal tears. Left, inflow aspect; right, outflow aspect.

Table 4
Clinical Utility of Retrieval Studies on Heart Valve Substitutes

Valve type	Knowledge gained/lessons learned
Caged ball valves	• Poppets fabricated from industrial silicone absorbed blood lipids and became swollen and brittle • Fragments of degraded heart valve material may embolize to other organs • Thrombosis can occur at stasis points downstream of the ball
Cloth-covered caged ball valves	• Cloth wear can cause hemolysis and cloth emboli • Cloth wear accompanied by silicone poppet wear can precipitate poppet escape • Healing of fabric may be more vigorous in animals than in humans • Quantitation of data (e.g., polymeric poppet weight and dimensions) may facilitate the understanding of a failure mode
Caged disk valves	• Teflon has poor wear resistance as a valve occluder • Poor design features may potentiate thrombosis
Tilting-disk valves	• Thrombosis may initiate downstream to the edge of a partially open disk at a region of stasis • A "minor" change in valve design can result in a new propensity toward failure • Animal implants instrumented with strain gauges can be used to test a hypothetical mechanical failure mechanism • Understanding a failure mode can lead to both new methods for noninvasive diagnosis (e.g., X-ray and acoustic) and modified patient management strategies (e.g., drugs to depress cardiac contractility)
Bileaflet tilting-disk valves	• Cavitation may cause critical materials damage in some valve designs • Thrombosis may be initiated in regions of microstasis at component junctions • Microscopic areas of stasis may be predicted by computer-assisted computation • Animal implant models may fail to predict vulnerability to thrombosis in humans

(continued)

Table 4 (Continued)

Bioprosthetic heart valves	• Tissue calcification is a major failure mode • Calcification is most pronounced in areas of leaflet flexion, where deformations are maximal • Calcification is accelerated in young recipients, especially children • Heart valve calcification can be studied outside of the circulation (e.g., subcutaneous implants in rats) • Calcification is initiated principally at cell remnants deep in the tissue
Cryopreserved allograft valves	• These valves are not viable and cannot grow • Failure is not immunologically mediated and, therefore, immunosuppression is inappropriate
Substitution of new materials	• Pyrolytic carbon has favorable clinical durability • Detailed examination of functional (not failed) prostheses may yield worthwhile data

Reproduced by permission from ref. *21.*

asymptomatic mitral Braunwald-Cutter prostheses need not be electively replaced. This case example demonstrates that (1) human trials and clinical use may reveal important complications not predicted by animal investigations, (2) the more vigorous healing that occurred in animals than humans essentially prevented preclinical prediction of the problem that occurred in people, (3) the propensity to and consequences of valve failure can depend on the specific valve replaced, (4) the use of quantitative data may facilitate the understanding of a failure mode, and (5) the results of implant retrieval studies can impact on patient management.

5.2. Björk-Shiley 60–70° Convexo-Concave Heart Valve

The Bjork-Shiley tilting disk mechanical valve prosthesis, widely used in the 1970s and evolving from a Delrin polymer to pyrolytic carbon disk was associated with an unacceptable late failure rate owing to thrombotic occlusion.[45] In this valve type, the disk is held in place by two wires called inflow and outflow struts. The inflow strut is an integral part of the valve base, and the outflow strut is welded to the base ring. To reduce thrombosis by improving flow through the valve, Shiley, Inc. redesigned the valve to achieve enhanced opening angles of the disk to 60° or 70° from the plane of the valve

Fig. 4. Braunwald-Cutter mitral heart valve prosthesis, demonstrating marked wear of the cloth covering at the ends of the struts.

base and changed the shape of the surfaces disk to a rounded shape (convexo-concave [C-C]). However, despite a seemingly insignificant design modification, unanticipated complications developed as a result of the new design. The mechanisms of this problem were elucidated through careful analysis using retrieved implant analysis, coupled with other methods.

The use of the redesigned Björk-Shiley C-C heart valve led to an unusually large cluster of cases in which the metallic outlet strut fractured, leading to disk escape (Fig. 5). The complication was fatal in the majority of patients in whom it occurred. By late 2003, strut fracture was reported in at least 663 of the more than 86,000 valves of this type implanted worldwide from beginning in 1978 (to 1983 for the 70° valves and 1986 for the 60° valves)[46]. Clinical studies identified large valve size, mitral site, recipient age younger than 50 years of age, and valve manufacture in 1981 and early 1982 as risk factors for this failure mode.[46,47]

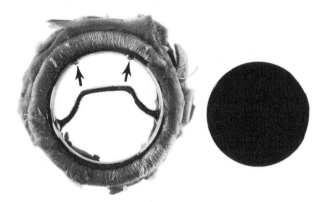

Fig. 5. Björk-Shiley heart valve prosthesis with fracture of the lesser (outflow) strut previously welded to the metal frame. Fracture sites are noted by arrows. The fractured strut could not be located at autopsy. (Reproduced with permission from ref. *48.*)

Pathological studies suggested that the underlying problem was metal failure, probably owing to over rotation of the disk during valve closure.[48] The design can cause an abnormally hard strike of the disk on the metallic outlet strut, leading to excessive bending stresses at or near the welds joining the outlet struts to the housing, potentially coupled with intrinsic weld flaws. In support of this failure mechanism, retrieved Bjork-Shiley valves often demonstrated a pronounced wear facet at the tip of the outlet struts, and some had localized pyrolytic carbon wear deposits at sites of apposition. Scanning electron microscopy of the fracture surfaces demonstrated that fractures began on the inlet side of the outlet strut and suggested that the first strut leg fracture typically initiated at or near the point of maximum bending stress. The initiation site was traced to a site of weld shrinkage porosity and/or inclusion in most cases. Occasionally, valves with only a single strut fracture were encountered.[49]

In subsequent animal studies in which Björk-Shiley 60–70° C-C valves were implanted in sheep and instrumented with strain gages showed that impact forces vary greatly with cardiac activity and that loads occurring during exercise were significantly elevated. Moreover, hyperdynamic cardiac activity was contributory to catastrophic failure, as supported by clinical analyses that showed higher risk of fracture with increased cardiac output.

This situation demonstrates that elucidation of a failure mode by detailed materials failure analysis and that effective use of carefully designed animal experiments can potentially affect patient management. Prophylactic valve replacement of susceptible prostheses may be indicated in some patients.[50] Understanding this mode of failure has justified development of noninvasive testing modalities (via high definition radiography[51] or acoustic characterization of strut status)[52] to establish when one strut has fractured before the onset of clinical failure and cautioning patients with mitral valves against vigorous exercise and the use of drugs (such as β-blockers) that reduce contractile failure.

5.3. Bioprosthetic Heart Valve Calcification

Calcification is an important pathological process contributing to failure of bioprostheses fabricated from porcine aortic and bovine pericardial bioprosthetic valves (*see* Fig. 3). Calcific failure occurs in 50% of valves within 12 to 15 years postoperatively in adults and occurs more frequently and earlier in children and adolescents. Studies of retrieved experimental and clinical implants have characterized calcification-induced failure modes, patterns and extent of mineral deposition,[53] the nature of the mineral phase,[54] and early calcification events.[55,56]

Further study directed to mechanisms and prevention have used bioprosthetic tissue implanted into both circulatory and subcutaneous experimental models, which calcify with morphological features similar to those observed in clinical specimens but show markedly accelerated calcification in experimental models. We have used the subcutaneous implantation model extensively as a technically convenient, economically advantageous and quantifiable model for investigating host and implant determinants and pathobiology of mineralization, as well as for screening and understanding the mechanisms of potential strategies for mineralization inhibition. Valves implanted as mitral replacements in sheep calcify extensively in 3 to 4 months and subcutaneous implants of bioprosthetic tissue in rats achieve calcium levels comparable to those of failed clinical explants in 8 weeks or less.[55]

Clinical and experimental studies indicate that calcification of bioprosthetic valves depends on host, implant, and mechanical factors.[37,55–57] For example, (1) the pretreatment of tissue with an aldehyde cross-linking agent (e.g., glutaraldehyde used in clinical implants) potentiates mineralization; (2) calcification is most pronounced in areas of leaflet flexion, where stress/deformations are maximal; and (3) neither nonspecific inflammation nor adaptive immunological responses appear to mediate bioprosthetic tissue calcification. Thus, the fundamental mechanisms of bioprosthetic tissue mineralization depend on specific biochemical modification of implant microstructural components induced by aldehyde pretreatments. The earliest mineral deposits in both clinical and experimental bioprosthetic tissue are localized to transplanted connective tissue cells; collagen involvement occurs later, possibly by an independent mechanism. Mineralization of connective tissue cells of bioprosthetic tissue appears to result from glutaraldehyde-induced cellular devitalization and the resulting disruption of cellular calcium regulation.

These studies of pathobiology have generated and provided the means to test approaches to reduce bioprosthetic valve failure by modifying host, implant, or mechanical influences. Most of the strategies for prevention of bioprosthetic tissue mineralization involve modifications of either valve design or preparation details or the local environment of the implant. Mechanisms of calcification inhibition by antimineralization treatments that have been investigated clinically and/or experimentally include (but are not limited to) extraction of calcifiable material, ionic and/or macromolecular binding to nucleation sites, and interference with calcium phosphate crystal growth.[58,59] Although durability of modified bioprosthetic heart valves can be assessed with certainty only by long-term clinical evaluation, an appropriate experimental testing program for the efficacy and safety of antimineralization treatments requires at least four components: (1) qualification (and whenever possible, assessment of mechanism) using subcutaneous implantation in rats; (2) hydrodynamic/durability testing to show lack of excessive obstruction or regurgitation, as well as verify that no new failure modes are evident; (3) morphological studies of unimplanted valve material to assess the potential for treatment-induced degradative changes;[60] and (4) valve replacement in the appropriate configuration and site in an animal model, usually sheep.

Thus, in the context of the present discussion, retrieval studies of bioprosthetic heart valves emphasize that biological failure mechanisms can be understood using specifically-designed animal models, guided by and correlated with the results of studies of retrieved clinical specimens. Furthermore, implant-retrieval studies can be used as a critical component of a thoughtful testing program to assure efficacy and safety of potential therapeutic modifications.

5.4. Beall Disk Heart Valve Prosthesis and the Durability of Pyrolytic Carbon

The Beall heart valve prosthesis, introduced in 1967, was a low-profile disk valve composed of a disk fabricated from extruded Teflon and metal struts coated with Teflon. This design and material were intended to minimize thromboresistance. Following realization of poor wear properties of the disk, with sequelae of disk abrasion including hemolysis and abnormal disk motion[61] (Fig. 6), its composition was changed to a denser compression-molded Teflon. Nevertheless, this prosthetic design continued to exhibit wear-related problems concentrated at the struts,[62] and a new model of the valve with disk and struts fabricated from pyrolytic carbon was introduced in the early 1970s.

Presently, as a result of favorable mechanical and biological properties of pyrolytic carbon, virtually all mechanical heart valve prostheses in use have pyrolytic carbon occluders and some have both carbon occluders and carbon cage components.[63] Pyrolytic carbon is strong, resistant to fatigue and abrasive wear, thromboresistant, and can be fabricated into a wide variety of shapes.

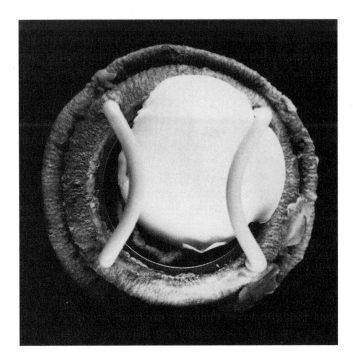

Fig. 6. Severe abrasive wear of disk poppet of Beall Teflon caged-disk mitral valve prosthesis.

In the early 1980s, clinical experience suggested that pyrolytic carbon had contributed to a major advancement in the durability of prosthetic heart valves; however, this had not been verified by direct valve observation. To confirm the anticipated favorable wear resistance of pyrolytic carbon in the clinical environment, we recovered at necropsy or surgery eight carbon-containing mechanical valve prostheses after long-term implantation and analyzed them by surface scanning electron microscopy and surface profilometry.[64] None of the prostheses had clinical or gross pathological malfunction or abrasive wear, but minimal strut wear was demonstrated by scanning electron microscopy. No appreciable wear on carbon valve occluders was demonstrated by analytical surface profilometry. Our study suggested that the use of pyrolytic carbon as an occluder and as a strut material for mechanical heart valve prostheses has minimized progressive abrasive wear as a long-term complication of cardiac valvular replacement. Favorable durability of pyrolytic carbon has subsequently been well documented.[65]

Analysis of implanted medical devices has traditionally concentrated primarily on those devices that failed in service and paid little attention to those

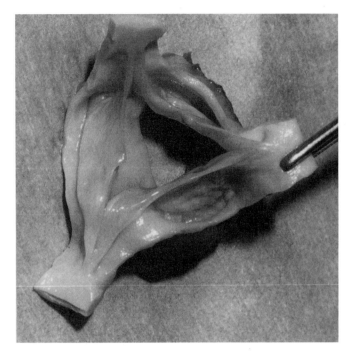

Fig. 7. Cryopreserved/thawed allograft heart valve. Removed from a patient who died, this aortic valve is ready for implantation in another as an aortic valve replacement (courtesy of CryoLife, Inc., Marietta, GA).

serving the patient until death or removal owing to unrelated causes. The study described above emphasizes that detailed examination of functional (not failed) prostheses recovered from patients following extended duration after implantation may yield worthwhile data to answer a focused question.

5.5. Cryopreserved Allograft Heart Valves

Allograft valves are removed at either autopsy or heart transplantation from one person and transplanted to another. Most such valves are cryopreserved (i.e., stored at −170° in dimethyl sulfoxide) until needed (Fig. 7). Aortic valve allografts have good hemodynamic profiles, low thromboembolic rates despite the absence of anticoagulation, low rates of infection, and low rates of degeneration.[66,67] There has been ongoing controversy regarding modes of failure, cellular viability, durability of the extracellular matrix, and whether immune responses contribute to failure.

We examined 20 explanted cryopreserved valves in place for several hours to 9 years, obtained by virtue of the author's role as core pathologist for a large clinical trial conducted under an investigational device exemption held by a consortium of five tissue banks. The original surgeries and explants were done in more than 100 hospitals. Arrangements were made by the consortium to have removed valves sent directly to the core pathology laboratory, accompanied by clinical data. We also examined thawed but unimplanted allografts (obtained via the same route) and 16 donor aortic valves obtained from heart transplants of recipients who later died.

Our studies indicated that removed cryopreserved allograft heart valves undergo progressively severe loss of the normal layered structure and stainable deep connective tissue cells with minimal inflammation.[68] Following either short-term or extended function, cryopreserved allograft heart valves have minimal, if any, viable cells but largely retain the original collagen network; it is this preservation of the autolysis-resistant collagenous skeleton that likely provides the structural basis for their generally favorable function. Because inflammation in these valves is minimal in almost all cases, we have found no evidence that immune responsiveness can impact late allograft function or degeneration. Interestingly, and in contrast, aortic valves of transplanted whole hearts maintained near-normal overall architecture, and cells.

Our studies of retrieved allograft valves emphasize that (1) not only failure modes but also mechanisms of successful function can be elucidated by careful study, (2) correlation of results from multiple types of implants can be exceedingly important in understanding structure–function correlations, and (3) a well-organized network can pool relatively unusual specimens obtained from many centers at a core pathology laboratory, lead to insights not readily obtainable from clinical material that accrues and thereby at a single institution.

5.6. Medtronic Parallel Valve

The Medtronic Parallel Bileaflet Heart Valve was an advanced product withdrawn from the market following clinical trials. In contrast to the 85° opening of the leaflets of the widely used St. Jude valve leaflets, this valve had a uniquely designed pivot mechanism that allowed the leaflets to open to fully parallel, thereby potentially providing better hemodynamics with a lower pressure drop across the valve. Mitral valve implants in sheep, performed and analyzed according to FDA guidelines, yielded no evidence of problems; however, clinical trials of the Parallel valve exhibited an excessive rate of thrombotic complications. Analysis of retrieved valves demonstrated that clinically important thrombi were likely initiated by microclot formation at the valve hinge pockets. Subsequent computational fluid dynamics analysis indicated a standing vortex and zone of blood stagnation at the hinge pockets and high shear stress near the hinge inflow channel wall capable of producing platelet and red cell damage.[69–71] This situation emphasizes that properly done preclinical investigation may fail to reveal critical problems that may occur in the clinical environment and pathological analysis coupled with computational flow studies can collaborate to rapidly identify the nature and mechanism of a valve complication.

5.7. The Carbomedics Photo-Fix Pericardial Heart Valve

Dye-mediated photo-oxidation is a method of glutaraldehyde-free preservation designed to prevent calcification in bioprosthetic valves. In a limited clinical trial, 10 aortic valves fabricated from photo-oxidized bovine pericardium were explanted for regurgitation after 8 to 23 months. Each valve had one to several commissural-basal tears (5 mm) to complete leaflet detachment (four valves). We studied the pathological findings in these valves, with emphasis on both mechanism of failure and histological appearance of tissue. The pattern of tearing was consistent among valves and suggested that a mechanism of design-related proximal-inflow surface cuspal abrasion against Dacron cloth was contributory.[72,73] Nevertheless, this study also suggested that photo-oxidized tissues would maintain biocompatibility and integrity during extended human in vivo function. This likely occurred as a result of this tissue being more compliant than glutaraldehyde-pretreated pericardium and its excursion during the closed phase were greater than in glutaraldehyde-treated tissue. This enhanced the contact of the tissue with and subsequent abrasion of the Dacron on the housing, and secondly, this problem had not been revealed by either conventional in vitro durability studies that were carried out at accelerated rates and did not permit full excursion of the tissues, and sheep mitral valve implants, in which the implant time was insufficient, did not reveal the problem. This situation demonstrated that implant retrieval and analysis can elucidate the key effect on valve function of the interactions of tissue mechanical properties with details of valve design.

6. Conclusions and Unresolved Issues

Pathological analysis of removed valve substitutes contributes to patient management and device development. Implant retrieval and analysis studies serve an important role in the evaluation of new/modified medical devices and may contribute to the management of individual patients. Optimal evaluation is driven by hypotheses and specific questions, uses technical steps appropriate for specific objectives, considers relevant failure modes, maximizes pertinent clinical data, and minimizes pitfalls. Nevertheless, difficulties remain, as summarized in Table 5, including: (1) difficulty in tracking implants; (2) autopsy or surgical removal required for pathological analysis; (3) legal/ethical issues, including ownership and confidentiality; (4) technical problems (e.g., removal-induced artifacts, overinterpretation of findings); (5) frequently inadequate

Table 5
Challenges in Clinical Device Retrieval

Need for implant registry (data on patients, prostheses types, complications)
Pathologist unaware that implant is in place
Autopsy/surgical removal of device from a potentially informative anatomic context
    before study commences
Decreasing autopsy rates overall
Induced artifacts
Overinterpretation of pathologic findings
Disposition of specimen (patient, manufacturer, Food and Drug Administration,
    attorney, pathology laboratory, research laboratory)
Inadequate clinical data
Potential biohazard/cleaning/packaging for transport/mode of discard
Priority of destructive vs nondestructive analyses
Inadequate funding for implant-retrieval activities
Legal/ethical concerns (informed patient consent, confusion concerning implant
    ownership, confidentiality of information)
Indentification of core laboratories with specific device expertise

clinical data; (6) disinterest of conventional funding sources; and (7) patient privacy issues. It is hoped that enhanced appreciation of the value of this activity will serve to facilitate the development of cooperation and collaborations that maximize the quantity and quality of the most useful specimens and relevant clinical data and stimulate appropriate funding mechanisms.

References

1. Rahimtoola, S. H. 1989. Perspective on valvular heart disease: An update. *J. Am. Coll. Cardiol.* 14:1–23.
2. Starr, A., Fessler C. L., Grunkemeier, G., and He, G. W. 2002. Heart valve replacement surgery: past, present and future. *Clin. Exp. Pharmacol. Physiol.* 29:735–738.
3. Silver, M. D. and Butany, J. 1987. Mechanical heart valves: Methods of examination, complications, modes of failure. *Hum. Pathol.* 18:577–585.
4. Anderson, J. M. 1993. Cardiovascular device retrieval and evaluation. *Cardiovasc. Pathol.* 2(Suppl):199S–208S.
5. Schoen, F. J. 1989. *Interventional and Surgical Cardiovascular Pathology: Clinical Correlations and Basic Principles.* Philadelphia: WB Saunders.
6. Butany, J., d'Amati, G., Fornasier, V., Silver, M. D., and Sanders, G. E. 1990. Detailed examination of complete bioprosthetic heart valves. *ASAIO Trans.* 36:M414–M417.
7. Schoen, F. J. 1995. Approach to the analysis of cardiac valve prostheses as surgical pathology or autopsy specimens. *Cardiovasc. Pathol.* 4:241–255.

8. Schoen, F. J. 1999. Evaluation of explanted cardiovascular prostheses. In: von Recum, A. F., ed., *Handbook of Biomaterials Evaluation: Scientific, Technical and Chemical Testing of Implant Material*, 2nd Edition. Washington: Taylor & Francis, pp. 671–686.

9. Schoen, F. J., Anderson, J. M., Didisheim, P., et al. 1990. Ventricular assist device (VAD) pathology analyses: guidelines for clinical studies. *J. Appl. Biomater.* 1:49–56.

10. Borovetz, H. S., Ramasamy, N., Zerbe, T. R., and Portner, P. M. 1995. Evaluation of an implantable ventricular assist system for humans with chronic refractory heart failure. Device explant protocol. *ASAIO J.* 41:42–48.

11. Wagner, W. R., Johnson, P. C., Kormos, R. L., and Griffith, B. P. 1993. Evaluation of bioprosthetic valve-associated thrombus in ventricular assist device patients. *Circulation* 88:203–209.

12. Fyfe, B. and Schoen, F. J. 1993. Pathologic analysis of 34 explanted Symbion ventricular assist devices and 10 explanted Jarvik-7 total artificial hearts. *Cardiovasc. Pathol.* 2:187–197.

13. Canizales, S., Charara, J., Gill, F., et al. 1982. Expanded polytetrafluoroethylene prostheses as secondary blood access sites for hemodialysis: Pathological findings in 29 excised grafts. *Can. J. Surg.* 54:17–26.

14. Downs, A. R., Guzman, R., Formichi, M., et al. 1991. Etiology of prosthetic anastomotic false aneurysms: Pathologic and structural evaluation in 26 cases. *Can. J. Surg.* 34:53–58.

15. Guidoin, R., Chakfe, N., Maurel, S., et al. 1993. Expanded polytetrafluoroethylene arterial prostheses in humans: histopathological study of 298 surgical excised grafts. *Biomaterials* 14:678–693.

16. Anderson, P. G., Bajaj, R. K., Baxley, W. A., and Roubin, G. S. 1992. Vascular pathology of balloon-expandable flexible coil stent in humans. *J. Am. Coll. Cardiol.* 19:372–381.

17. Van Beusekom, H. M. M., van der Giessen, W. J., van Suylen, R. J., Bos, E., Bosman, F.T., and Serruys, P. W. 1993. Histology after stenting of human saphenous vein bypass grafts: Observations from surgically excised grafts 3 to 320 days after stent implantation. *J. Am. Coll. Cardiol.* 21:45–54.

18. Farb, A., Sangioregi, G., Carter, A. J., et al. 2002. Pathology of acute and chronic coronary stenting in humans. *Circulation* 105:2932–2933.

19. Schwartz, R. S., Edelman, E. R., Carter, A., et al. 2002. Drug-eluting stents in preclinical studies: Recommended evaluation from a consensus group. *Circulation* 106:1867–1873.

20. Jabcobs, T. S., Won, J., Gravereaux, E. C., et al. 2003. Mechanical failure of prosthetic human implants: A 10-year experience with aortic stent graft devices. *J. Vasc. Surg.* 37:16–26.

21. Anderson, J. M., Shoen, F. J., Brown, S. A., and Merritt, K. 2004. Implant retrieval and evaluation. In: Ratner, B. D., Hoffman, A. S., Shoen, F. J., and Lemons, J. E., eds., *Biomaterials Science*, 2nd Edition. New York: Elsevier, pp. 771–782.

22. Savage, R. A. 1991. New law to require medical device injury report. *CAP Today*, July, pp. 40.
23. Kahan, J. S. 1991. The Safe Medical Devices Act of 1990. *Med. Dev. Diagn. Ind.*, Jan, p. 67.
24. Atkins, C. W. 1995. Results with mechanical cardiac valvular prostheses. *Ann. Thorac. Surg.* 60:1836–1844.
25. Turina, J., Hess, O. M., Turina, M., and Krayenbuehl, H. P. 1993. Cardiac bioprostheses in the 1990s. *Circulation* 88:775–781.
26. Bloomfield, P., Wheatley, D. J., Prescott, R. J., and Miller, H. C. 1991. Twelve-year comparison of a Bjork-Shiley mechanical heart valve with porcine bioprostheses. *N. Engl. J. Med.* 324:573–579.
27. Hammermeister, K. E., Sethi, G. K., Henderson, W. G., Oprian, C., Kim, T., and Rahimtoola, S. 1993. A comparison of outcomes in men 11 years after heart-valve replacement mechanical valve or bioprosthesis. *N. Engl. J. Med.* 328:1289–1296.
28. Schoen, F. J. 2001. Pathology of heart valve substitution with mechanical and tissue prostheses. In: *Cardiovascular Pathology*, 3rd Ed., Silver, M. D., Gotlieb, A. I., Schoen, F. J., eds., WB Saunders pp. 629–667.
29. Davis, P. F. and Bowyer, D. E. 1975. Scanning electron microscopy: Arterial endothelial integrity after fixation at physiological pressure. *Atherosclerosis* 21:463–69.
30. Clowes, A. W., Gown, A. M., Hanson, S. R., and Reidy, M. A. 1985. Mechanisms of arterial graft failure. 1. Role of cellular proliferation in early healing of PTFE prostheses. *Am. J. Pathol.* 118:43–54.
31. Palatianos, G. M., Dewanjee, M. K., Panoutsopoulos, G., Kapadvanjwala, M., Novak, S., and Sfakianakis, G. N. 1994. Comparative thrombogenicity of pacemaker leads. *Pacing Clin. Electrophysiol.* 17:141–145.
32. Reidy, M. A., Chao, S. S., Kirkman, T. R., and Clowes, A. W. 1986. Endothelial regeneration. VI. Chronic nondenuding injury in baboon vascular grafts. *Am. J. Pathol.* 123:432–439.
33. Levy, R. J., Schoen, F. J., Levy, J. T., Nelson, A. C., Howard, S. L., and Oshry, L. J. 1993. Biologic determinants of dystrophic calcification and osteocalcin deposition in glutaraldehyde-preserved porcine aortic valve leaflets implanted subcutaneously in rats. *Am. J. Pathol.* 113:143–155.
34. Schoen, F. J., Hirsch, D., Bianco, R. W., and Levy, R. J. 1994. Onset and progression of calcification in porcine aortic bioprosthetic valves implanted as orthotopic mitral valve replacements in juvenile sheep. *J. Thorac. Cardiovasc. Surg.* 108:880–887.
35. Schoen, F. J., Titus, J. L., and Lawrie, G. M. 1983. Autopsy-determined causes of death after cardiac valve replacement. *J. Am. Med. Assoc.* 249:899–902.
36. Schoen, F. J. and Hobson, C. E. 1985. Anatomic analysis of removed prosthetic heart valves: causes of failure of 33 mechanical valves and 58 bioprostheses, 1980–1983. *Human Pathol*. 16:549–559.
37. Schoen, F. J. and Levy, R. J. 1999. Tissue heart valves: current challenges and future research perspectives. *J. Biomed. Mater. Res.* 47:439–465.

38. Butany, J. and Leask, R. 2001. The failure modes of biological prosthetic heart valves. *J. Long Term Eff. Med. Implants* 11:115–135.
39. Schoen, F. J. 1999. Future directions in tissue heart valves: Impact of recent insights from biology and pathology. *J. Heart Valve Dis.* 8:350–358.
40. O'Rourke, R. A., Peterson, K. L., and Braunwald, N. S. 1973. Postoperative hemodynamic evaluation of a new fabric-covered ball-valve prosthesis. *Circulation* 47,48(Suppl 3):74–79.
41. Braunwald, N. S. and Bonchek, L. I. 1967. Prevention of thrombus on rigid prosthetic heart valves by the ingrowth of autogenous tissue. *J. Thorac. Cardiovasc. Surg.* 54:630–638.
42. Braunwald, N. S. and Morrow, A. G. 1968. Tissue ingrowth and the rigid heart valve. *J. Thorac. Cardiovasc. Surg.* 56:307–319.
43. Blackstone, E. H., Kirklin, J. W., Pluth, J. R., Turner, M. E., and Parr, G. V. S. The performance of the Braunwald-Cutter aortic prosthetic valve.
44. Schoen, F. J., Goodenough, S. H., Ionescu, M. I., and Braunwald, N. S. 1984. Implications of late morphology of Braunwald-Cutter mitral heart valve prostheses. *J. Thorac. Cardiovasc. Surg.* 88:208–216.
45. Murphy, D. A., Levine, F. H., Buckley, M. J., et al. 1983. Mechanical valves: a comparative analysis of the Starr-Edwards and Bjork-Shiley prostheses. *J. Thorac. Cardiovasc. Surg.* 86:746–752.
46. Blot, W. J., Ibrahim, M. A., Ivey, T. D., et al. 2005. Twenty-five-year experience with the Björk-Shiley convexoconcove heart valve. A continuing clinical concern. *Circulation* 111:2850–2857.
47. Orszulak, T. A., Schaff, H. V., DeSmet, J. -M., Danielson, G. K., Pluth, J. R., and Puga, F. J. 1993. Late results of valve replacement with Björk-Shiley valve (1973–1982). *J. Thorac. Cardiovasc. Surg.* 105:302–312.
48. Shoen, F. J., Levy, R. J., and Piehler H. R. 1992. Pathological considerations in replacement cardiac valves. *Cardiovasc. Pathol.* 1:29–52.
49. de Mol, B. A. J. M., Koornneef, F., and van Gaalen, G. L. 1995. What can be done to improve the safety of heart valves? *Intl. J. Risk Safety Med.* 6:157–168.
50. Van Gorp, M. J., Steyerberg, E. W., and van der Graff, Y. 2004. Decision guidelines for prophylactic replacement of Bjork-Shiley convexo-concave heart valves: impact on clinical practice. *Circulation* 109:2092–2096.
51. O'Neill, W. W., Chandler, J. G., Gordon, R. E., et al. 1995. Radiographic detection of strut separations in Bjork-Shiley convexo-concave mitral valves. *N. Engl. J. Med.* 333:414–419.
52. Plemons, T. D. and Hovenga, M. 1995. Acoustic classification of the state of artificial heart valves. *J. Acoust. Soc. Am.* 97:2326–2333.
53. Schoen, F. J., Kujovich, J. L., Webb, C. L., and Levy, R. J. 1987. Chemically determined mineral content of explanted porcine aortic valve bioprostheses: Correlation with radiographic assessment of calcification and clinical data. *Circulation* 76:1061–1066.
54. Tomazic, B. B., Edwards, W. D., and Schoen, F. J. 1995. Physicochemical characterization of natural and bioprosthetic heart valve calcific deposits. Implications for prevention. *Ann. Thorac. Surg.* 60(2 Suppl):S322–S327.

55. Schoen, F. J., Levy, R. J., Nelson, A. C., Bernhard, W. F., Nashef, A., and Hawley, M. A. 1985. Onset and progression of experimental bioprosthetic heart valve calcification. *Lab. Invest.* 52:523–532.

56. Schoen, F. J., Tsao, W., and Levy, R. J. 1986. Calcification of bovine pericardium used in cardiac valve bioprostheses: implications for the mechanisms of bioprosthetic tissue mineralization. *Am. J. Pathol.* 123:134–145.

57. Schoen, F. J., Kujovich, J. L., Levy, R. J., et al. 1988. Bioprosthetic heart valve pathology: clinicopathologic features of valve failure and pathobiology of calcification. *Cardiovasc. Clin.* 18:289–317.

58. Schoen, F. J., Levy, R. J., Hilbert, S. L., and Bianco, R. W. 1992. Antimineralization treatments for bioprosthetic heart valves: assessment of efficacy and safety. *J. Thorac. Cardiovasc. Surg.* 104:1285–1288.

59. Shoen, F. J. and Levy, R. J. 2005. Calcification of tissue heart valve substitutes: progress toward understanding and prevention. *Ann. Thorac. Surg.* 79:1072–1080.

60. Flomenbaum, M. A. and Schoen, F. J. 1993. Effects of fixation back-pressure and antimineralization treatment on the morphology of porcine aortic bioprosthetic valves. *J. Thorac. Cardiovasc. Surg.* 105:154–164.

61. Robinson, M. J., Hildner, F. J., and Greenberg, J. J. 1971. Disc variance of Beall mitral valve. *Ann. Thor. Surg.* 11:11–17.

62. Silver, M. D. and Wilson, G. J. 1977. The pathology of wear in the Beall Model 104 heart valve prosthesis. *Circulation* 56:617–622.

63. More, R. B., Haubold, A. D., and Bokros, J. C. 2004. Pyrolytic carbon for long-term medical implants. In: Ratner, B. D., Hoffman, A. S, Schoen, F. J., Lemons, E., eds., *Biomaterials Science: An Introduction to Materials in Medicine*, 2nd Ed. New York: Elsevier, pp.170–182.

64. Schoen, F. J., Titus, J. L., and Lawrie, G. M. 1982. Durability of pyrolytic carbon-containing heart valve prostheses. *J. Biomed. Mater. Res.* 16:559–570.

65. Haubold, A. D. 1994. On the durability of pyrolytic carbon in-vivo. *Med. Prog. Tech.* 20:201–208.

66. Kirklin, J., Smith, D., Novick, W., et al. 1993. Long-term function of cryopreserved aortic homografts. A ten-year study. *J. Thorac. Cardiovasc. Surg.* 106:154–166.

67. Grunkemeier, G. L. and Bodnar, E. 1994. Comparison of structural valve failure among different "models" of homograft valves *J. Heart Valve Dis.* 3:556–560.

68. Mitchell, R. N., Jonas, R. A., Schoen, F. J. 1998. Pathology of explanted cryopreserved allograft heart valves: comparison with aortic valves from orthotopic heart transplants. *J. Thorac. Cardiovasc. Surg.* 115:118–127.

69. Gross, J. M., Shu, M. C., Dai, F. F., Ellis, J., and Yoganathan, A. P. 1996. A microstructural flow analysis within a bileaflet mechanical heart valve hinge. *J. Heart Valve Dis.* 5:581–590.

70. Ellis, J. T., Healy, T. M., Fontaine, A. A., Saxena, R., and Yoganathan. A. P. 1996. Velocity measurements and flow patterns within the hinge region of Medtronic Parallel bileaflet mechanical valve with clear housing. *J. Heart Valve Dis.* 5:572–573.

71. Gao, Z. B., Hosein, N., Dai, F. F., and Hwang, N. H. 1999. Pressure and flow fields in the hinge region of bileaflet mechanical heart valves. *J . Heart Valve Dis.* 8:197–205.

72. Schoen, F. J. 1998. Pathologic findings in explanted clinical bioprosthetic valves fabricated from photooxidized bovine pericardium. J. *Heart Valve Dis.* 7:174–179.

73. Butterfield, M. and Fisher, J. 2000. Fatigue analysis of clinical bioprosthetic heart valves manufactured using photooxidized bovine pericardium. *J . Heart Valve Dis.* 9:161–166.

19

The Use of Surrogate Outcome Measures

A Case Study: Home Prothrombin Monitors

John J. Whyte

1. Introduction

One of the most important aspects of study design is the selection of outcome measures. The choice of the most appropriate primary and secondary outcome measures can be a complicated decision and represents one of the most complex issues in the design of a clinical trial. Appropriate outcome measures will determine both safety and clinical effectiveness, as well as potential reimbursement, based on the application of the rigorous evidence-based approach taken by the Centers for Medicare and Medicaid Services (CMS) and other payor/providers.[1] Therefore, companies should devote considerable time to making certain the outcome measures chosen will be viewed as appropriate by reviewers and clinicians.

Ideally, one should choose an outcome measure that is directly impacted by the intervention and is the specific clinically relevant outcome one seeks to determine. For example, one measures cardiac deaths in a study by looking at effectiveness of acetyl cholinesterase inhibitors in patients with congestive heart failure. For some types of medical devices, however, the intervention being studied may be dependent on an infrequent or extremely harmful event. As a result, primary or direct outcome measures may not be practical or readily usable. Therefore, surrogate outcome measures are increasingly used. The implications of using surrogate measures needs to be sufficiently understood before one considers their use.

2. Surrogate Outcome Measures

A *surrogate outcome measure* or *surrogate endpoint* can be defined as a "variable which is relatively easily measured and which predicts a rare or distant outcome of a therapeutic intervention but which is not itself a direct mea-

From: *Clinical Evaluation of Medical Devices: Principles and Case Studies, Second Edition*
Edited by: K. M. Becker and J. J. Whyte © Humana Press Inc., Totowa, NJ

sure of either harm or benefit."[2] The surrogate measures are typically selected for medical device trials in cases in which there are rare events, since the sample size of a study must be rather large to detect a difference in events between the device and a control. A negative study result could consequently be misleading owing to a lack of power to detect a statistically significant difference. In addition, the time course of a trial with infrequent events can be quite lengthy, requiring an impractical length of time for a sufficient quantity of rare events to occur. Both of these variables, the occurrence of rare events and the number of events as a function of time, can prevent an important trial from being conducted owing to logistical or practical (namely cost) constraints. There are also some infrequent circumstances when the primary outcome measure could be rather harmful, and therefore, the risks to the clinical participants do not outweigh the benefits of using such an outcome. In these instances, a surrogate outcome measure may be more practical.

There are potential drawbacks, however, of using a surrogate measure. Most importantly, it does not directly answer the primary question. A commonly cited example is the use of elevations in blood pressure in many clinical trials as a surrogate for myocardial injury. The elevations in blood pressure are extrapolated to determine effects on myocardium, although the actual injury on myocardium, which may be the specific outcome that is sought, would not be measured in the study.

With respect to diagnostic tests, the ideal, randomized controlled trial for effectiveness occurs when the results of the diagnostic test alter a treatment intervention and the health outcomes associated with such treatment are included within the boundaries of that same study. However, there are other trial designs in which appropriate linkages of randomized controlled trials can demonstrate a similar, strong relationship between testing and health outcomes. In this instance, there is a coupling of surrogate outcomes with eventual health outcomes.

3. Case Study: Home Prothrombin Monitors

A good case study on the use of surrogate endpoint is the Medicare National Coverage Determination on Home Prothrombin Time International Normalized Ratio (INR) Monitoring for Anticoagulation Management.[3]

This case relates to the use of home prothrombin INR monitors for patients with mechanical heart valves being anticoagulated with warfarin. Patients who require anticoagulation must have their prothrombin time (PT) measured to determine whether they are in the correct range. The INR was developed to ensure consistency of different laboratory measurements. In this method, the ratio of the patient's PT is compared to the mean PT for a group of normal individuals. The ratio is adjusted for the sensitivity of the laboratory's thromboplastin, resulting in the INR.

Proper anticoagulation has been a problem for many patients, especially Medicare beneficiaries. Many patients are either over- or under-anticoagulated, or they are not anticoagulated at all owing to safety concerns. Improper anticoagulation has significant health effects, with an increased risk of stroke, myocardial injury, and clot formation.

Warfarin has a narrow therapeutic index. As a result, treatment of each anticoagulation therapy patient can be highly individualized. This variability necessitates frequent testing. The use of a home prothrombin INR monitor could potentially allow patients to do more frequent testing, manage dosing better, and therefore potentially reduce complications of anticoagulation therapy.

Home prothrombin monitors arrived on the market after being cleared by the Food and Drug Administration under a 510(k)-clearance process in the late 1990s. They were granted waived status under the Clinical Laboratory Improvement Act.

In 2001, CMS used two analytical frameworks in making a decision to cover and therefore pay for the use of these devices. Up to this time, they were not covered or reimbursed by Medicare owing to the lack of evidence in support of beneficial outcomes. The first analytical framework considered in the evidence-based method was the typical study design comparing a home INR test to a standard care control group (values taken at a laboratory) and used bleeding episodes and thromboembolism as the main outcome measures. Although there were two positive studies with these outcome measures, both were small in size and had some methodological issues (e.g., patient-selection criteria, adequate controls, improper statistical analysis) that made generalizability difficult.[4,5]

Because it was believed that there were valuable data suggesting the usefulness of this intervention (especially for patients with mechanical heart valves who need to be anticoagulated for life and at a higher value than other patients being anticoagulated for conditions such as atrial fibrillation or congestive heart failure), CMS decided to consider another analytic framework: efficacy.[6] The second framework allowed a study design that compared a home INR test group to a standard care control group but used the time-in-therapeutic range (TTR) as a surrogate measure of primary events. This was a more achievable outcome, given the much larger sample size needed to adequately compare the primary event rates.

The use of this surrogate measure, TTR, was closely scrutinized. A body of literature clearly demonstrated that use of these INR devices in the home, with or without patient self-management, significantly increased TTR. To account for potential frequency bias, the metric of percentage of tests out of range or patient-days out of range was also used. It was concluded that a sufficient number of studies using different statistical frameworks showed these devices increased TTR. This increase in TTR likely results from the fact that more information allows patients/physicians to make more timely decisions regarding dosage adjustments and therefore reduces variability of the INR.

However, the connection between TTR and adverse events still needed to be demonstrated. A specific question in the analysis was the following: "Given that the incidence of thromboembolic and/or hemorrhagic events are small, and some studies may be underpowered to detect a difference in incidence amongst various management methods, is TTR an adequate surrogate for reduction in thromboembolic/hemorrhagic events?"

The decision memorandum acknowledges that because the incidence of thromboembolic and hemorrhagic events is small (1–5%), a study would need to enroll hundreds to thousands of patients to have enough power to detect a statistically significant difference in events. Therefore, the use of TTR as a valid surrogate outcome measure was considered.

For TTR, there was a considerable volume of literature evaluating its use as a surrogate. Of note, the Managing Anticoagulation Service Trial, a randomized, controlled trial comparing anticoagulation services to usual care, used TTR as its primary outcome. An Agency for Healthcare Research and Quality (AHRQ) external review panel approved the study's design.[7] Additionally, a literature review by Samsa and Matchar noted that 20 studies demonstrated that increased TTR leads to a reduction in thromboembolic and hemorrhagic events.[8] The authors concluded that there was a strong relationship between TTR and event rate that is supported by a large literature.

Based on both of these analytic frameworks, Medicare reversed its noncoverage policy and issued a national coverage policy that allowed the use of these devices for patients with mechanical heart valves undergoing anticoagulation with warfarin.

4. Conclusion

Surrogate outcome measures have an important role in clinical trial design. The key is to choose a surrogate that will be viewed as a validated (reliable as well as reproducible) measure of treatment efficacy or diagnostic accuracy. As noted in case above, one of the most important characteristics of a surrogate is being a true predictor of disease, such as elevated blood pressure predicting heart disease and TTR predicting hemorrhagic and thromboembolic events. This requires a body of literature supporting the relationship. Additionally, the treatment must affect the desired primary outcome via the surrogate or a biological mechanism reliably correlated with the surrogate, and this relationship needs to be clearly supported by valid evidence.

When chosen correctly, the use of surrogate outcome measures allows important trials to be conducted while minimizing cost and time burden when primary events are infrequent and harmful.

References

1. Burken, M. I. and Whyte, J. J. 2002. Home international normalized ratio monitoring: Where evidence-based medicine is exemplified in the Medicare coverage process. *J. Thromb. Thrombolysis* 13:5–7.
2. Greenhalgh, T. 2001. *How to Read a Paper*. London: BMJ Books.
3. Decision can be accessed at http://www.cms.hhs.gov/mcd/viewnca.asp?from= ncd&nca_id=72
4. Ansell, J. E., Patel, N., Ostrovsky, D., Nozzolillo, E., Peterson, A. M., and Fish, L. 1995. Long-term patient self-management of oral anticoagulation. *Arch. Intern. Med.* 155:2185–2189.
5. White, R. H., McCurdy, S. A., von Marensdorff, H., Woodruff, D. E. Jr., and Leftgoff, L. 1989. Home prothrombin time monitoring after the initiation of warfarin therapy. A randomized, prospective study. *Ann. Intern. Med.* 111:730–737.
6. Over 400,000 patients currently have mechanical heart valves.
7. Matchar, D. B., Samsa, G. P., Cohen, S. J., Oddone, E. Z., and Jurgelski, A. E. 2002. Improving the quality of anticoagulation of patients with atrial fibrillation in managed care organizations: Results of the managing anticoagulation services trial. *Am. J. Med.* 113:42–51.
8. Samsa, G. P. and Matchar, D. B. 2000. Relationship between test frequency and outcomes of anticoagulation: a literature review and commentary with implications for the design of randomized trials of patient self-management. *J. Thromb. Thrombolysis* 9:283–292.

INDEX